AF606185

WHEN CONTINENTS COLLIDE: GEODYNAMICS AND GEOCHEMISTRY OF ULTRAHIGH-PRESSURE ROCKS

Petrology and Structural Geology

VOLUME 10

Series Editor:

ADOLPHE NICOLAS

Department of Earth and Space Sciences,
University of Montpellier, France

The titles published in this series are listed at the end of this volume.

When Continents Collide: Geodynamics and Geochemistry of Ultrahigh-Pressure Rocks

Edited by

BRADLEY R. HACKER

Geological Sciences,
University of California, Santa Barbara, U.S.A.

JUHN G. LIOU

Geological and Environmental Sciences,
Stanford University, Stanford, U.S.A.

KLUWER ACADEMIC PUBLISHERS

DORDRECHT / BOSTON / LONDON

552.4
W

A C.I.P. Catalogue record for this book is available from the Library of Congress.

ISBN 0-4128-2420-5

Published by Kluwer Academic Publishers,
P.O. Box 17, 3300 AA Dordrecht, The Netherlands.

Sold and distributed in North, Central and South America
by Kluwer Academic Publishers,
101 Philip Drive, Norwell, MA 02061, U.S.A.

In all other countries, sold and distributed
by Kluwer Academic Publishers Group,
P.O. Box 322, 3300 AH Dordrecht, The Netherlands.

Cover figure:

Locations of ultrahigh-pressure rocks, UHP rocks with diamonds (italic) and high-pressure rocks formed at >2 GPa (parentheses). NNW section through the Hindu Kush showing 1977 seismicity recorded by (Chatelain et al., 1980). Eclogitic garnet containing a 250 mm coesite grain partially transformed to quartz (Zhang and Liou, 1996).

Chatelain, J.L., Roecker, S.W., Hatzfeld, D., and Molnar, P. (1980) Microearthquake activity and fault plane solutions in the Hindu Kush region and their tectonic implications, Journal of Geophysical Research 85, 1365-1387.

Zhang, R.Y. and Liou, J.G. (1996) Coesite inclusions in dolomite from eclogite in the southern Dabie Mountains, China: The significance of carbonate minerals in UHPM rocks, American Mineralogist 81, 181-186.

Printed on acid-free paper

First published 1998
Reprinted with corrections 1999

All Rights Reserved
© 1998 Kluwer Academic Publishers
No part of the material protected by this copyright notice may be reproduced or
utilized in any form or by any means, electronic or mechanical,
including photocopying, recording or by any information storage and
retrieval system, without written permission from the copyright owners.

Printed in the Netherlands

Contents

Acknowledgments

While the need for a volume such as this has been apparent for some time, it was Ian Francis, formerly of Chapman–Hall, who had the vision to start the project. Petra van Steenbergen shepherded the tome through to its timely conclusion. Dave Root graciously proofed the entire volume. Tremendous thanks are due the reviewers of the chapters, whose only reward was to see the news before it broke:

Geoff Abers	Gray Bebout
Cong Bolin	Andy Calvert
Christian Chopin	Jean Crespi
W.G. Ernst	John M. Ferry
Gayle Gleason	Trevor Ireland
John Holloway	Shaocheng Ji
Jeff Lee	H.-J. Massonne
Alan Matthews	James M. Mattinson
Jed Mosenfelder	Craig Nicholson
Simon Peacock	Lothar Ratschbacher
Douglas Rumble III	Tracy Rushmer
Werner Schreyer	Jan Tullis
J. Douglas Walker	An Yin
Ye Kai	T.Z. Yui

Preface

It was once easy for most geologists to dismiss the couple of small coesite-bearing outcrops discovered by Chopin (1984) and Smith (1984) as rare tectonic freaks worthy of only idle speculation. In the last 10 years, however, the number of orogenic belts world-wide that contain diamond, coesite, or other indications of metamorphic pressures >2 GPa, has exploded to more than 15! It is time for all geologists to consider the manifold ramifications of this proof that continental blocks as large as 5 x 50 x 100 km were subducted to depths of 100–150 km commonly during the history of the Earth and may have played a significant role in the formation of most mountain belts.

The best-known modern examples of continental crust at UHP in the mantle today are Australia–Timor and the Pamir–Hindu Kush. In the first chapter of this volume, **Lin and Roecker** use seismic tomography, earthquake data, geodesy, and leveling to argue that Taiwan also belongs to this select group, showing that the title of this volume, "When Continents Collide" should not be taken too literally. **Patiño Douce and McCarthy** use an experimental perspective to show that the bulk of continental crust is unlikely to melt during subduction to UHP depths, but instead should undergo dehydration melting during decompression. Eclogites have often been thought of as strong rocks, but **Stöckhert and Renner** argue that eclogites are weak and that continental material at UHP is too soft to be exhumed as a coherent slab. **Davies and von Blanckenburg** study the detachment of oceanic lithosphere from continental lithosphere during continental collision, and show that the depth of breakoff depends strongly on the strength of the subducted continental lithosphere; breakoff in old lithosphere occurs along the passive margin, but breakoff in young lithosphere occurs within the continent. **Grasemann et al.** present simpler thermal calculations and suggest that UHP rocks should undergo two-stage cooling histories. They emphasize that the slowest cooling corresponds to the most rapid period of exhumation and that the fastest cooling corresponds to the slowest period of exhumation. **Blythe** notes that few modern exhumation rates are as fast as exhumation rates commonly argued for exhumed high-pressure rocks, and suggests that this implies either special precipitation or glaciation patterns or that high-pressure rocks are exhumed more slowly than commonly assumed.

Excess ^{40}Ar has been a long-standing problem in high-pressure rocks, but **Scaillet** reveals how this can be turned to advantage to reveal fluid–rock interactions at UHP conditions. Eclogites and ultramafic rocks in UHP orogens might plausibly come from subducted oceanic lithosphere, but **Jahn** shows for the Dabie–Su–Lu orogen that the eclogites and ultramafic rocks were derived from continental intrusions and subcontinental upper mantle, respectively. UHP rocks in China have yielded astoundingly low stable isotopic ratios, which **Rumble** interprets as evidence that the protolith represents a high-latitude geothermal system. The spatial extent of the fossil geothermal system indicates that UHP metamorphism, without a pervasive free fluid phase, occurred over a length scale of >100 km. **Tabata et al.** report on Raman spectroscopy on mineral inclusions in paragneiss that show that the entire southern Dabie Shan of China was subducted to mantle depths. The rate and means by which volatiles are subducted control many facets of subduction zones. **Ernst et al.** note that mineral assemblages in exhumed metamorphic rocks conflict with conclusions drawn from recent experimental studies, and that mafic lithosphere carries more volatiles into the mantle than continental lithosphere. Realization of the important geodynamical effects of slow phase transformations is growing, and **Austrheim** shows that the rates of metamorphic reactions in the deep crust are controlled by fluid availability and can be sluggish at temperatures as high as 800°C. In aggregate, these papers demark the frontier of our understanding of the creation, preservation, and exhumation of ultrahigh-pressure rocks.

Bradley R. Hacker and J.G. Liou

Chapter 1

Active Crustal Subduction and Exhumation in Taiwan

C. H. Lin
Inst. of Earth Sciences, Academia Sinica, P. O. Box 1-55, Nankang, Taipei, Taiwan, ROC, lin@earth.sinica.edu.tw

S. W. Roecker
Dept. of Earth and Environmental Sciences, Rensselaer Polytechnic Institute, Troy, NY 12180, USA, roecker@gretchen.geo.rpi.edu

Abstract: The evolution of the Taiwan orogeny is often described in terms of deformation of rocks above a shallow decollement, either in a continuously deforming wedge or a discretely deforming stack of thrust sheets. Recently acquired evidence from seismic tomography, earthquake locations and focal mechanisms, GPS geodesy, leveling, and exhumed shear zones suggest that such models are inadequate. We propose that these observations are better explained by a model of continental subduction followed by crustal exhumation. In the context of this model, the Eastern Central Range is the locus of active crustal exhumation. Simple thermal modeling shows that the observed seismicity patterns across Taiwan and the high heat flow in the Eastern Central Range are well explained by this crustal exhumation model.

1. INTRODUCTION

Despite its being less dense than the underlying mantle, a variety of geological evidence suggests that continental crust is at least occasionally subducted to depths of 150 km (e.g., Chopin, 1984; Smith, 1984; Wang *et al.*, 1989; Dewey *et al.*, 1993). The significance of continental subduction is often dismissed, however, because a number of simple observations argue for the persistence of continental crust at the Earth's surface. These include the longevity of the continents and the shortening (rather than disappearance) of continental crust during collision. The buoyancy of continental crust implies that continental subduction must be aided by some other factor such as the concurrent subduction

B. R. Hacker and J. G. Liou (eds.),
When Continents Collide: Geodynamics and Geochemistry of Ultrahigh-Pressure Rocks, 1–25.
© 1998 *Kluwer Academic Publishers. Printed in the Netherlands.*

of some thickness of continental mantle lithosphere, or by the precedence of subducted oceanic lithosphere, both of which act to counterbalance the density deficit. Such possibilitics were considered by Molnar and Gray (1979), who inferred that the subduction of continental crust can occur under physically reasonable circumstances.

The existence at the surface of geologic evidence for crustal subduction implies some mechanism of exhumation as well, but the dynamics of any such mechanism are not well understood. Platt (1987), for example, presented a simple physical model for crustal subduction which, because of the coupling required to take the crust down in the first place, suggested that the return of the subducted crust to the surface was implausible. On the other hand, Ernst (1983) appealed to buoyancy-driven exhumation of subducted crust after detachment from the descending slab, and a recent series of simulations (Chemenda, 1993; Chemenda *et al.*, 1995; 1996) suggests that the buoyancy of subducted continental crust will eventually lead to failure seaward of the subduction front and the subsequent return of the crust to the surface. According to their models, continental crust may be subducted to depths of ~200 km before such failure and exhumation occurs.

Most of the postulated examples of crustal exhumation are from long-extinct systems, and therefore dynamical models of this process are constrained only by surface geology and end results. Tomographic images in two active orogens, the Hindu Kush (Roecker, 1982) and Taiwan (Roecker *et al.*, 1987), were interpreted to indicate active subduction of continental crust, but until recently, evidence for active exhumation has been missing.

This chapter reviews recent observations that suggest that continental subduction and crustal exhumation not only has occurred, but is presently occurring in eastern Taiwan. We discuss these observations in light of prevailing models for orogenic processes in Taiwan, and argue that the best explanation of the geophysical and geological data from this area is the active exhumation of previously subducted continental crust.

2. GEOLOGIC AND TECTONIC OVERVIEW OF TAIWAN

The island of Taiwan is located at a complex boundary between the Philippine Sea and Eurasian plates (Fig. 1), and has evolved as a consequence of the collision between an island arc on the Philippine Sea plate, represented on Taiwan by the Coastal Range, and the continental shelf of the Eurasian plate, which makes up the rest of the island. East of Taiwan, the Philippine Sea plate subducts northward beneath the Ryukyu arc, while south of the island, Eurasian plate oceanic lithosphere beneath the south China Sea subducts eastward beneath the Philippine Sea plate. The collision that produced the island initiated in the northern part of the island at 4 Ma and propagated to the south, where it is now in an incipient stage (e.g., Chi *et al.*, 1981; Suppe, 1981).

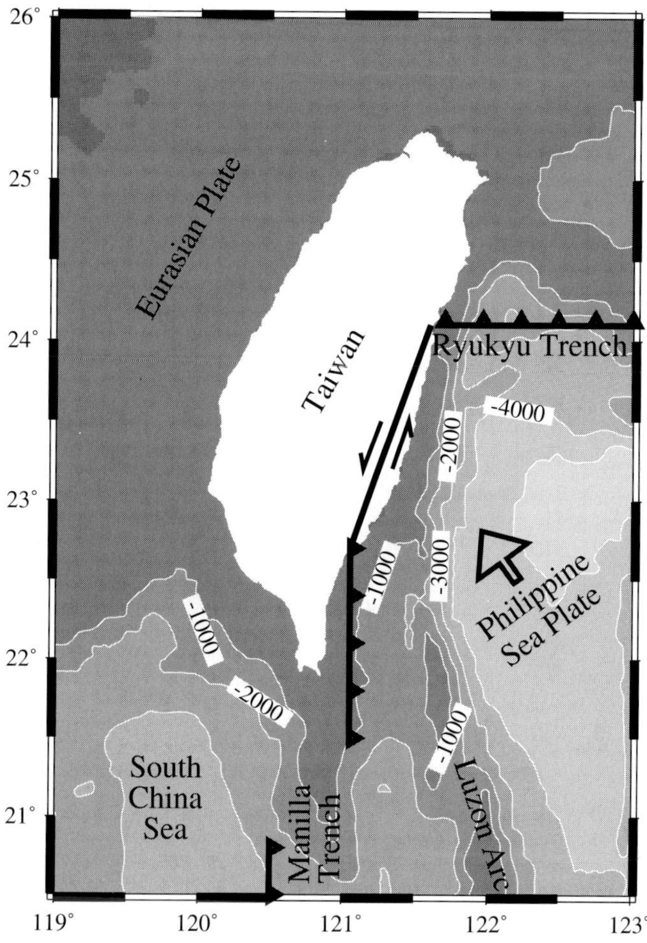

Figure 1. Generalized tectonics in the Taiwan area (after Tsai *et al.*, 1977). Heavy lines mark major convergent boundaries between the Philippine Sea and Eurasian plates. Triangles along one side of the boundary indicate the direction of subduction. The large arrow shows the direction of motions of the Philippine Sea plate with respect to the Eurasian plate. Bathymetry is shown with a contour interval of 1000 m.

The principal geologic units of Taiwan generally strike NNE-SSW and are divided by a number of subparallel faults (Fig. 2). From west to east these units are the Coastal Plains, the Western Foothills, the Western Central Range, the Eastern Central Range, the Longitudinal Valley, and the Coastal Range. The Coastal Plains are underlain by Pleistocene rocks and sediments, while the

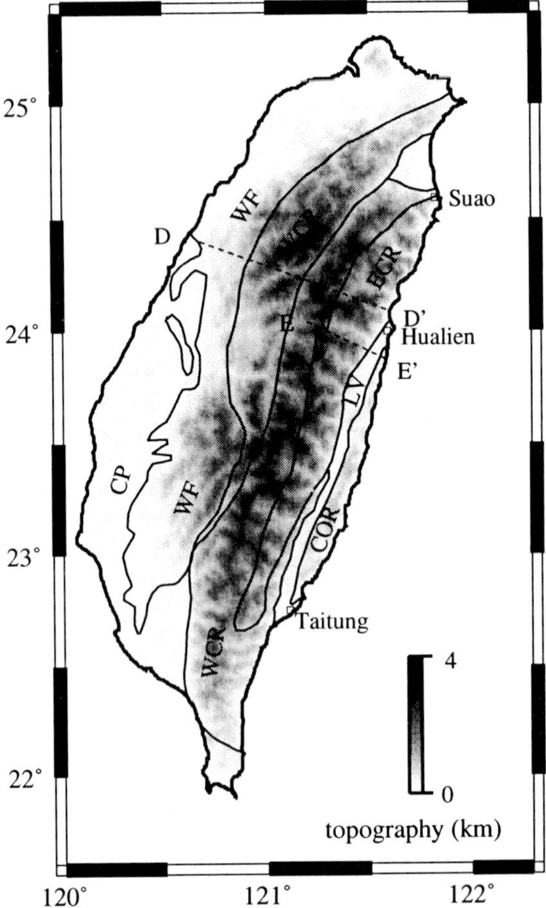

Figure 2. Topography (shaded) and major geologic provinces in Taiwan (after Ho, 1988), including the Coastal Plain (CP), Western Foothills (WF), Western Central Range (WCR), Eastern Central Range (ECR), Longitudinal Valley (LV) and Coastal Range (COR). D-D' and E-E' are the two P-wave velocity profiles of figures 3 and 4.

Western Foothills unit is an unmetamorphosed to slightly metamorphosed (zeolite facies) Oligocene to Neogene fold-and-thrust belt. The Western Foothills are bounded by the Chuchih fault, east of which is the Western Central Range, a slate belt deformed in Plio-Pleistocene times (Ho, 1988). The Eastern Central Range is a pre-Tertiary metamorphic complex composed largely of the "Tananao Schist", which is an assortment of marble, dolostone, and metavolcanics with minor metaigneous rocks. The Eastern Central Range is separated from the Coastal Range to the east by the Longitudinal Valley. The Coastal Range is a manifestation on Taiwan of the Luzon volcanic arc of the Philippine Sea plate, and thus the Longitudinal Valley generally is considered to be the suture.

The Tananao Schist of the Eastern Central Range, of particular importance to the present study, are the oldest rocks in Taiwan (Yen, 1954). This unit extends ~250 km along strike, decreasing in width from about 30 km in the north to about 10 km in the south. The southern end is marked by a distinct change in both the elastic wave velocity structure and the level and type of seismic activity (Roecker *et al.*, 1987). The evolution of the belt has been documented in paleontological (Yen *et al.*, 1951; Yen, 1953; Lee, 1984) and isotopic (Jahn *et al.*, 1984; Yui, 1987) investigations, which suggest an origin of clastic and carbonate deposition during Permo-Triassic time (200-255 Ma). The belt experienced three significant thermal events (Lan *et al.*, 1990) at 90 Ma, 40 Ma, and since 12 Ma. The cooling ages for the last event, associated with the current arc-continent collision, increase westward, which Lan *et al.* (1990) explained as incomplete resetting due to westward decreasing peak temperatures.

3. Recent Observations Related to the Arc-Continent Collision

Geological and geophysical studies in the past several years bear on the evolution of the Eastern Central Range. These include imaging with seismic tomography, estimates of stress orientation though earthquake mechanisms, strain rates from geodetic observations, uplift and erosion from a variety of techniques, and temperature gradients from heat flow.

3.1 Tomographic Imaging

Over the past ten years images of the subsurface of Taiwan have been constructed using seismic tomography (e.g., Roecker *et al.*, 1987; Lin *et al.*, 1989; Yeh *et al.*, 1989; Yeh and Chen, 1994; Rau and Wu, 1995; Ma *et al.*, 1996; Lin *et al.*, 1998). Studies of the large-scale structure of central Taiwan consistently reveal a low velocity horizon that thickens eastward beneath the Coastal Plains and Eastern Central Range (Fig. 3). The depth of this low velocity region is greater than expected from isostatic crustal thickening, and was interpreted by Roecker *et al.* (1987) as continental crust carried down with the subducting Eurasian lithosphere. An east-dipping zone associated with ambient earthquake activity in the southern part of the island was similarly interpreted as evidence of ongoing subduction of the Eurasian continental shelf.

The low velocity regions determined in each of the large-scale tomographic studies terminate beneath the Eastern Central Range against a higher velocity feature that generally has been interpreted as oceanic lithosphere of the Philippine Sea plate. Recently, however, Lin *et al.* (1998) carried out a high resolution study of the region between the Eastern Central Range and the Coastal

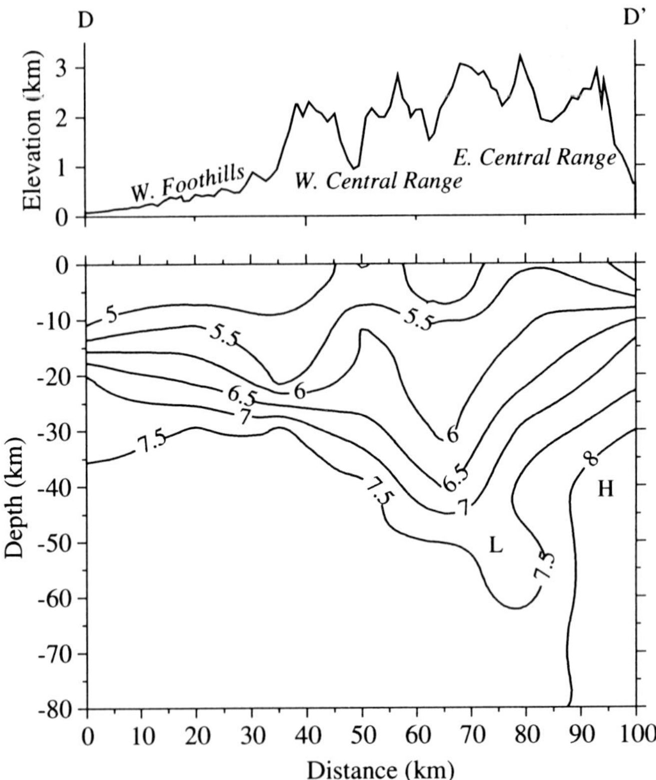

Figure 3. P-wave velocity structure and topography along profile D-D' in figure 2 (after Rau and Wu, 1995). Regions of high and low velocities are marked "H" and "L".

Range using data from a dense deployment of the PANDA II network in the Hualien region in the east-central part of the island (Fig. 2). Images derived from these data (Fig. 4) reveal that this higher velocity region does not extend as far east as the Coastal Range, but is confined to the Eastern Central Range by another region of low velocities beneath the Longitudinal Valley. This surprising result suggests that the Eastern Central Range is not the result of crustal thickening or underplating, nor a simple accumulation of crust above a shallow decollement. Rather, it appears to be related to more deep-seated (>40 km depth) phenomena.

3.2 Styles of Deformation

Sitting on an active plate boundary, Taiwan has abundant local earthquake activity that can be used to infer both the rheology of the subsurface through the mapping of brittle deformation, and the styles of deformation through the construction of focal mechanisms. The general picture derived from earthquake locations (Fig. 5) is that most of the activity is associated with the Philippine Sea plate east of Taiwan subducting to the north, and the Eurasian plate south of about 23°N subducting to the east. The remainder of the seismicity is scattered throughout the Western Foothills, and is related to shortening in the fold-and-thrust belt. For the most part, focal mechanisms from this belt show thrust faulting with slip orthogonal to the strike of the major faults on the island. Horizontal compression in a NW-SE direction thus dominates the region.

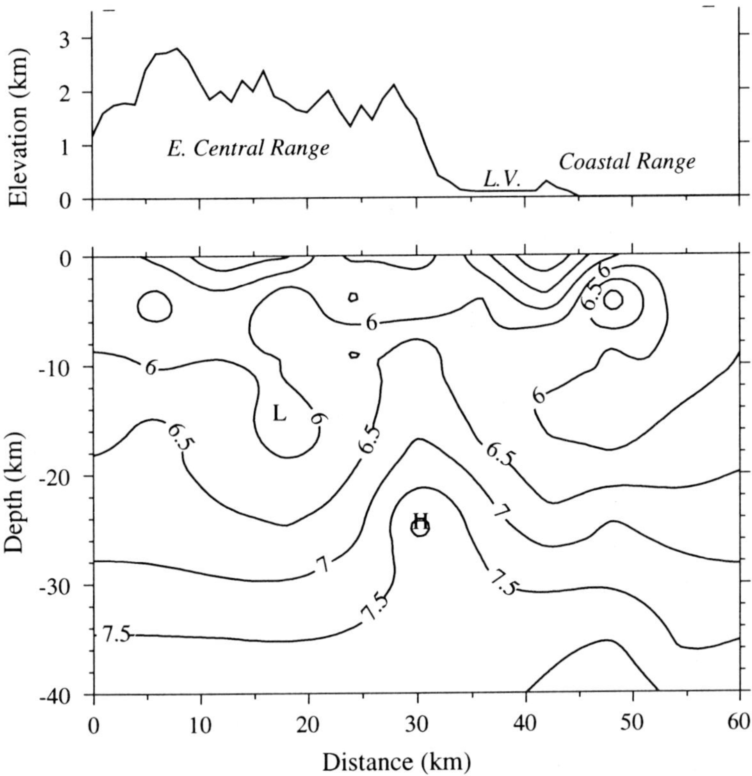

Figure 4. P-wave velocity structure and topography along profile E-E' in figure 2 (after Lin *et al.*, 1998). Regions of high and low velocities are marked "H" and "L".

While such features are the first-order effects of the collision between Eurasia and the Philippine Sea plate, closer examination requires modifications to this simple picture. For example, Lin and Roecker (1993) showed that a zone of anomalously deep (20-100 km) earthquakes beneath the central part of the island at about 24°N are too deep to be associated with a decollement or other feature related solely to crustal deformation. Lin and Roecker suggested that these earthquakes result from an offset in the mantle lithosphere which has juxtaposed the upper mantle with subducted crust and cooled it to temperatures permitting brittle deformation. Equally surprising is the observation that the Eastern Central Range, despite its location at the eastern front of deformation, is nearly aseismic (section CC' in Fig. 5). Moreover, in contrast to the rest of the island, many of the earthquakes in this region have normal faulting mechanisms, which suggests a predominance of vertically, rather than horizontally, oriented maximum principal stress. For example, Cheng (1995) reported 24 such events with $M_d > 4$ located throughout the Eastern Central Range. Lin *et al.* (1998) noted a similar persistence of normal faulting in the Eastern Central Range in the mechanisms of the microearthquakes recorded by the PANDA II network (Table 1 and Fig. 6).

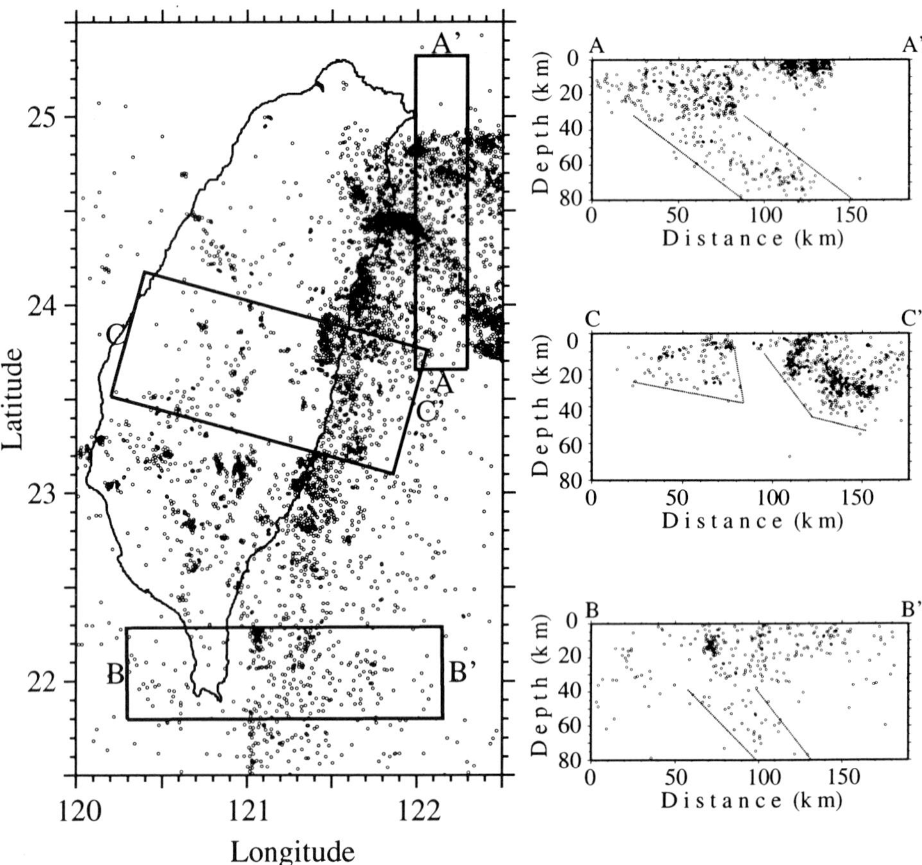

Figure 5. Earthquakes located by the Central Weather Bureau Seismograph Network (CWBSN) between 1992 and 1995 with magnitude of greater than 3.0. Uncertainties in locations are less than 5 km in epicenter and 10 km in depth. Hypocenter projections on three profiles (A-A', B-B' and C-C') are shown to the right.

The inference of extension in the Eastern Central Range is also supported by recent geological studies of rock deformation in the area. In particular, Lee

(1995) interpreted drag folds and quartz veins as evidence of normal faulting. Crespi et al. (1996), in a study of late-stage brittle normal faults and semibrittle to ductile normal-sense shear zones, suggested that normal faulting in the Eastern Central Range extends at least to the depths of the brittle-ductile transition.

Table 1. Earthquakes in the Eastern Central Range.

No.	Date yr/mo/day	Time hr:min	Lat °N	Lon °E	Depth km	Mag
1	1983/05/10	0:15	24.419	121.560	28	6.0
2	1987/06/27	7:38	24.290	121.646	27	5.1
3	1987/11/10	4:33	24.418	121.724	34	4.9
4	1992/08/06	16:49	24.338	121.752	48	4.5
5	1990/12/19	0:08	23.662	121.366	1	5.2
6	1990/12/26	0:20	23.816	121.433	1	4.1
7	1991/01/04	13:08	23.661	121.364	1	4.0
8	1991/01/18	1:36	23.692	121.346	11	5.4
9	1991/01/18	1:39	23.671	121.288	1	4.4
10	1991/01/18	10:49	23.689	121.291	1	5.0
11	1991/01/18	10:54	23.671	121.306	2	4.8
12	1991/01/18	12:40	23.731	121.234	3	4.5
13	1991/01/19	15:59	23.702	121.274	1	4.5
14	1991/01/21	18:02	23.639	121.359	3	5.1
15	1991/01/21	18:19	23.658	121.353	4	4.0
16	1991/01/21	22:13	23.640	121.304	1	4.3
17	1991/01/24	1:17	23.671	121.327	0	4.2
18	1991/01/24	4:01	23.657	121.308	2	4.2
19	1991/02/16	18:13	23.674	121.326	4	4.0
20	1991/07/19	18:42	23.647	121.401	4	4.7
21	1991/07/27	18:17	23.641	121.400	5	4.5
22	1988/02/28	12:40	22.739	120.938	7	4.2
23	1990/07/06	9:16	22.712	120.896	1	4.3
24	1991/09/09	20:57	22.726	120.924	1	4.8
25	1993/10/10	0:0	23.809	121.443	11	<4.0
26	1993/09/01	16:53	23.868	121.404	19	<4.0
27	1993/08/12	8:43	23.872	121.406	19	<4.0
28	1993/08/30	4:54	23.917	121.445	16	<4.0
29	1993/11/10	16:15	23.973	121.452	20	<4.0
30	1993/09/20	3:15	24.070	121.505	34	<4.0
31	1993/11/26	23:04	24.139	121.617	12	<4.0
32	1993/11/02	17:40	24.186	121.583	15	<4.0

Source parameters of the 32 fault-plane solutions reported by Cheng (1995) (1-24) and Lin *et al.* (1998) (25-32). Mechanisms of the first two events determined by waveform modeling, the remainder by first-motion polarities. Earthquakes reported by Lin et al (1998) were all microearthquakes recorded by a local network and had no magnitude reported other than that they were less than 4.0.

Finally, results from GPS surveys of Taiwan (Yu and Chen, 1996), during which 26 subnets were occupied 6-7 times over a period of 6 years, show that while Taiwan is in general under horizontal contraction, most of the Eastern Central Range is extending (Fig. 7). The extensional strain rates in most of the Eastern Central Range from 0.2-0.62 µstrain/a, although in central Taiwan there is evidence of slight contraction at 0.02 µstrain/a.

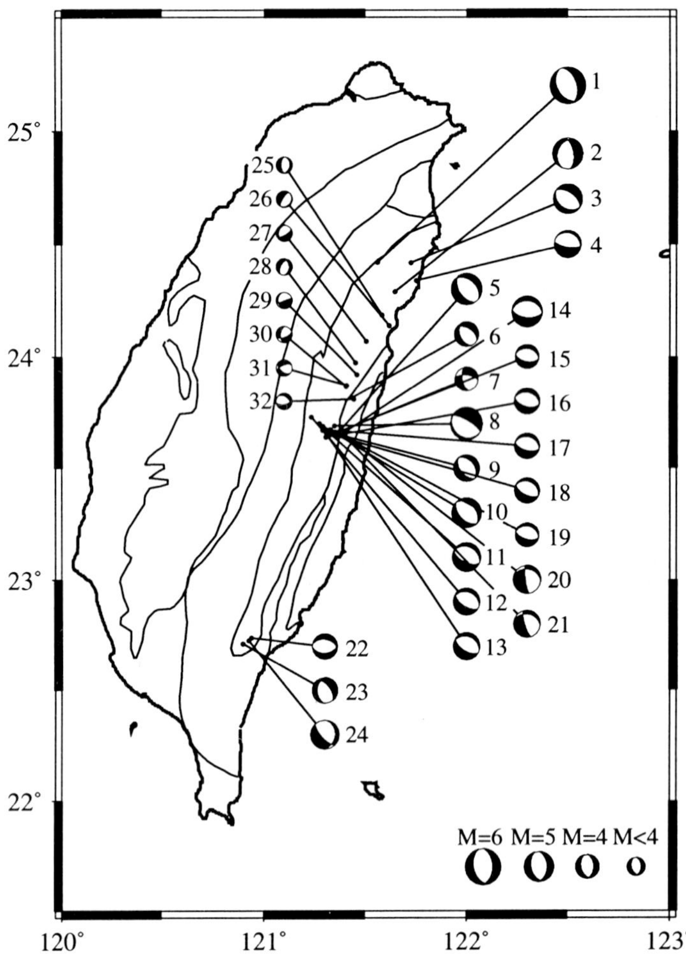

Figure 6. Earthquakes in the Eastern Central Range reported by Cheng (1995) and Lin *et al.* (1998) to have normal fault plane solutions. Solutions are lower hemisphere projections. Numbers next to the mechanisms correspond to those in the first column of Table 1.

3.3 Uplift and Erosion

Uplift and erosion rates in Taiwan are much higher than average. Based on long-term geological evidence (Li, 1975; Lee, 1977; Peng *et al.*, 1977; Liu, 1982; Lan *et al.*, 1990), these rates have been considered to be approximately equal and on the order of 5 mm/a. This has contributed to an interpretation of Taiwan as a steady state wedge, albeit one with quite rapid turnover. A recent study of zircon fission track ages of rocks in the Central Range by Tsao (1996), however, suggests that the rate of uplift increased from 7-16 mm/a at 1.5 Ma, and recent leveling measurements (Liu, 1995) show rates of uplift of 36-42 mm/a over the past decade (Fig. 8). Thus, it seems that uplift of much of the Central Range is recent and has accelerated, rather than being uniform over the past several million years.

3.4 Heat Flow

Heat flow in Taiwan varies from about 30-254 mW/m^2 (Lee and Cheng, 1986), and to first order the contours of heat flow correlate with topographic relief (Fig. 9). North of 23°N, contours of heat flow increase eastward and parallel the general NNE trend of the principal geologic units, with the highest values obtained in the Eastern Central Range. Farther east, in the Coastal Range, the heat flow returns to the low values found in the western Coastal Plains. South of the termination of the Eastern Central Range at about 23°N, the trend of the heat flow contours shifts to NNW. High heat flow in the Eastern Central Range implies a localized elevated temperature gradient in the area, which in turn suggests higher than normal temperatures in the crust.

3.5 Metamorphism

Useful boundary conditions on temperature and pressure can be derived from the metamorphic facies observed at the surface. To review briefly, the metamorphic grade of rocks in Taiwan increases eastward, from unmetamorphosed to slightly metamorphosed (zeolite facies) in the Western Foothills, through prehnite-pumpellyite facies in the Western Central Range, to greenschist facies in the Eastern Central Range, indicating a general increase in the maximum temperatures and (probably) pressures endured by these rocks. For much of the Central Range, temperatures range from 250-450°C and pressures from 300-400 MPa (e.g., Liou, 1981a; 1981b; Lee *et al.*, 1982; Ernst, 1983; Ernst and Harnish, 1983; Warneke and Ernst, 1983; Ernst, 1984), although temperatures and pressures as high as 550-650°C and 500 MPa have been reported in the Nanao-Suao area north of Hualien (Ernst and Jahn, 1987). A somewhat unusual aspect of the metamorphic terrain in Taiwan is that the higher grade (prehnite-pumpellyite to greenschist) facies cover a rather broad area, extending in places over more than half the width of the island.

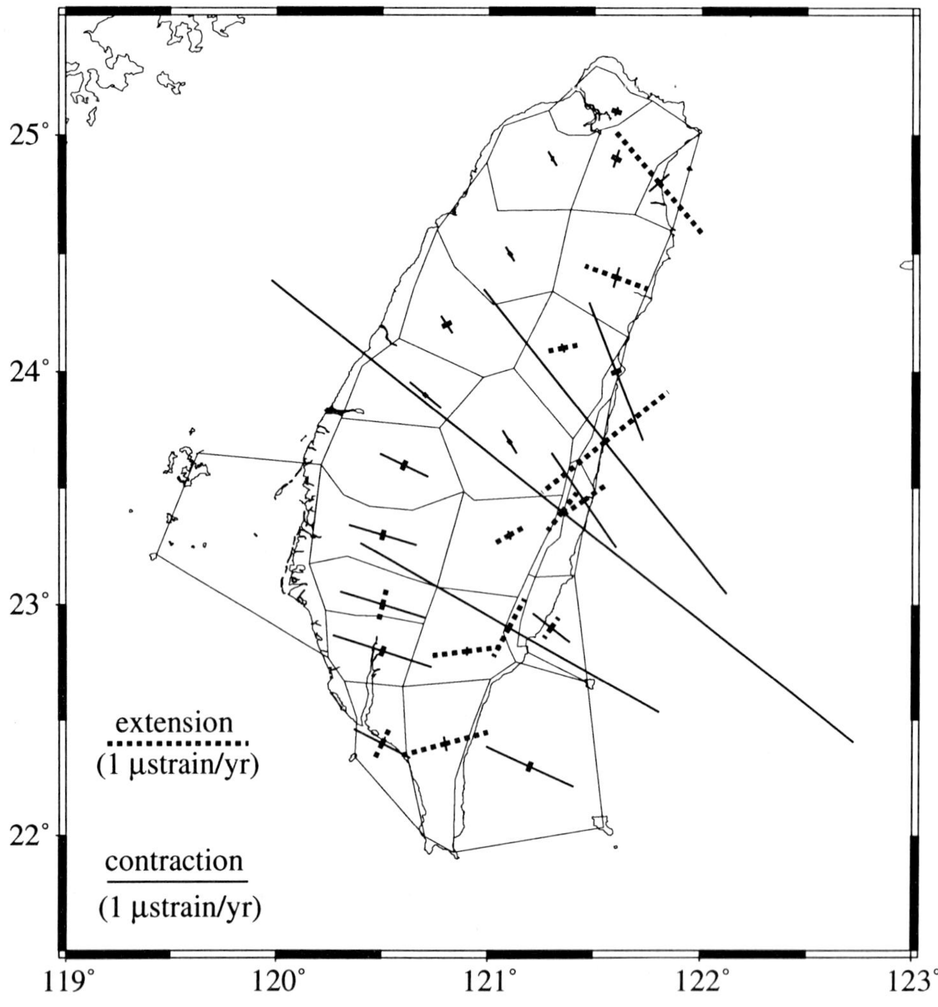

Figure 7. Results of GPS measurements of strain rate in Taiwan, adapted from Yu and Chen (1996). Note predominance of EW extension in the Central Range.

4. IMPLICATIONS FOR GEODYNAMICS

Existing models for the evolution of Taiwan differ in detail but generally assign all the important activity to deformation in the upper 15-20 km, above or near a shallow dipping decollement. These include a wedge model (e.g., Barr and Dahlen, 1989; Dahlen and Barr, 1989; Barr *et al.*, 1991), a thrust model (Hwang and Wang, 1993), and an underplating model (e.g., Platt *et al.*, 1985; Platt, 1987). In the former two models, all the relevant activity takes place above the decollement. In the latter model, material is allowed to be incorporated into the wedge from below the decollement, but again most of the activity is shallow.

Taiwan is frequently characterized as an archetypal steady-state accretionary wedge. A series of analytical investigations into the properties of such wedges (cited above) has shown this model to be consistent with a variety of surface observations in Taiwan; in particular topography and heat flow appear well explained. According to this model, as crustal material enters the collisional wedge, it decouples along a shallow (7-15 km depth) decollement. The material above the decollement is incorporated into the mountain-building wedge, while the remainder disappears into the mantle. The depth at which material enters the wedge controls the depth to which it descends and the distance it penetrates into

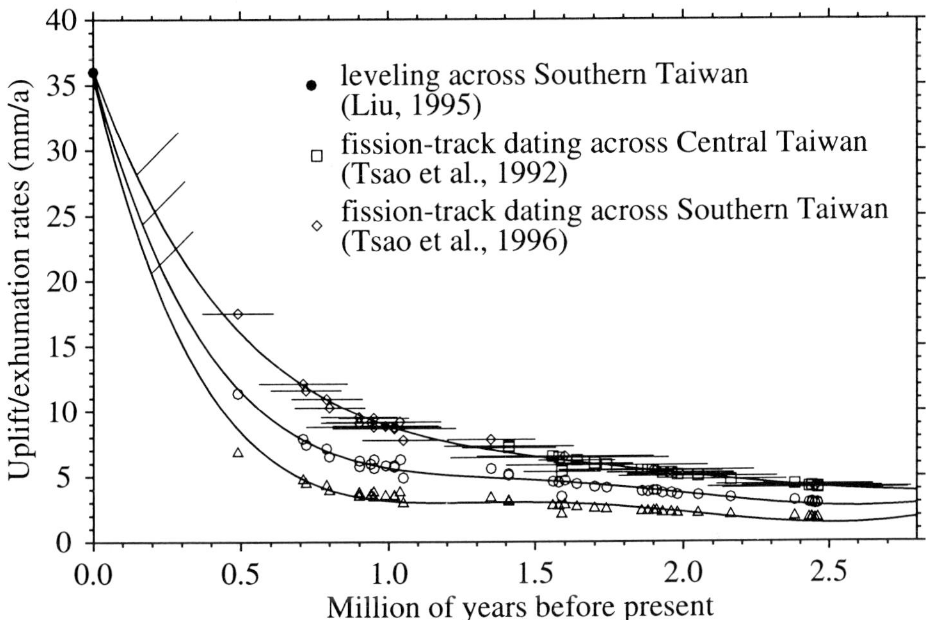

Figure 8. Uplift (from leveling) and exhumation (from fission track) rates for the Central Range, adapted from Liu (1995), Tsao *et al.* (1992) and Tsao (1996), which suggest a dramatic acceleration of mountain building since 1 Ma. The fission track observations were made from zircons with closure temperatures of ~235°C. Because the inference of exhumation rate from fission-track observations requires some assumptions about thermal gradient and exhumation history, results of 3 different models are shown. The uppermost points, from Tsao (1996), assume a thermal gradient of 30°/km with all the exhumation occurring after the most recent measurement. The open circles correspond to a model with the same thermal gradient of 30°/km but with all the exhumation before the oldest datum. The triangles correspond to a model where all the exhumation occurs before the oldest datum and assumes a very high thermal gradient of 50°/km. Solid lines are 5th-order polynomial fits. Note that even when a true "steady state" condition is assumed (*i.e.*, high elevations and thermal gradient since 2.5 Ma), the exhumation rate at 0.5 Ma is still seven times the rate at 2.5 Ma.

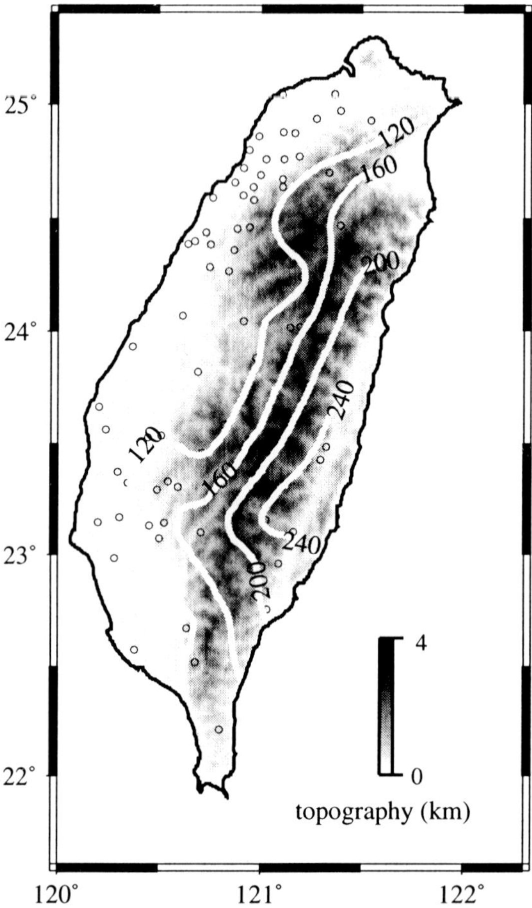

Figure 9. Surface heat flow and topography in Taiwan (after Lee and Cheng, 1986). Circles show the locations of heat flow measurements and contours are in mW/m^2. These contours do not include heat flow measurements in the Coastal Range, which are similar to the low values in the western part of the island (~120 mW/m^2).

the wedge. The rocks in the Eastern Central Range, therefore, would have remained close to the decollement until being forced to the surface at their present location by the "backstop" provided by the Coastal Range.

Barr *et al.* (1991) attempted to model the metamorphic facies in Taiwan with the wedge model, but while they were able to qualitatively reproduce the west-to-east increasing grade of metamorphism, they were unable to account for the substantial width of the prehnite-pumpellyite to greenschist belts. Underplating, which can modify the stress and temperature fields within the wedge by allowing material beneath the decollement to enter the wedge, was invoked by Barr *et al.* (1991) to increase the width of these zones, but they found the amount of underplating required (more than 50%) to be excessive and impossible to accommodate in a steady state.

Similar topography and heat flow profiles were generated by Hwang and Wang (1993) using a discrete thrust model with much of the heat being produced by friction on thrust faults, but it is unclear if this type of mechanism can explain the metamorphic facies better than the wedge model. Moreover, the high temperatures necessary to produce the heat flow observed in the Central Range imply a ductile rheology, and so differences between the continuously deforming wedge model and the discrete thrusting model are perhaps less important in this region.

The high heat flow observed in the Central Range indicates a higher than normal thermal gradient, and thus may be considered as evidence for high temperatures in the crust. As a surface observation, heat flow does not allow deduction of a unique thermal profile, but patterns of earthquake activity in Taiwan provide additional clues to temperature distributions at depth. For example, the occurrence of earthquakes at depths greater than 20 km beneath the Western Foothills and into the Western Central Range (CC' in Fig. 5) suggest active deformation below the decollement, and imply that not all of the crustal deformation takes place within the wedge.

The level of seismic activity drops off sharply beneath the Eastern Central Range (section CC' in Fig. 5). This relative lack of earthquake activity could be explained either by a lack of deviatoric stress, by an ability of the rocks to withstand or transmit stress without deforming internally (*i.e.*, they are relatively strong), or by an inability of the rocks in this region to accumulate strain in a way that leads to brittle failure (*i.e.*, they are relatively weak). Given the level of activity in the regions surrounding the Central Range, a lack of deviatoric stress seems unlikely. Similarly, the extensive shearing of rocks exposed at the surface tends to discount the idea that they are exceptionally strong, or in any event are capable of transmitting stress without themselves deforming. It therefore seems most likely that the rocks are deforming ductilely as a consequence of higher than normal temperatures throughout the upper 20 km or so of crust.

As discussed above, any of the previously proposed mechanisms could be expected to create elevated temperatures beneath the Eastern Central Range, but the sharp lateral variation in temperature suggested by the gap in seismic activity seems difficult to explain in the context of any of them; one would instead expect a gradual decrease in the depth of activity.

Beyond temperature profiles, the existing models do not seem appropriate for many of the other observations discussed. In particular, the wedge and thrust models are both steady state models constrained by long-term estimates of uplift and erosion over the past 4-5 m.y. Although a modest amount of acceleration is predicted by the particle paths of the wedge model, these models appear at odds with the recent observations from leveling and fission-track dating showing that uplift and exhumation have accelerated dramatically within the past million years. Moreover, this acceleration is not uniform throughout the island, but is concentrated in the Central Range. Nor do existing models allow for normal faulting, particularly at depths of 10's of kilometers, as observed both in the earthquake mechanisms and in rock deformation. It is also difficult to model observed gravity anomalies with only a simple wedge model of low density rock. In particular, Ellwood *et al.* (1996) and Yen *et al.* (1998) reported that the observed gravity anomalies require some basement involvement in the mountain building. Finally, it is difficult to explain the tomographic images of the subsurface, particularly those of Lin *et al.* (1998) which suggest a connection between the Eastern Central Range with structures at depths of several 10's of km, in the context of any model that restricts its attention to the region above a shallow decollement.

Given the shortcomings of the existing geodynamical models for Taiwan, we propose a different explanation for the evolution of the Central Range. Specifically, our view is that the observations reviewed above are well explained by a process of subduction and exhumation of continental crust. The remainder of this chapter discusses this idea and how it fits the observations.

5. THE EXHUMATION MODEL

In the context of the proposed model of crustal subduction and exhumation the evolution of the Central Range of Taiwan is as follows (e.g., Ernst, 1983; Pelletier and Stephan, 1986). The current arc-continent collision began after subduction of the Oligocene-Miocene South China Sea along the Manila Trench, initiating at the northern part of Taiwan and propagating south. Analogous to the simulations conducted by Chemenda *et al.* (1995; 1996), we envision (Fig. 10) that as collision progressed, the continental shelf of the Eurasian plate was dragged down into the mantle, carrying with it some thickness of continental crust. As more crust was consumed in this fashion, the resistive buoyancy of the lighter material eventually increased to the point where decoupling occurred near the base the crust, forming a crustal slice which, through the combined effects of buoyancy and erosion, made its way back up to the surface, even while the lithospheric mantle continued to subduct. The exhumed crustal slice now is exposed at the surface in the Eastern Central Range.

To first order, the existence of exhumed crust can explain the observed high temperatures and the aseismicity of the Central Range simply by the ascension of hot, ductile crustal material to shallow depth. Moreover, the exhumation of a slice of crust explains the prevalence of normal faulting and extension observed. Because the development of the crustal slice would occur only after an extended period of continental subduction, the evolution of the range would not be a steady state process, but would be characterized by an acceleration of uplift following decoupling. Thus, this type of orogenic evolution is consistent with the observed acceleration of uplift over the past 1 m.y. Because it is a body with thermal and lithological characteristics distinctively different from surrounding rocks at the same depth, an exhumed slice of crust could explain the lateral variations in seismicity and the connection of the Eastern Central Range with deeper structure observed in the tomographic images. Finally, a wide slice of exhumed crust could

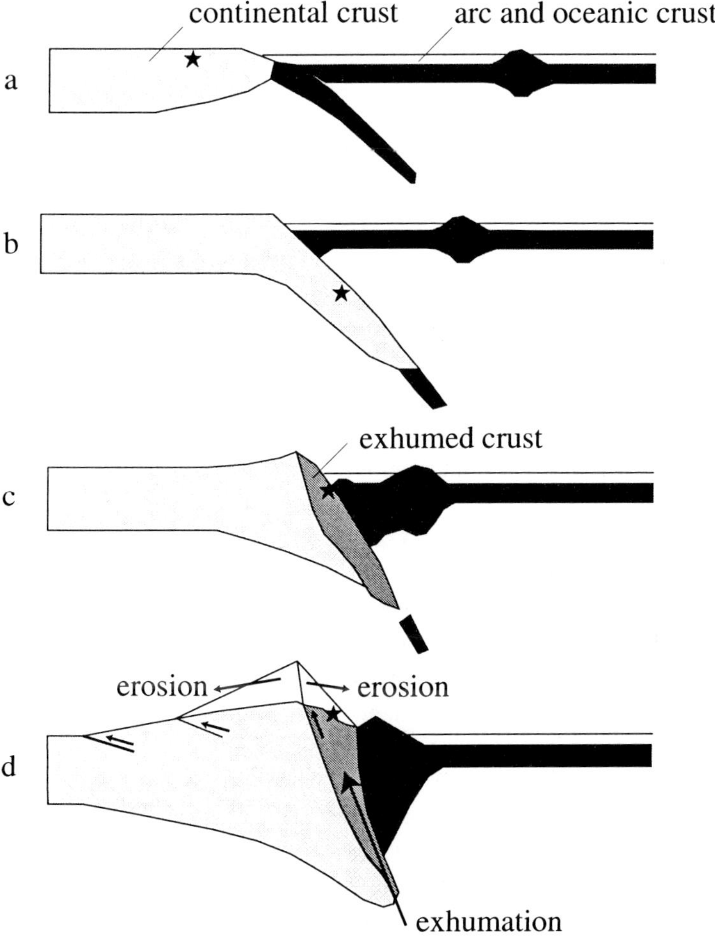

Figure 10. Schematic diagrams summarizing the model of continental subduction followed by crustal exhumation. a) Subduction of oceanic crust. b) Subduction of continental margin and failure along the leading edge of subducting crust. c) Exhumation initially due to buoyancy. (d) Acceleration of exhumation due to rapid erosion, and deformation of both the exhuming and overlying crusts. The dark arrows indicate buoyancy forces. Stars are markers corresponding to 15 km depth at stage b. Thin horizontal line shows approximate location of sea level for reference.

explain the extended width of the greenschist-facies metamorphism in the Eastern Central Range.

6. Thermal Modelling as a Quantitative Test

The fact that an exhumation model can provide general explanations for a variety of observations is all well and good, but admittedly appears to be somewhat ad hoc at this level of detail. In order to be more quantitative, we derived a thermal history using a simple model and reasonable boundary conditions to test whether this model makes physical sense within the context of tectonics in Taiwan, and to see if crustal subduction and exhumation can explain any observations better than previous models.

We employ a finite-difference method to compute a two-dimensional thermal model using a method initially developed by Minear and Toksöz (1970). The computational scheme consists of translating temperatures following the motion field and then allowing thermal diffusion over the time interval until the next movement. For our two-dimensional modeling we chose a representative cross section of Taiwan perpendicular to the strike of the general patterns of surface geology, heat flow, and gravity. The calculation of the thermal structures follows the scenario discussed above—crustal exhumation subsequent to the subduction of continental margin in Taiwan—taking as a guide the simulations of crustal exhumation conducted by Chemenda *et al.* (1995; 1996).

6.1 Boundary Conditions and Parameters

The boundary conditions and parameters for the thermal modeling are taken either from existing evidence or from reasonable assumptions. In particular, the temperatures at the surface and at 100 km are initially set at 0°C and 1000°C, respectively. This implies an initial heat flow of about 95 mW/m^2, which agrees with that observed near the Coastal Plains, far from the deformation front. Thermal conductivity is taken to be 4 W/MK, and shear strain heating is not considered. The subduction progresses at a rate of 7 cm/a in accordance with the convergence rate between the Eurasian and Philippine Sea plates (Seno, 1977). The exhumation of a portion of subducted crust within a region 30 km long and 20 km thick takes place during the last 1 m.y., with an exhumation rate of 3 cm/a as suggested by the results of fission-track, K-Ar and $^{40}Ar/^{39}Ar$ dating combined with petrographic textures (Tsao *et al.*, 1992; Tsao, 1996) and recent leveling surveys (Liu, 1995), which indicate that uplift of the Eastern Central Range accelerated during the last 1 m.y.

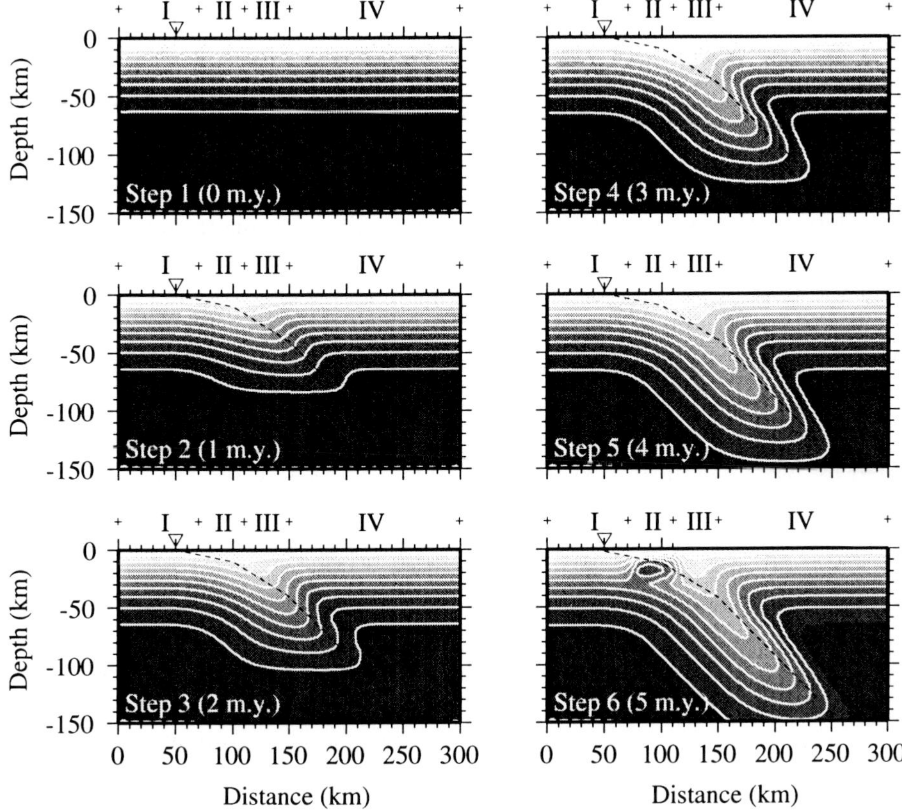

Figure 11. Six thermal regimes calculated for the model of continental subduction and crustal exhumation. Triangles mark the beginning of subduction on the surface, and the upper boundary of the subducted Eurasian continental shelf is indicated by a solid black line that cuts across the isotherms. Vertical markers at the top of each section indicate the distance sections discussed in the text. Subduction of crust continues from the first step until step 6, when exhumation takes place.

The calculations proceed over 6 stages, beginning at 5 Ma and separated by times steps of 1 m.y. (Fig. 11). To facilitate discussion of the results, we divide the distance along the profile into four sections: (I) west of Central Range, 0-70 km; (II) Central Range, 70-110 km; (III) east of the Central Range, 110-150 km; and (IV) the far east, 150-200 km. Subduction initiates at 50 km distance along the profile, as indicated by the triangles in Fig. 11.

6.2 Results of Geothermal Evolution

Generally, the results of the modeling show that a thermal depression is produced by subduction in stages 1 to 5, while a temperature increase is caused by crustal exhumation at the last stage. At the end of 5 m.y. of continental subduction and exhumation, the thermal structures change significantly from west to east across the profile. The thermal gradients in sections I and IV are not disturbed by the effects of continental subduction and exhumation, and maintain a value of about 20°C/km. In section II the gradient in the upper crust increases significantly as a result of crustal exhumation. The thermal gradient near the surface is more than 50°C/km, and thus temperatures in the upper crust increase rapidly with depth to a few hundred degrees. The gradient in section III decreases due to the subduction of cold continental crust. Temperatures are depressed along the top of the subducted slab, and the gradient is less than 10°C/km.

6.3 Relation Between Thermal Gradients and Seismicity

If earthquakes in the quartz-dominant lithologies of the crust are approximately limited by the 350°C isotherm (e.g., Chen and Molnar, 1983), we can compare the observed pattern of seismicity to our estimate of the brittle-ductile transition (Fig. 5 and 12). In section I, the normal thermal gradient of about 20°C/km has not been disturbed by the processes assumed in the model. This result generally is consistent with earthquake depths in this section being limited to <25 km. In section II, the higher temperatures within the crust outline the aseismic zone observed in the Central Range. Because of the high thermal gradient near the surface, there are no earthquakes at depths >10 km. In section III, there is a diffuse distribution of earthquakes from the surface to a depth of ~50 km. The occurrence of those earthquakes can be attributed to a low thermal gradient caused by the subduction of cold crust into the mantle. Additionally, these earthquakes are delimited by a boundary dipping east at about 40°, which mimics the isotherms we calculated. In section IV, earthquakes are again restricted to depths <25 km. The pattern of seismicity is similar to that in Section I, because neither of those geothermal gradients were disturbed by continent subduction and crustal exhumation. The geothermal model derived here thus is consistent with the seismicity pattern.

6.4 Comparison to Observed Heat Flow

The range of heat flow predicted by our model (Fig. 12) is between 100 and 250 mW/m^2, with the highest values in the Eastern Central Range and the lower values in the Coastal Plains and the Coastal Range. Generally, this pattern is consistent with the heat flow observed in Taiwan (Lee and Cheng, 1986). A comparison of the contours in Fig. 9 with the calculated heat flow in Fig. 12 suggests that the gradient in heat flow to the west is significantly less than we predict, but these contours are constrained by few observations (the circles in Fig. 9) and the actual gradient is not well known. These contours, and the ones modeled in other investigations, are significantly smoothed representations of the actual observations.

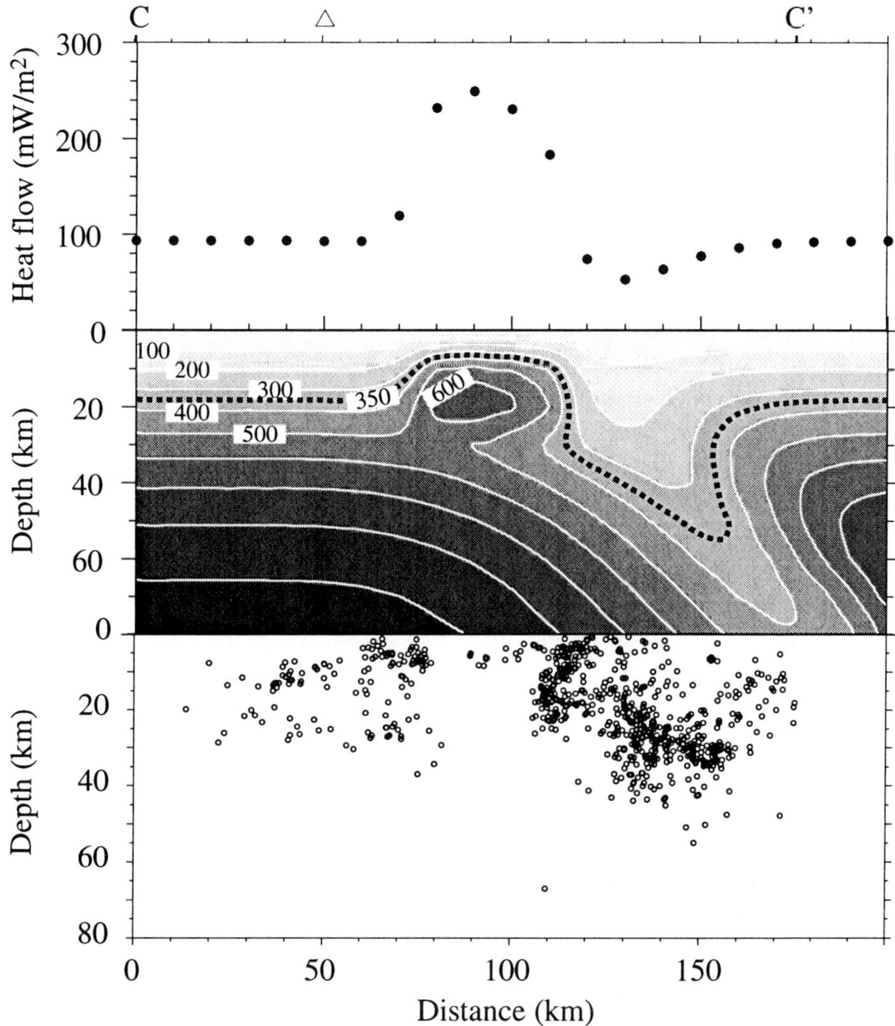

Figure 12. (Top) Heat flow predicted by the crustal exhumation model. (Middle) The final temperature profile at step 6 with the 350°C isotherm indicated by a dashed line. (Bottom) Seismicity projected on C-C' profile of fig. 5.

6.5 Discussion

Constrained mostly by surface and long-term observations related primarily to topography and heat flow, each of the geodynamic models proposed for the evolution of the Central Range appear to be equally likely. Without evidence showing that the rocks at the surface once resided at depths greater than about 15 km, the difference between the supra-decollement models and the exhumation model is one of stacking order: does Taiwan evolve in a "first-in, first-out", or a "first-in, last-out" fashion?

When evidence of subsurface structure and the nonstationary nature of uplift are considered, however, the model of continental subduction and crustal exhumation appears more likely. As discussed above, a variety of observations, and simple quantitative modeling of these observations, may be marshaled in support of exhumation. However, perhaps the most significant piece of evidence comes from the tomographic imaging of Lin *et al.* (1998) in which the structure that makes up the Eastern Central Range appears to continue to depths of several tens of kilometers. Thus, while it might be possible to adjust the variables of supra-decollement models to allow for accelerated exhumation of extended metamorphic terrains, the tomographic evidence implies that some more deep-seated structure is involved.

We note that the crustal exhumation model may pertain only to the evolution of the Central Range; we do not use this model to explain the structure of the remainder of the island. Indeed, the wedge model may still provide the best explanation of processes in the Western Foothills, and thus a global model for the evolution of Taiwan might be some hybrid of the two. Nevertheless, the available evidence strongly supports the idea that the Central Range is part of an ongoing process of subduction and exhumation of continental crust, and provides the best example known so far of this process occurring at present.

ACKNOWLEDGMENTS

We would like to thank Geoff Abers, Jean Crespi, Bradley Hacker, and Craig Nicholson for their insightful reviews.

REFERENCES

Barr, T.D. and Dahlen, F.A. (1989) Brittle frictional mountain building, 2, Thermal structure and heat budget, *Journal of Geophysical Research* **94**, 3923–3947.

Barr, T.D., Dahlen, F.A., and McPhail, D.C. (1991) Brittle frictional mountain building, 3, Low-grade metamorphism, *Journal of Geophysical Research* **96**, 10319–10338.

Chemenda, A.I. (1993) Subduction of lithosphere and back-arc dynamics: insights from physical modeling, *Journal of Geophysical Research* **98**, 16167–16185.

Chemenda, A.I., Mattauer, M., and Bokun, A.N. (1996) Continental subduction and a mechanism for exhumation of high-pressure metamorphic rocks: new modeling and field data from Oman, *Earth and Planetary Science Letters* **143**, 173–182.

Chemenda, A.I., Mattauer, M., Malavieille, J., and Bokun, A.N. (1995) A mechanism for syn-collisional rock exhumation and associated normal faulting: Results from physical modelling, *Earth and Planetary Science Letters* **132**, 225–232.

Chen, W.P. and Molnar, P. (1983) The depth distribution of intracontinental and intraplate earthquakes and its implications for the thermal and mechanical properties of the lithosphere, *Journal of Geophysical Research* **88**, 4183–4214.

Cheng, S.N. (1995) The study of stress distribution in and around Taiwan. Chungli, p. 215. National Central University, Taiwan.

Chi, W.R., Namson, J., and Suppe, J. (1981) Stratigraphic record of plate interactions in the Coastal Range of eastern Taiwan, *Memoir of the Geological Society of China* **4**, 155–194.

Chopin, C. (1984) Coesite and pure pyrope in high-grade blueschists of the western Alps: a first record and some consequences, *Contributions to Mineralogy and Petrology* **86**, 107–118.

Crespi, J., Chan, Y.C., and Swaim, M. (1996) Synorogenic extension and exhumation of the Taiwan hinterland, *Geology* **24**, 247–250.

Dahlen, F.A. and Barr, T.D. (1989) Brittle friction mountain building, 1, Deformation and mechanical energy budget, *Journal of Geophysical Research* **94**, 3906–3922.

Dewey, J.F., Ryan, P.D., and Andersen, T.B. (1993) Orogenic uplift and collapse, crustal thickness, fabrics and metamorphic phase changes; the role of eclogites, *Geological Society Special Publications* **76**, 325–343.

Ellwood, A., Wang, C.Y., Teng, L.S., and Yen, H.Y. (1996) Gravimetric examination of thin-skinned detachment vs. basement-involved models for the Taiwan Orogen, *Journal of the Geological Society of China* **39**, 209–221.

Ernst, W.G. (1983) Mineral parageneses in metamorphic rocks exposed along Tailuko Gorge, Central Mountain Range, Taiwan, *Journal of Metamorphic Geology* **1**, 305–329.

Ernst, W.G. (1984) The nature of the biotite isograd in the Tailuko Gorge transect, Central Mountain Range, Taiwan, *Proceedings of the Geological Society of China* **27**, 25–40.

Ernst, W.G. and Harnish, D. (1983) Mineralogy of some Tananao greenschists facies rocks, Mu-Kua Chi area, eastern Taiwan, *Proceedings of the Geological Society of China* **26**, 99–110.

Ernst, W.G. and Jahn, B.M. (1987) Crustal accretion and metamorphism in Taiwan, a post-Palaeozoic mobile belt, *Philosophical Transactions of the Royal Society of London* **A321**, 129–161.

Ho, C.S. (1988) *An introduction to the geology of Taiwan (second Edition): explanatory text of the geologic map of Taiwan*, Ministry of Economic Affairs, Taipei.

Hwang, W.T. and Wang, C.Y. (1993) Sequential thrusting model for mountain building: constraints from geology and heat flow of Taiwan, *Journal of Geophysical Research* **98**, 9963–9973.

Jahn, B.M., Martineau, F., and Cornichet, J. (1984) Chronological significance of Sr isotopic compositions in the crystalline limestones of the Central Range, Taiwan, *Memoir of the Geological Society of China* **6**, 295–301.

Lan, C.Y., Lee, T., and Wang, C. (1990) The Rb-Sr isotopic record in Taiwan gneisses and its tectonic implication, *Tectonophysics* **183**, 129–143.

Lee, C.R. and Cheng, W.T. (1986) Preliminary heat flow measurements in Taiwan, in M.K. Horn (ed.), *Transactions of the Fourth Circum-Pacific Energy and Mineral Resources Conference*, Circum-Pacific Council for Energy and Mineral Resources, Tulsa.

Lee, C.S. (1984) Stratigraphic study of the Tananao Group in the region north of the Liwuchi, Taiwan, *Central Geological Survey of Taiwan Special Publication* **3**, 1–10.

Lee, C.W., Wang, C.Y., Yen, T.P., and Lo, C.H. (1982) Polymetamorphism in some gneiss bodies, Hoping-Chipan area, Hualien, eastern Taiwan, *Acta Geologica Taiwanica* **21**, 122–139.

Lee, P.Y. (1977) Rate of the Early Pleistocene uplift in Taiwan, *Memoir of the Geological Society of China* **2**, 71–76.

Lee, Y.H. (1995) A study of the stress field system evolution of the Taiwan mountain belt, central cross-island highway and surrounding regions, *Bulletin of the Central Geological Survey* **10**, 51–89.

Li, Y.H. (1975) Denudation of Taiwan Island since the Pliocene Epoch, *Geology* **4**, 105–107.

Lin, C.H. and Roecker, S.W. (1993) Deep earthquakes beneath central Taiwan; mantle shearing in an arc-continent collision, *Tectonics* **12**, 745–755.

Lin, C.H., Yeh, Y.H., and Roecker, S.W. (1989) Seismic velocity structures beneath the Sanyi-Fengyuan area, central Taiwan and their tectonic implications, *Proceedings of the Geological Society of China* **32**, 101–120.

Lin, C.H., Yeh, Y.H., Yen, H.Y., Chen, K.C., Huang, B.S., Roecker, S.W., and Chiu, J.M. (1998) Three-dimensional elastic velocity structure of the Hualien region of Taiwan: evidence of active crustal exhumation, *Tectonics* **17**, 89–103.

Liou, J.G. (1981a) Petrology of metamorphosed oceanic rocks in the Central Range of Taiwan, *Memoir of the Geological Society of China* **4**, 291–342.

Liou, J.G. (1981b) Recent high CO_2 activity and Cenozoic progressive metamorphism in Taiwan, *Memoir of the Geological Society of China* **4**, 551–582.

Liu, C.H. (1995) Geodetic monitoring of mountain building in Taiwan, *Transactions of the American Geophysical Union, Eos* **76**, 636.

Liu, T.K. (1982) Tectonic implication of fission track ages from the Central Range, Taiwan, *Proceedings of the Geological Society of China* **25**, 22–37.

Ma, K.F., Wang., J.H., and Zhao, D. (1996) Three-dimensional seismic velocity structure of the crust and uppermost mantle beneath Taiwan, *Journal of Physics of the Earth* **44**, 85–105.

Minear, J.W. and Toksöz, M.N. (1970) Thermal regime of a downgoing slab and new global tectonics, *Journal of Geophysical Research* **75**, 1397–1419.

Molnar, P. and Gray, D. (1979) Subduction of continental lithosphere: some constraints and uncertainties, *Geology* **7**, 58–62.

Pelletier, B. and Stephan, J.F. (1986) Middle Miocene obduction and late Miocene beginning of collision registered in the Hengchun Peninsula: geodynamic implications for the evolution of Taiwan, *Memoir of the Geological Society of China* **7**, 301–324.

Peng, T.H., Li, Y.H., and Wu, F.T. (1977) Tectonic uplift of the Taiwan island since the early Holocene, *Memoir of the Geological Society of China* **2**, 57–69.

Platt, J., Legget, J., Young, J., Razza, H., and Alam, S. (1985) Large-scale sediment underplating in the Makran accretionary prism, southwest Pakistan, *Geology* **13**, 507–511.

Platt, J.P. (1987) The uplift of high-pressure low-temperature metamorphic rocks, *Philosophical Transactions of the Royal Society of London* **A321**, 87–103.

Rau, R.J. and Wu, F.T. (1995) Tomographic imaging of lithospheric structures under Taiwan, *Earth and Planetary Science Letters* **133**, 517–532.

Roecker, S.W. (1982) Velocity structure of the Pamir-Hindu Kush region: possible evidence of subducted crust, *Journal of Geophysical Research* **87**, 945–959.

Roecker, S.W., Yeh, Y.H., and Tsai, Y.B. (1987) Three-dimensional P and S wave velocity structures beneath Taiwan: deep structure beneath an arc-continent collision, *Journal of Geophysical Research* **92**, 10547–10570.

Seno, T. (1977) The instantaneous rotation vector of the Philippine Sea Plate relative to the Eurasian Plate, *Tectonophysics* **42**, 209–226.

Smith, D.C. (1984) Coesite in clinopyroxene in the Caledonides and its implications for geodynamics, *Nature* **310**, 641–644.

Suppe, J. (1981) Mechanics of mountain building in Taiwan, *Memoir of the Geological Society of China* **4**, 67–89.

Tsai, Y.B., Ten, T.L., Chiu, J.M., and Liu, H.L. (1977) Tectonics implications of the seismicity in the Taiwan region, *Memoir of the Geological Society of China* **2**, 13–41.

Tsao, S.H. (1996) The geological significance of illite crystallinity, zircon fission-track ages, and K-Ar ages of metasedimentary rocks of the Central Range of Taiwan. Taipei, p. 272. National Taiwan University, Taiwan.

Tsao, S.H., Li, T.C., Tien, J.L., Chen, C.H., Liu, T.K., and Chen, C.H. (1992) Illite crystallinity and fission-track ages along east central cross-island highway of Taiwan, *Acta Geologica Taiwanica* **30**, 45–64.

Wang, X., Liou, J.G., and Mao, H.K. (1989) Coesite-bearing eclogite from the Dabie Mountains in central China, *Geology* **17**, 1085–1088.

Warneke, L.F. and Ernst, W.G. (1983) Progressive Cenozoic metamorphism of rocks cropping out along the southern East-West Cross-Island Highway, Taiwan, *Memoir of the Geological Society of China* **6**, 21–48.

Yeh, Y.H. and Chen, C.H. (1994) Three-dimensional P and S wave velocity structures beneath Taiwan from regional three-component digital data, *Transactions of the American Geophysical Union, Eos* **75**, 243.

Yeh, Y.H., Lin, C.H., and Roecker, S.W. (1989) A study of crustal structures beneath northeastern Taiwan: possible evidence of the western extension of Okinawa Trough, *Proceedings of the Geological Society of China* **32**, 139–156.

Yen, H.Y., Yeh, Y.H., and Wu, F.T. (1998) Two-dimensional crustal structures of Taiwan from gravity data, *Tectonics* **17**, 104–111.

Yen, T.P. (1953) On the occurrence of the late Paleozoic fossils in the metamorphic complex of Taiwan, *Bulletin of the Geological Survey of Taiwan* **4**, 23–26.

Yen, T.P. (1954) The gneisses of Taiwan, *Bulletin of the Geological Survey of Taiwan* **5**, 1–100.

Yen, T.P., Sheng, C.C., and Keng, W.P. (1951) The discovery of fusuline limestone in the metamorphic complex of Taiwan, *Bulletin of the Geological Survey of Taiwan* **3**, 23–25.

Yu, S.B. and Chen, H.Y. (1996) Spatial variations of crustal strain in the Taiwan area, *The 6th Taiwan symposium on Geophysics, Chiayi, Taiwan*, 659–668.

Yui, T.F. (1987) Carbon isotope composition of marble: a possible criterion for geochronologic/stratigraphic correlation in the Tananao Group, Taiwan, *Memoir of the Geological Society of China* **8**, 123–133.

Chapter 2

Melting of Crustal Rocks During Continental Collision and Subduction

Alberto E. Patiño Douce and T.C. McCarthy
Department of Geology, University of Georgia, Athens GA 30602, USA, alpatino@uga.cc.uga.edu and burger@arches.uga.edu

Abstract: The continental crust can partially melt to generate silicic magmas. Compositional diversity among these magmas is determined by the pressure of melting, by the availability of free aqueous fluids, and by the composition of the protolith. Because crustal rocks are subject to extreme pressure and temperature conditions during continental collisions, collisional orogens potentially are environments for magma generation. This chapter discusses the nature of the melts and solid residues likely to be formed in response to continental collision.

In the absence of H_2O-rich fluids, melting of metamorphic rocks is triggered by the breakdown of hydrous minerals. These incongruent dehydration-melting reactions give rise to H_2O-bearing melts and to anhydrous (or less hydrous) solid residues. With the exception of muscovite-rich metapelitic rocks at pressures of less than ~1 GPa, dehydration melting of common crustal rocks requires temperatures in excess of 850°C. Heating of crustal rocks to such temperatures in collisional orogens takes place only if the lithospheric mantle becomes detached from the crust, if the crust is invaded by mantle-derived magmas, or if subduction transports rocks to the upper mantle. Melts generated by dehydration melting of a wide range of quartzofeldspathic rocks at temperatures of 850-1100°C are granitic, and become less ferromagnesian and richer in total alkalis and alumina with increasing pressure. The solid residues are more variable and depend on source composition, but are generally granulitic at P ≤1 GPa and eclogitic at P ≥2 GPa, with a transitional interval of garnet granulite. In thickened continental crust underlain by a lid of lithospheric mantle, and in slices of continental crust that are buried by subduction and exhumed rapidly, temperatures are unlikely to exceed 800-850°C. Under these conditions only muscovite-rich metapelitic schists can undergo dehydration melting, yielding peraluminous leucogranites. The rather flat dP/dT slope of the muscovite dehydration-melting reaction means that melting most likely takes place at relatively shallow depth (<0.8 GPa), during decompression caused by tectonic exhumation of deep-seated rocks. Deeper melting of any quartzofeldspathic rock in these relatively "cold" environments requires influx of H_2O-rich fluids. Melts formed in this manner

B. R. Hacker and J. G. Liou (eds.),
When Continents Collide: Geodynamics and Geochemistry of Ultrahigh-Pressure Rocks, 27–55.
© 1998 *Kluwer Academic Publishers. Printed in the Netherlands.*

are more sodic than the hotter melts formed by dehydration melting, and the residues are rich in micas.

1. INTRODUCTION

Magma generation during continental collisions requires the convergence of appropriate source rocks and thermal conditions. Quartzofeldspathic rocks of sedimentary and igneous origin are the most abundant rocks in the continental crust, and can partially melt to yield silicic magmas. Continental collision belts thus are fertile environments for magma generation on a large scale. Whether this potential is realized depends on the thermal structure of the collisional orogen. Processes such as thermal blanketing resulting from tectonic thickening, delamination of the subcrustal lithosphere, and extensional collapse of thickened crust favor the attainment of temperatures hot enough to cause widespread crustal melting. On the other hand, subduction of cold supracrustal rocks may refrigerate the collisional orogen and inhibit melting. Because all of these processes are likely to operate to varying extents during continental collisions, the timing, location and extent of crustal anatexis may be different in each collisional orogen.

This chapter links the melting behavior of crustal rocks (determined experimentally) with the thermal structure of collisional orogens (determined from field observations and numerical models). Discussion focuses on three rocks that are representative of the most abundant material in the continental crust: i) a muscovite schist, which corresponds to mature clastic sediments (pelites) typical of passive margins; ii) a model metagreywacke (biotite + plagioclase + quartz), representative of immature clastic sediments abundant in active continental margins; and iii) a calc-alkaline tonalite, representative of the average crustal igneous composition. We discuss the melting behavior of these three rock types at pressures between 0.4 and 3.0 GPa. Source composition, pressure, and availability of externally derived H_2O all have strong effects on melting temperatures and on the compositions of melts and solid residues. Thus, besides the issue of whether crustal anatexis takes place during continental collisions, we also discuss how the outcome of crustal anatexis is likely to vary as a function of depth.

2. THERMAL EVOLUTION DURING CONTINENTAL COLLISION AND SUBDUCTION

Perhaps the most dramatic effect of continental collisions is that they routinely transport large volumes of supracrustal rocks to depths of at least 60 km, and locally as much as 120 km. This transport is accomplished by a combination of thrusting, homogeneous shortening, and subduction. Exhumation mechanisms entail both tectonic and erosional processes. In addition, the mantle lithosphere that underlies a zone of continental collision can be either thickened (e.g., by continued subduction) or thinned (by gravitational instability or thermal erosion), strongly affecting the heat flux into the base of the orogen. A wide range of thermal structures of thickened continental crust is therefore possible, depending on how supracrustal rocks are buried and exhumed, and on the behavior of the underlying mantle lithosphere. In order to explore whether, where, and how metamorphic rocks of supracrustal origin can partially melt during continental convergence, we start with an overview of the possible thermal regimes in collisional orogens.

Model P-T paths for rocks buried at the base of the continental crust after doubling of the crustal thickness (~70 km, Fig. 1, paths A and B) show a range in maximum predicted temperatures of almost 200°C. This range arises from different assumed distributions of heat-producing elements in the crust, and from different assumed thickness of the subcrustal lithosphere. In addition, these thermal models do not account for two factors that may cause the actual P-T evolution of collisional belts to differ from the model results. First, both models assume that erosion exhumes the rocks relatively slowly, allowing heating and then cooling. Following a continental collision, however, rocks may undergo much faster, near-adiabatic exhumation during extensional collapse of the orogen (*e.g.*, C in Fig. 1). Second, the lithospheric mantle underlying the collisional orogen might detach and sink (e.g., Houseman *et al.*, 1981), raising the temperature of the base of the crust to values characteristic of asthenospheric mantle (>1200°C). A wide range of thermal regimes thus appears possible for the crustal section of collisional orogens. Crustal rocks may undergo yet more extreme P-T histories if subduction transports them to greater depths.

The best known occurrences of supracrustal rocks buried to depths of ~100 km are in the western Alps (e.g., Chopin *et al.*, 1991), the Dabie-Su-Lu region of eastern China (Liou *et al.*, 1996), the Kokchetav Massif of northern Kazakhstan (Shatsky *et al.*, 1995), and the Western Gneiss region of Norway (Wain, 1997). These rocks, characterized by the presence of coesite ± diamond, define coesite-eclogite facies metamorphism (Fig. 1), at ultrahigh pressure (2.5–4.0 GPa) and relatively low temperature (600-870°C). In order for metamorphism to take place at such conditions, supracrustal rocks must be buried rapidly by subduction. Moreover, in order for mineralogic indicators of coesite-eclogite facies metamorphism to survive exhumation, it is necessary that the lithosphere be cooled by continued subduction while exhumation takes place (Hacker and Peacock, 1994). Simultaneous subduction and exhumation can give rise to decompression paths that cool steadily, such as D in Fig. 1. In the absence of the

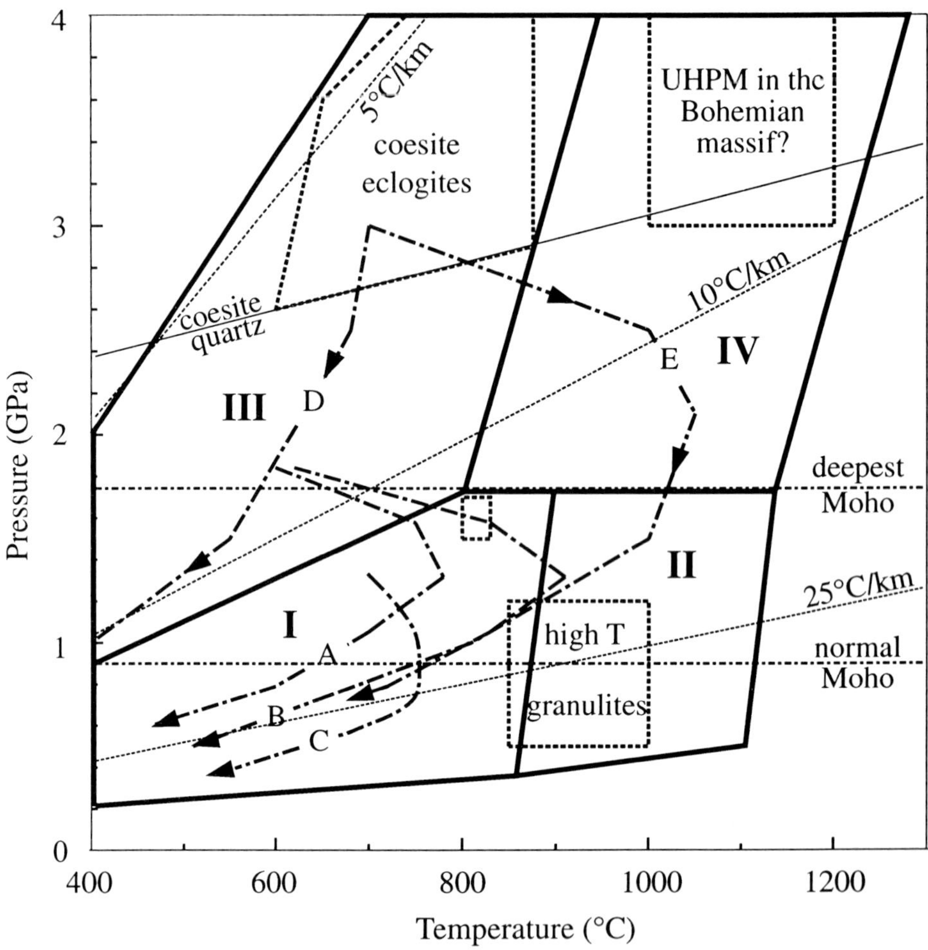

Figure 1. Constraints on the thermal structure of continental collision belts. A and B are model P-T paths for continental collisions followed by erosion of thickened crust (England and Thompson, 1986; Patiño Douce *et al.*, 1990). C is a decompression path for Himalayan rocks, inferred by Butler *et al.* (1997) from thermobarometry. D and E are possible P-T paths during continental subduction, after Hacker and Peacock (1994). "Coesite eclogites" corresponds to conditions for UHP rocks from Dora Maira (Chopin *et al.*, 1991), Northern Kazakhstan (Shatsky *et al.*, 1995), and Eastern China (Liou *et al.*, 1996). UHPM in the Bohemian massif was inferred by Becker & Altherr (1992). "High-temperature granulites" shows P-T conditions of high-grade terranes from Antarctica (Harley, 1985), India (Sengupta *et al.*, 1991), Brazil (Ackermand *et al.*, 1987), and Africa (Harris and Holland, 1984). P-T ranges in four distinct tectonic environments are: I) continental collision; II) continental collision followed by detachment of the subcontinental lithospheric mantle; III) "cold" continental subduction; and IV) "hot" continental subduction.

refrigerating effect of subduction, rocks that were initially buried to coesite-eclogite facies conditions would heat on their return to the surface, such as path E in Fig. 1. Mineralogical indicators of coesite-eclogite facies conditions would not be preserved along these hot decompression P-T paths, which implies that recognized UHP coesite-eclogite facies terranes did not follow such paths. However, there may be other instances in which supracrustal rocks buried to depths of 100 km or more have heated significantly during, or prior to, decompression. An example of this may be present in the Bohemian Massif (Fig. 1), where there is evidence, from K-feldspar exsolution lamellae in clinopyroxene, of calc-silicate marbles metamorphosed at pressures of 3.0-4.0 GPa and temperatures greater than 1100°C (Becker and Altherr, 1992). Preservation of metamorphic rocks from such extreme conditions is likely to be rare, however, because most supracrustal rocks will undergo extensive melting (marbles being a notable exception).

In order to discuss anatexis during continental collisions, we distinguish four P-T fields that represent different thermal regimes in collisional orogens (Fig. 1). The boundaries among the different fields are necessarily arbitrary, but each of the fields corresponds to a specific tectonic environment. The diagram thus provides a useful framework within which to discuss the melting behavior of supracrustal rocks during continental convergence.

Field I corresponds to tectonic thickening of continental crust overlying lithospheric mantle that remains attached to the base of the crust. This is the classical environment of Barrovian metamorphism. Fast adiabatic decompression by tectonic collapse of an orogen can give rise to Buchan overprinting of Barrovian assemblages. Buchan P-T conditions are thus used to define the low pressure boundary of field I. If the lithospheric mantle detaches from the bottom of the collisional orogen, temperatures in the thickened crust will increase (field II in Fig. 1). High-temperature granulites such as those from Antarctica (Harley, 1985), India (Sengupta *et al.*, 1991), Brazil (Ackermand *et al.*, 1987), and Africa (Harris and Holland, 1984) may have formed in thermal regimes such as this. In field III, crustal rocks are buried by subduction and are exhumed while subduction is still active. The "cold" conditions that ensue (e.g., Hacker and Peacock, 1994) are characterized by the formation of eclogites and coesite eclogites. If continental rocks are subducted to depths of 100 km or more, but subduction ceases before the rocks are exhumed, the rocks may heat and be subjected to ultrahigh pressures and temperatures such as those inferred for the Bohemian Massif (field IV in Fig. 1). In most cases, however, metamorphic evidence that rocks have traversed this field will be obliterated by large-scale anatexis.

3. REPRESENTATIVE SUPRACRUSTAL LITHOLOGIES

This chapter discusses the melting behavior of three protoliths representative of rocks that are abundant in the continental crust (Table 1). The muscovite schist

(MS) is a model for mature clastic sediments. Sample MS is a kyanite-zone metapelite from the High Himalayan Crystalline Sequence, with 43 wt% quartz, 28 % plagioclase, 22 % muscovite, 5 % garnet, and 2 % biotite. Melting of this rock was studied experimentally at pressures of 0.6–1.0 GPa, with and without added H_2O, by Patiño Douce and Harris (1998). The model metagreywacke (SBG) represents immature clastic sediments and is a mechanical mixture of 37 wt% natural biotite (Mg# 55), 27 % plagioclase (An_{38}), 34 % quartz and 2 % ilmenite. Melting of this assemblage at pressures of 0.3–1.5 GPa was studied experimentally by Patiño Douce and Beard (1995) without added H_2O and by Patiño Douce (1996) with added H_2O. The calc-alkaline tonalite (CAT) models the average crustal igneous composition. It is an intrusive rock from Lee Vining Canyon in the Sierra Nevada of California that contains 45 wt% plagioclase, 20 % quartz, 13 % biotite, 13 % hornblende, 7 % K-feldspar, and 2 % oxides. This rock was studied experimentally at 0.4 and 0.8 GPa by Patiño Douce (1997).

Table 1. Bulk chemical compositions of starting materials.

	SiO_2	Al_2O_3	TiO_2	$FeO^{*\dagger}$	MgO	MnO	CaO	Na_2O	K_2O	Total
MS	75.28	14.29	0.36	2.40	0.66	0.13	0.94	2.77	2.40	99.23
SBG	63.40	12.30	2.50	7.80	4.70	0.10	2.10	2.00	3.60	98.50
CAT	60.76	16.87	0.89	5.85	2.65	0.11	5.33	3.83	2.54	98.83

† Total Fe as FeO.

We performed additional experiments on all three starting materials at pressures up to 3 GPa. Most of these new experiments were performed without added H_2O (dehydration melting), but a few experiments contained small amounts of added H_2O. The experimental and analytical procedures used in the new experiments were identical to those used in previous studies (Patiño Douce, 1995; Patiño Douce and Beard, 1995; Patiño Douce and Harris, 1998). The phase assemblages in all the experimental products, both new and previously published, are given in Table 2.

Because each of these starting materials was studied over a wide P-T range, focusing discussion on them makes it possible to have a continuous view of the changing melting behaviors of common quartzofeldspathic rocks, from the shallow crust to the ultradeep environment of continental subduction. Other studies of similar starting materials are available in the literature, some of which are used to complement the discussion in the following sections.

4. ROLE OF H_2O IN CRUSTAL ANATEXIS

Dehydration melting is a limiting case in which all the H_2O required to lower the solidus of a rock is provided by incongruent breakdown of hydrous minerals (e.g., Thompson and Algor, 1977; Thompson, 1982). The other limiting case for crustal melting is that in which there is sufficient H_2O to saturate the melt (e.g., Tuttle and Bowen, 1958; Wyllie, 1977; Puziewicz and Johannes, 1990). Because a pervasive free fluid phase is unlikely to exist at depths beyond the uppermost few km of crust (Yardley and Valley, 1997), and because the solubility of H_2O in silicate melts increases with pressure (e.g., Burnham, 1979a), dehydration melting is the process that best explains the generation of large volumes of magmas in the continental crust (Clemens and Vielzeuf, 1987). Infiltration of aqueous fluids into high-grade rocks may be locally important, for example, when cold and wet low-grade rocks are rapidly buried under higher grade rocks. We call this situation, in which melting takes place with some added H_2O but not enough for saturation, H_2O-fluxed melting. In the presence of even small amounts of added H_2O ($\leq$1 wt%), the solidus temperature of the rock decreases, and the melt compositions change significantly from those generated by dehydration melting (Conrad *et al.*, 1988; Patiño Douce, 1996; Patiño Douce and Harris, 1998).

Although we discuss in greater detail dehydration melting of crustal rocks, the effects of H_2O infiltration on crustal anatexis are also addressed, by means of experiments at selected P-T-X_{H_2O} conditions. Water-saturated melting is not discussed here; see Wyllie (1977), Stern and Wyllie (1981) and Huang and Wyllie (1981) for discussion of the H_2O-saturated melting behavior of crustal rocks at pressures of up to 3.5 GPa.

5. BEHAVIOR OF BIOTITE DURING CRUSTAL ANATEXIS

Biotite is the most abundant H_2O reservoir in medium- to high-grade crustal rocks. It is therefore important to understand how biotite compositions change as a function of pressure, temperature, and coexisting mineral assemblage. Biotites in the muscovite schist and in the metagreywacke depart from stoichiometric phlogopite-annite solid solutions, toward more aluminous compositions (Fig 2a). Al enrichment is stronger in the muscovite schist than in the metagreywacke, because only the former rock contains an aluminumsilicate. No single exchange can account for the total Al content in biotite. Rather, at least two aluminous components must be considered, a tschermak exchange (eastonite-siderophyllite) and a dioctahedral (muscovite) exchange (Fig. 2a, see also Patiño Douce *et al.*, 1993). The dioctahedral component becomes increasingly important with increasing pressure, as shown by the contents of octahedral Al (Fig. 2a) and octahedral vacancies (Fig. 2b). The strong effect of pressure on the dioctahedral component of biotite in assemblages containing aluminumsilicate and garnet was demonstrated by Patiño Douce *et al.* (1993). Remarkably, however, this behavior is also observed in the metagreywacke, which lacks aluminumsilicate and

Table 2. Phase assemblage in partial melting experiments.

P GPa	T °C	H_2O wt%	Phase assemblage								
MS, muscovite schist											
0.6	700	4	Qtz	Pl	Bt	Ms	Grt				
0.6	750	0	Qtz	Pl	Bt	Ms	Grt	Melt			
0.6	750	1	Qtz	Pl	Bt	Ms	Grt	Melt			
0.6	750	2	Qtz	Pl	Bt	Ms	Grt	Melt			
0.6	775	0	Qtz	Pl	Kfs	Bt	Ms	Grt	Als	Melt	
0.6	775	2	Qtz	Pl	Bt	Ms	Grt	Als	Rt	Melt	
0.6	800	0	Qtz	Pl	Kfs	Bt	Ms	Grt	Als	Melt	
0.6	820	0	Qtz	Pl	Kfs	Bt	Grt	Als	Melt		
0.6	850	0	Qtz	Pl	Kfs	Bt	Grt	Als	Melt		
0.6	900	0	Qtz	Pl	Kfs	Bt	Grt	Als	Melt		
0.8	800	0	Qtz	Pl	Kfs	Bt	Ms	Grt	Als	Melt	
1	700	4	Qtz	Pl	Ms	Grt	Rt	Melt			
1	750	1	Qtz	Pl	Bt	Ms	Grt	Melt			
1	750	2	Qtz	Pl	Bt	Ms	Grt	Als	Melt		
1	775	4	Qtz	Pl	Ms	Grt	Melt				
1	820	0	Qtz	Pl	Bt	Ms	Grt				
1	835	0	Qtz	Pl	Kfs	Bt	Ms	Grt	Als	Melt	
1	850	0	Qtz	Pl	Kfs	Bt	Ms	Grt	Als	Melt	
1	900	0	Qtz	Pl	Kfs	Bt	Grt	Als	Melt		
1.5	900	0	Qtz	Pl	Kfs	Bt	Ms	Grt	Als	Rt	Melt
1.5	950	0	Qtz	Pl	Kfs	Bt	Ms	Grt	Als	Rt	Melt
2.1	700	2	Qtz	Ms	Grt	Cpx	Rt	Melt			
2.1	940	0	Qtz	Kfs	Bt	Ms	Grt	Als	Rt	Melt	
2.1	960	0	Qtz	Kfs	Grt	Als	Rt	Melt			
2.1	975	0	Qtz	Kfs	Bt	Grt	Als	Rt	Melt		
2.1	1025	0	Qtz	Kfs	Bt	Grt	Als	Rt	Melt		
2.1	1060	0	Qtz	Pl	Grt	Als	Rt	Melt			
2.7	950	1	Qtz	Ms	Grt	Cpx	Als	Melt			
3	725	2	Qtz	Bt	Grt	Cpx					
3	975	0	Qtz	Ms	Grt	Cpx	Melt				
3	1010	0	Qtz	Kfs	Ms	Grt	Cpx	Als	Melt		
3	1075	0	Qtz	Kfs	Ms	Grt	Cpx	Als	Rt	Melt	
SBG, model graywacke											
0.3	840	0	Qtz	Pl	Bt	Ilm					
0.3	875	0	Qtz	Pl	Bt	Opx	Ilm	Melt			
0.3	900	0	Qtz	Pl	Bt	Opx	Ilm	Melt			
0.3	925	0	Qtz	Pl	Bt	Opx	Ilm	Melt			
0.5	840	0	Qtz	Pl	Bt	Ilm					
0.5	875	0	Qtz	Pl	Bt	Opx	Ilm	Melt			
0.5	900	0	Qtz	Pl	Bt	Opx	Ilm	Melt			
0.5	925	0	Qtz	Pl	Bt	Opx	Ilm	Melt			
0.5	950	0	Qtz	Pl	Opx	Ilm	Melt				
0.7	875	0	Qtz	Pl	Bt	Opx	Ilm	Melt			
0.7	900	0	Qtz	Pl	Bt	Opx	Ilm	Melt			
0.7	925	0	Qtz	Pl	Opx	Ilm	Melt				
0.7	925	1	Qtz	Pl	Bt	Opx	Ilm	Melt			

P GPa	T °C	H_2O wt%	Phase assemblage								
0.7	938	0	Qtz	Pl	Opx	Ilm	Rt	Melt			
0.7	950	0	Qtz	Pl	Opx	Ilm	Melt				
1	875	0	Qtz	Pl	Bt	Opx	Ilm				
1	900	0	Qtz	Pl	Bt	Opx	Ilm	Melt			
1	925	0	Qtz	Pl	Bt	Opx	Ilm	Melt			
1	925	1	Qtz	Pl	Bt	Grt	Opx	Ilm	Melt		
1	925	2	Opx	Ilm	Melt						
1	950	0	Qtz	Pl	Opx	Ilm	Rt	Melt			
1	975	0	Qtz	Pl	Opx	Ilm	Rt	Melt			
1	1000	0	Qtz	Pl	Opx	Ilm	Rt	Melt			
1.25	930	0	Qtz	Pl	Bt	Grt	Opx	Ilm	Rt	Melt	
1.25	960	0	Qtz	Pl	Kfs	Bt	Grt	Opx	Rt	Melt	
1.5	925	0	Qtz	Pl	Kfs	Bt	Grt	Opx	Ilm		
1.5	925	1	Qtz	Pl	Bt	Grt	Hbl	Rt	Melt		
1.5	925	6	Bt	Grt	Opx	Ilm	Rt	Melt			
1.5	925	8	Bt	Opx	Ilm	Melt					
1.5	950	0	Qtz	Pl	Kfs	Bt	Grt	Cpx	Opx	Rt	Melt
1.5	975	0	Qtz	Kfs	Grt	Cpx	Opx	Rt	Melt		
1.5	1000	0	Qtz	Kfs	Grt	Cpx	Opx	Rt	Melt		
2.1	940	0	Qtz	Kfs	Bt	Grt	Cpx	Rt	Melt		
2.1	960	0	Qtz	Kfs	Bt	Grt	Cpx	Rt	Melt		
2.1	975	0	Qtz	Kfs	Bt	Grt	Cpx	Rt	Melt		
2.1	1025	0	Qtz	Kfs	Bt	Grt	Cpx	Rt	Melt		
2.1	1060	0	Qtz	Grt	Cpx	Opx	Rt	Melt			
3	975	0	Qtz	Bt	Grt	Cpx	Rt	Melt			
3	1010	0	Qtz	Bt	Phn	Grt	Cpx	Rt	Melt		
3	1075	0	Qtz	Bt	Grt	Cpx	Rt	Melt			
CAT, tonalite											
0.4	950	0	Pl	Bt	Cpx	Opx	Ilm	Melt			
0.8	950	0	Qtz	Pl	Bt	Cpx	Opx	Rt	Melt		
1.5	900	0	Qtz	Pl	Kfs	Bt	Grt	Cpx	Rt	Melt	
1.5	950	0	Qtz	Pl	Kfs	Bt	Grt	Cpx	Rt	Melt	
2.1	700	2	Qtz	Phn	Cpx	Hbl	Ilm	Melt			
2.1	940	0	Qtz	Kfs	Grt	Cpx	Rt	Melt			
2.1	960	0	Qtz	Kfs	Grt	Cpx	Rt	Melt			
2.1	975	0	Qtz	Kfs	Grt	Cpx	Rt	Melt			
2.1	1025	0	Qtz	Kfs	Grt	Cpx	Rt	Melt			
2.1	1060	0	Pl	Grt	Cpx	Rt	Melt				
3	725	2	Qtz	Phn	Grt	Cpx	Amp				
3	975	0	Qtz	Kfs	Cym	Zo	Phn	Grt	Cpx	Rt	Melt
3	1010	0	Qtz	Kfs	Cym	Zo	Phn	Grt	Cpx	Rt	Melt
3	1075	0	Qtz	Kfs	Zo	Grt	Cpx	Rt	Melt		

MS experiments at 0.6 and 1.0 GPa from Patiño Douce & Harris (1998). SBG experiments at 0.3–1.5 GPa from Patiño Douce & Beard (1995) and Patiño Douce (1996). CAT experiments at 0.4 and 0.8 GPa from Patiño Douce (1997). Mineral abbreviations after Kretz (1983), except: Als = aluminumsilicate, Cym = K-cymrite.

muscovite. At 3 GPa, biotite in the metagreywacke contains ~0.5 octahedral vacancies p.f.u., suggesting that micas with intermediate trioctahedral–dioctahedral compositions (Green, 1981) may be common in UHP metamorphic rocks. The other notable change in biotite composition with pressure is the increase in the phengite component, which becomes especially

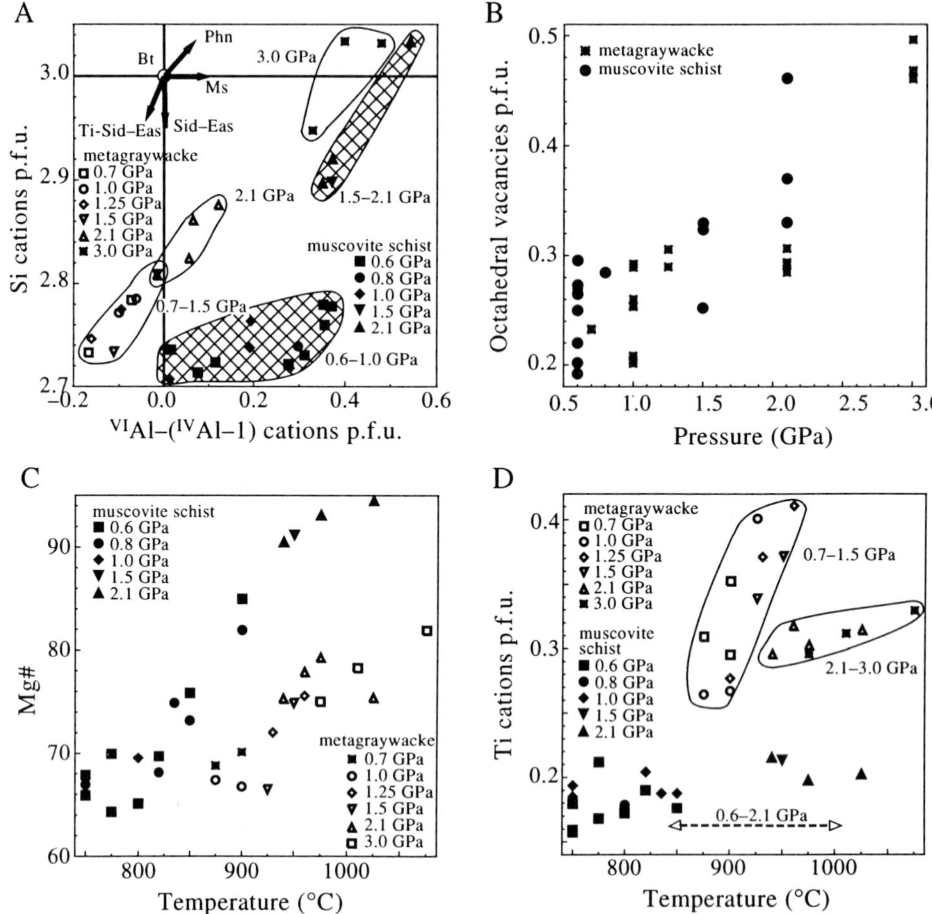

Figure 2. Changes in biotite composition with P, T, and bulk composition (formulas calculated on the basis of 12 oxygen anions; see table 2). a) Si vs. "excess" octahedral Al (*i.e.*, ^{VI}Al in excess of that accounted for by tschermak substitution). Arrows point toward endmembers: phengite, $KAlR^{2+}Si_4O_{10}(OH)_2$; muscovite, $KAl_2AlSi_3O_{10}(OH)_2$; siderophyllite-eastonite, $KR^{2+}{}_2AlAl_2Si_2O_{10}(OH)_2$; and Ti-siderophyllite-eastonite, $KR^{2+}{}_{2.5}Ti_{0.5}Al_2Si_2O_{10}(OH)_2$. b) Octahedral vacancies (= 7-total cations in tetrahedral and octahedral coordinations) as a function of P. c) Mg# = 100*Mg/(Mg+Fe), in molar proportions) as a function of T. d) Ti as a function of T. A Ti-saturating phase (ilmenite and/or rutile) is always present in the metagreywacke, but is absent from the muscovite schist.

significant at pressures of ~3.0 GPa (Fig. 2a). Incorporation of phengite component into trioctahedral micas at high pressure was also demonstrated experimentally by Massone and Schreyer (1987).

Temperature has a strong effect on the Mg# of biotite, which in the muscovite schist approaches the Mg endmember at temperatures of ~1000°C (Fig. 2c). Biotite in the metagreywacke coexists with a Ti-saturating phase, and becomes enriched in Ti with increasing temperature (Fig. 2d; see also Patiño Douce *et al.*, 1993). There is a sharp discontinuity in the behavior of Ti at pressures of ~1.5–2.0 GPa. The effect of temperature is stronger at lower pressures, where Ti can be considered to enter biotite as a Ti-tschermak exchange (Fig. 2a). At P >2 GPa, Ti contents at a given temperature are markedly lower than at lower pressures (Fig. 2d), and there is no compelling evidence for a Ti-tschermak exchange (Fig. 2a). This discontinuity in the behavior of Ti may reflect either a change in the coexisting assemblage (*e.g.*, the disappearance of orthopyroxene, which is part of the Ti-buffering assemblage together with ilmenite, or rutile, and quartz), or a crystallochemical change in biotite. In contrast to the metagreywacke, Ti contents in biotite in the ilmenite- and rutile-free muscovite schist tend to remain constant (Fig. 2d).

6. MELTING PHASE RELATIONS OF COMMON CRUSTAL ROCKS

6.1 Muscovite Schist

The dehydration-melting solidus of muscovite schist has a positive and rather flat dP/dT slope of ~5 MPa/°C at pressures less than ~1.0-1.5 GPa (Fig. 3). Incipient melting (≤5 vol% melt) is observed in experiments at 1.5 GPa and 900°C, at 2.1 GPa and 940°C, and at 3.0 GPa and 975°C (Table 2). These data require that the solidus steepen at high pressures.

Incongruent melting at P ≤1 GPa consumes plagioclase and quartz in addition to muscovite, and produces aluminumsilicate, K-feldspar and biotite. The dehydration-melting reaction is therefore of the form:

$$\text{ms} + \text{pl} + \text{qtz} \rightarrow \text{melt} + \text{kfs} + \text{als} + \text{bt}$$

This incongruent melting reaction consumes more muscovite than plagioclase (Patiño Douce and Harris, 1998), leading to a low pressure (P ≤1 GPa) residue rich in plagioclase and aluminumsilicate. Melting at temperatures exceeding the muscovite-out boundary is controlled by dissolution of potassium feldspar and incongruent breakdown of biotite + aluminumsilicate, with crystallization of

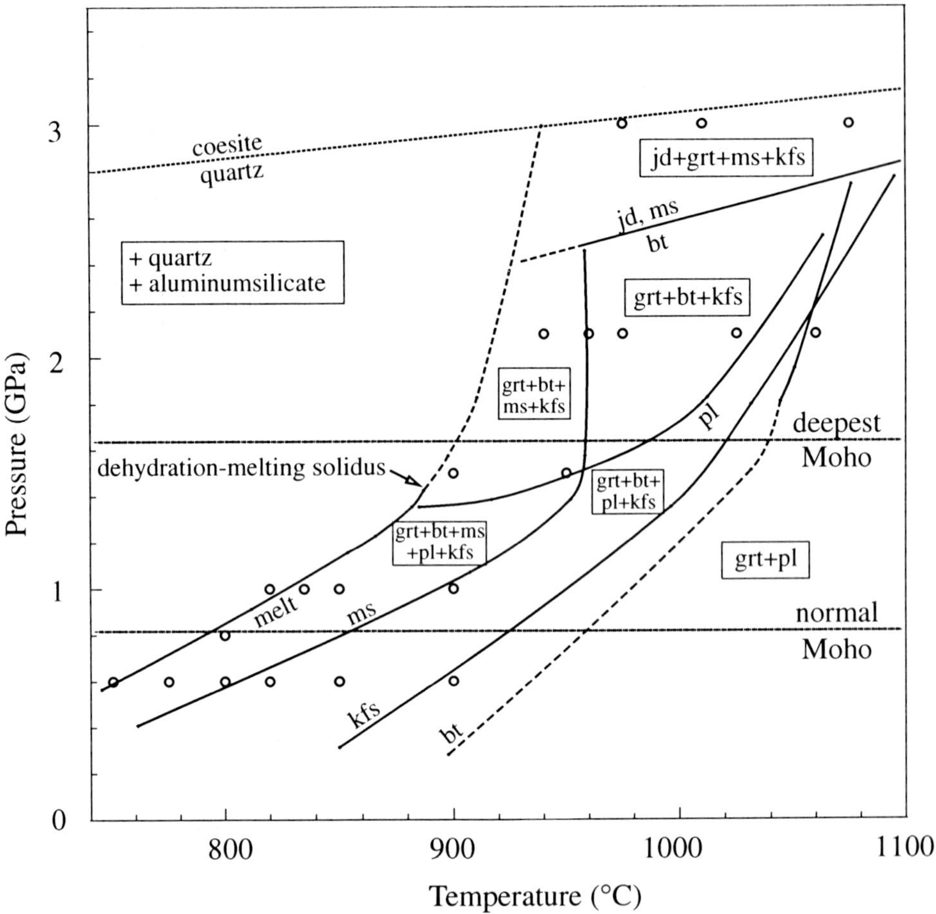

Figure 3. Dehydration-melting phase relations of muscovite schist. Circles show P-T conditions of constraining experiments (Table 2). Phases are labeled on the side of the phase boundary on which the phase is present. The phase boundary between biotite and jadeite + muscovite is schematic. Transition between these assemblages likely occurs over a multivariant field. This field is bracketed bt the 2.1 and 3.0 GPa experiments and almost certainly has a positive dP/dT slope, but its location is unknown.

garnet (Vielzeuf and Holloway, 1988; Patiño Douce and Johnston, 1991; Patiño Douce and Harris, 1998) . Plagioclase disappears from the solidus at P ~1.5 GPa, but, K-feldspar is formed by the incongruent melting reaction to at least 3 GPa, and always persists over a wide temperature interval (~100°C) above the solidus. Jadeite (Fig. 4) becomes stable on the dehydration-melting solidus at pressures of ~2.5 GPa (Table 2, Fig. 3). Concomitant with the appearance of jadeite, biotite disappears and there seems to be an expansion of the muscovite stability field (Fig. 3). The residual assemblage formed by incongruent melting of muscovite schist at high pressure (P ≥2.5 GPa) is dominated by jadeite and quartz, accompanied by minor garnet, K-feldspar and kyanite. It is uncertain whether muscovite in 3 GPa experiments at temperatures >1000°C is a stable phase (Fig. 3, see also Patiño Douce and Harris, 1998). If so, some H_2O and incompatible elements such as K, Rb, and Ba, could be locked in muscovite deep in collisional orogens, as well as in subducted slices of continental crust, even in rocks that have undergone partial melt extraction.

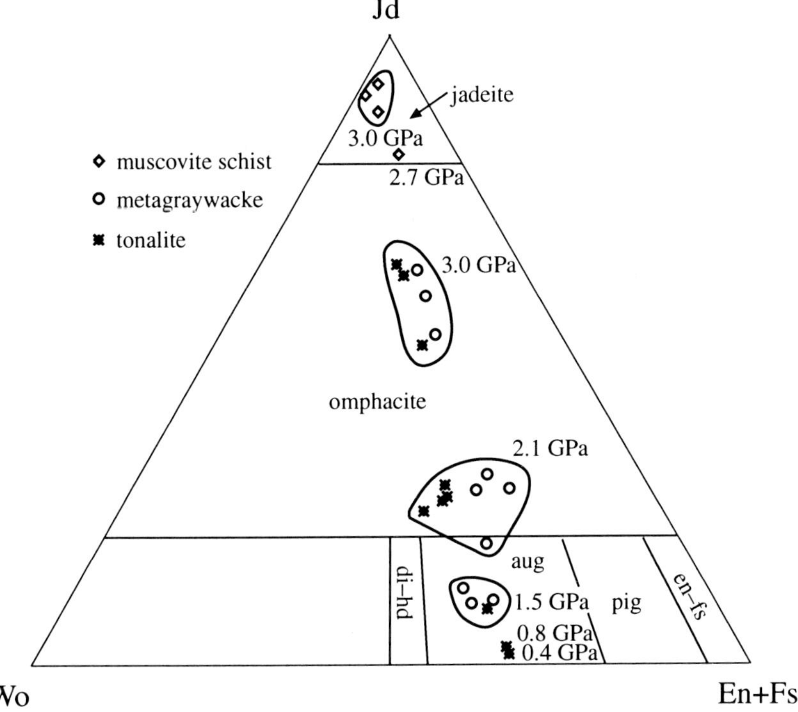

Figure 4. Clinopyroxene compositions in residues of partial melting of crustal protoliths. Classification scheme after Morimoto (1988).

The melting reaction of muscovite schist with added H_2O (H_2O-fluxed melting) is different from the dehydration-melting reaction. During dehydration melting, H_2O is supplied by incongruent breakdown of muscovite, so that the ratio of muscovite to plagioclase consumed by the melting reaction is approximately 3:1 (Patiño Douce and Harris, 1998). The addition of H_2O depresses the plagioclase + quartz solidus but has only a minor effect on the stability of muscovite (Huang and Wyllie, 1981; Patiño Douce and Harris, 1998). As a consequence, H_2O-fluxed melting within the stability field of plagioclase occurs via a congruent reaction of the form:

$$pl + qtz + ms + H_2O \rightarrow melt$$

that can take place at temperatures more than 100°C lower than the incongruent dehydration-melting reaction. This reaction consumes up to twice as much plagioclase as muscovite (Patiño Douce and Harris, 1998). The residue formed by H_2O-fluxed melting is thus enriched in muscovite and depleted in plagioclase (or jadeite at P ≥2 GPa, Huang and Wyllie, 1981), K-feldspar and aluminumsilicate, compared to the residue left by dehydration melting (see also Patiño Douce and Harris, 1998).

6.2 Metagreywacke

The dehydration-melting phase relations of metagreywackes are affected strongly by pressure, Mg/Fe ratio, and oxygen fugacity (Patiño Douce and Beard, 1996).

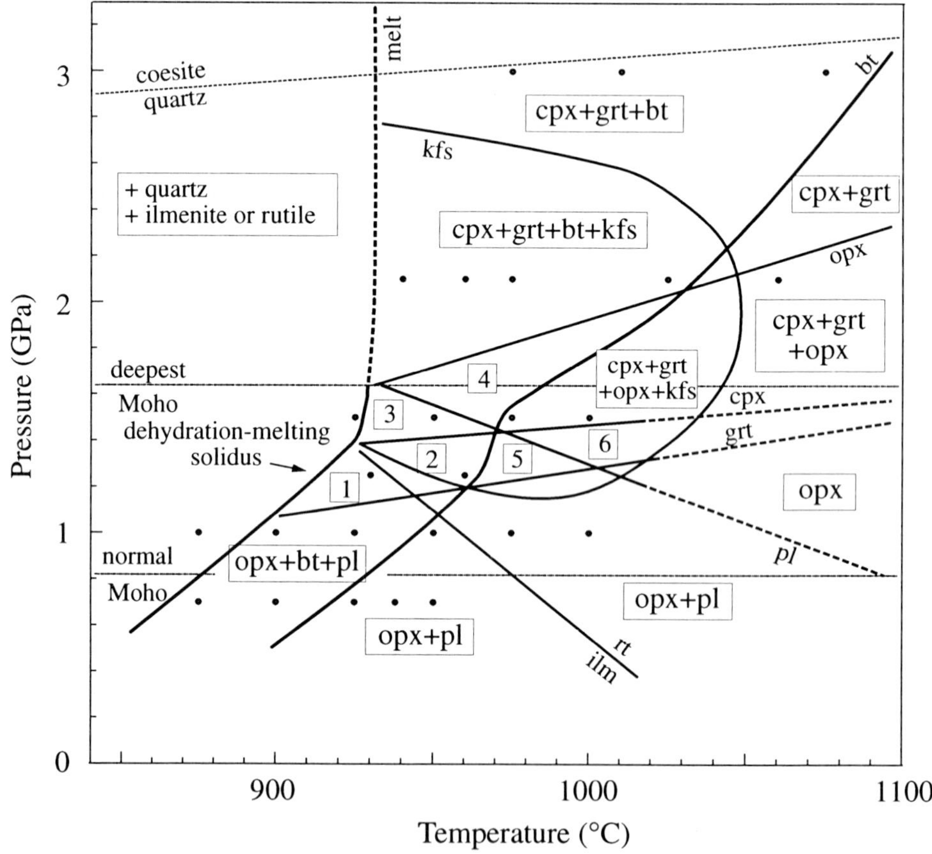

Figure 5. Dehydration-melting phase relations of metagreywacke. Circles and phase boundaries as in Fig. 3. Numbered fields are: 1: opx + bt + pl + grt; 2: opx + bt + pl + grt + kfs; 3: opx + cpx + bt + pl + grt + kfs; 4: opx + cpx + bt + grt + kfs; 5: opx + pl + grt + kfs; 6: opx + grt + kfs.

The behavior of rocks of intermediate Mg/Fe ratio is shown in Fig. 5. The dehydration-melting solidus has positive dP/dT slope up to 1.5 GPa. At higher pressures the solidus must steepen, because melting is observed at 2.1 GPa and 940°C and at 3.0 GPa and 975°C. It remains to be determined experimentally whether the dehydration-melting solidus bends back at P >1.5 GPa, as suggested, for example, by Lambert and Wyllie (1972), Burnham (1979b), and Le Breton and Thompson (1988).

There are two fields, at high and low pressure, well-defined by distinct assemblages of residual solid phases (Fig. 5). At pressure ≤1 GPa, the solid assemblage formed by incongruent melting is dominated by orthopyroxene + plagioclase. At pressure ≥1.5 GPa these phases are replaced by garnet + clinopyroxene ± K-feldspar. Clinopyroxene compositions (Fig. 4) vary from aluminous augite at 1.5 GPa, through low-Na omphacite at 2.1 GPa, to omphacite at 3.0 GPa. The transition from granulitic residues (plagioclase + orthopyroxene) to eclogitic residues (garnet + clinopyroxene) takes place between 1–2 GPa (Fig. 5), which roughly corresponds to the lower crust in thickened collisional orogens. Several assemblages, consisting of different combinations of plagioclase, orthopyroxene, clinopyroxene and garnet, are stable within restricted P-T intervals throughout the transitional interval (numbered fields in Fig. 5). Garnet appears at the dehydration-melting solidus at pressures of 1.0-1.2 GPa, and clinopyroxene at pressures of 1.2–1.5 GPa (Fig. 5). Plagioclase and orthopyroxene disappear from the solidus at pressures between 1.5 and 2.1 GPa, with orthopyroxene persisting to 2.1 GPa only at temperatures >1050°C.

The role of plagioclase in the melting reaction changes with pressure (Patiño Douce and Beard, 1995; 1996). At pressures of less than 1 GPa, plagioclase dissolves congruently in the melt, leading to the melting reaction:

$$bt + pl + qtz \rightarrow melt + opx$$

At P >1 GPa, and up to the maximum pressure at which it is a stable solidus phase (P <2.1 GPa), plagioclase reacts incongruently as follows:

$$bt + pl + qtz \rightarrow melt + grt + cpx \pm opx$$

Breakdown of anorthite to grossular ± diopside provides the Al for crystallization of garnet, and the albite component of plagioclase partitions into the melt (see also Patiño Douce and Beard, 1996).

As the protolith becomes richer in Fe/Mg, the stability fields of both pyroxenes shrink and the residual assemblage becomes more sensitive to redox conditions (Patiño Douce and Beard, 1996). Garnet is stable in reducing environments (f_{O_2} ≤QFM). Under such conditions, garnet is a residual phase at P ≥0.7 GPa in metagreywackes of Mg# 44 (Vielzeuf and Montel, 1994), and a dominant residual phase at pressures as low as 0.5 GPa in metagreywackes of Mg # 20 (Patiño Douce and Beard, 1996). If melting takes place in an oxidizing environment (f_{O_2} ≥Ni-NiO), garnet is not stable and is replaced by magnetite + plagioclase (Patiño Douce and Beard, 1996).

As in the case of muscovite schist, the addition of H_2O to metagreywacke lowers the plagioclase + quartz solidus but has less effect on the stability of biotite (Conrad *et al.*, 1988). The contrast between the behaviors of plagioclase and biotite becomes stronger with increasing pressure (cf. Patiño Douce, 1996). The addition of small amounts of H_2O (≤2 wt%) to metagreywacke expands the stability field of garnet at the expense of orthopyroxene and, at P = 1.5 GPa, also stabilizes hornblende (Patiño Douce, 1996). Crystallization of garnet or hornblende requires incongruent breakdown of plagioclase. Therefore, residual assemblages formed by H_2O-fluxed melting of metagreywackes will generally be enriched in biotite + garnet (± hornblende if the temperature is low enough) relative to plagioclase + orthopyroxene, compared to the assemblages formed by dehydration melting at similar P-T conditions (Patiño Douce, 1996).

6.3 Tonalite

The dehydration melting relations of the hornblende + biotite-bearing calc-alkaline tonalite (CAT) are shown in Fig. 6. Incongruent breakdown of hornblende in quartz-bearing assemblages can produce either plagioclase or clinopyroxene, depending on pressure (Patiño Douce and Beard, 1995; Patiño Douce, 1997). Dehydration melting at 0.4 GPa produces a solid assemblage dominated by Ca-rich plagioclase (An_{77}) and orthopyroxene, with subordinate clinopyroxene (Patiño Douce, 1997). Peritectic crystallization of plagioclase is suppressed by increasing pressure (Patiño Douce and Beard, 1995). At 0.8 GPa, the solid assemblage formed by the melting reaction is dominated by clinopyroxene, and orthopyroxene + plagioclase are formed in much smaller proportions than at 0.4 GPa (Patiño Douce, 1997). Orthopyroxene is replaced by garnet at P ~1.2 GPa, plagioclase disappears at pressures between 1.5 and 2.1 GPa, and zoisite becomes a peritectic phase at pressures between 2.1 and 3.0 GPa (Fig. 6, Table 3). Clinopyroxene compositions change from low-Al augite at 0.4 GPa to omphacite at 3.0 GPa (Fig. 4). Biotite is present above the dehydration-melting solidus of the tonalite at P ≤1.5 GPa, but is absent from dehydration-melting experiments at higher pressures. These relationships suggest that the biotite-out boundary has negative dP/dT slope (Fig. 6). Phengite appears in dehydration-melting experiments at 3.0 GPa, persisting to temperatures >1000 °C, where it becomes notably enriched in Ti (Tables 2 and 3, Fig. 6). As in the metagreywacke, the pressure interval 1–2 GPa spans the transition from granulitic residues (plagioclase + one or two pyroxenes ± garnet) to eclogitic residues (garnet + omphacite ± zoisite ± phengite). The peritectic assemblage formed by dehydration melting of tonalite at 3 GPa is hydrous, as it contains zoisite and phengite. As noted by Schmidt (1993), this shows that hydrous

Table 3. Zoisite and phengite compositions in tonalite experiments at 3.0 GPa.

T°C	phase	SiO_2	Al_2O_3	TiO_2	$FeO^{*\dagger}$	MgO	MnO	CaO	Na_2O	K_2O	F	Total
975	Zo	40.58	33.83	0.06	0.44	0.00	0.00	23.11	0.14	0.28	0.01	98.45
975	Phn	50.75	30.40	1.25	2.45	1.93	0.01	0.10	0.35	9.27	0.25	96.66
1010	Zo	40.43	32.74	0.06	0.60	0.01	0.06	23.21	0.15	0.29	0.04	97.56
1010	Phn	48.83	29.87	2.24	2.09	2.04	0.06	0.05	0.30	10.50	0.23	96.08
1075	Zo	40.19	32.99	0.09	1.26	0.06	0.14	22.70	0.16	0.35	0.19	98.05

Values are averages of 6 analyses of different crystals. F equivalent subtracted from listed totals.

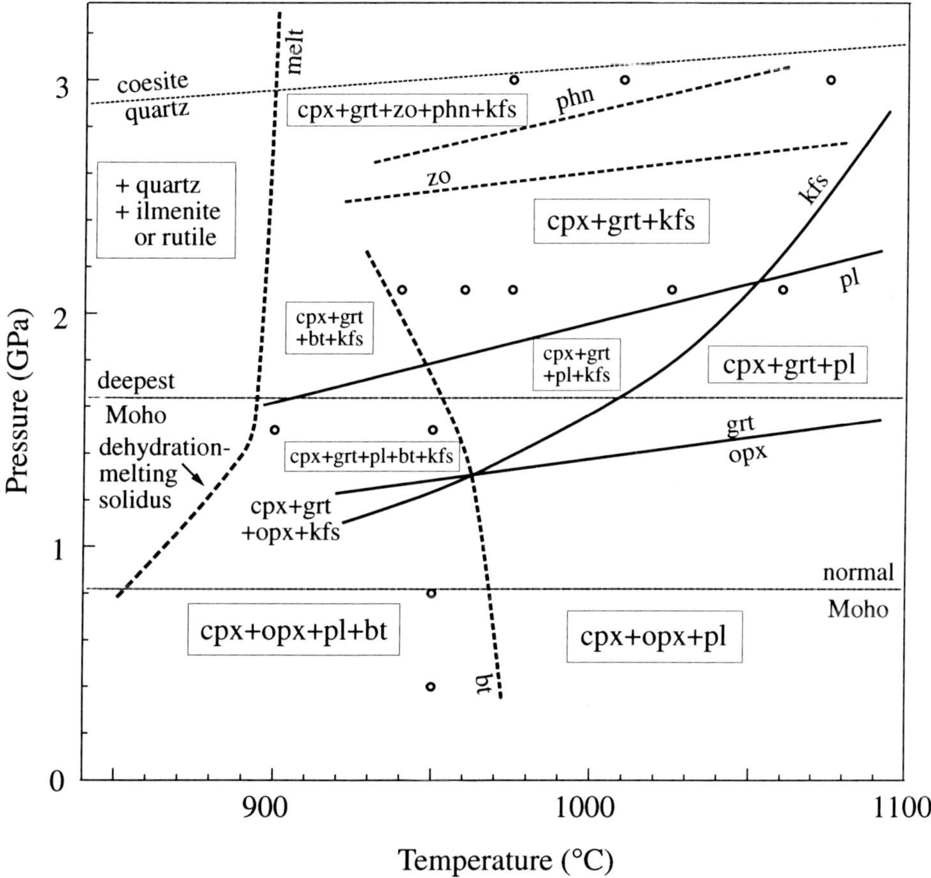

Figure 6. Dehydration-melting phase relations of calc-alkaline tonalite. Circles and phase boundaries as in Fig. 3. The positions and slopes of the phase boundaries for zoisite and phengite are schematic. Zoisite is present in all 3.0 GPa experiments, but not at 2.1 GPa. Phengite is only present in the two lowest temperature 3.0 GPa experiments. The dP/dT slopes of both phase boundaries are almost certainly positive, but their locations are unknown.

minerals in eclogites are not necessarily retrograde—they can form at or near the peak temperature attained by the rock.

We have not determined the dehydration-melting solidus of the tonalite. The presence of melt in the experiment at 900°C and 1.5 GPa shows that the solidus must lie at lower temperature than that of the metagreywacke (compare Figs. 5 and 6). We tentatively show a solidus with the same shape as that of the metagreywacke but displaced to lower temperature, because the breakdown of biotite plays an important role in the melting of both rocks. This inferred solidus is broadly consistent with the experimental results of Skjerlie and Johnston (1993; 1996) on rocks of tonalitic bulk compositions. It is possible, however, that the tonalite dehydration-melting solidus might bend back at pressures between 1 and 2 GPa, in response to the breakdown of hornblende to garnet (Wyllie and Wolf, 1993).

Melting of tonalite with added H_2O has been studied by Lambert and Wyllie (1974), Carroll and Wyllie (1990), and Schmidt (1993), among others. As in other quartzofeldspathic lithologies, the addition of H_2O promotes melting of plagioclase + quartz at temperature lower than that of biotite breakdown, and thus stabilizes micas in the residue. The mica present in tonalites close to the H_2O-saturated solidus (650-700°C) is biotite at P <1.4 GPa and phengite at higher pressures (Schmidt, 1993). Phengite is present in our experiments on the tonalite with 2 wt% added H_2O at 700°C and 2.1 GPa and at 725°C and 3.0 GPa (Table 2). At P >0.6 GPa, epidote is also an important near-solidus phase (Schmidt, 1993), but it disappears at temperatures of ~800°C (Carroll and Wyllie, 1990; Schmidt and Thompson, 1996).

6.4 Melt Compositions

Representative melt compositions are shown in Tables 4 and 5. For dehydration melting (Table 4), the compositions shown correspond to the temperature at which the dominant hydrous phase disappears at each pressure. Choosing terminal conditions for the hydrous phase to compare melt compositions across pressure ensures that the dehydration-melting reaction has gone to completion, and thus brings out most clearly the way in which pressure-induced changes in the nature and stoichiometry of the melting reaction affect melt compositions.

Melts from the three starting materials are silica rich at all pressures and temperatures. FeO + MgO + TiO_2 decreases with increasing pressure (Table 4, Fig. 7a), so that melts formed at pressures greater than 1.5 GPa, and even at temperatures as high as 1075°C, are leucogranitic (FeO + MgO + TiO_2 <2 wt%). Lime contents are lowest at high pressure (Table 4). Total alkali contents increase with pressure, from ~8 wt% at P <1 GPa to ~10 wt% at P = 3 GPa (Fig. 6b). These compositional trends reflect the fixed H_2O budget of dehydration melting, as a result of which the activity of H_2O decreases with increasing

Table 4. *Representative melt compositions: dehydration melting.*

P GPa	T °C	SiO_2	Al_2O_3	TiO_2	$FeO^{*†}$	MgO	MnO	CaO	Na_2O	K_2O	Total
Muscovite schist (MS)											
0.6	800-820	74.51	15.41	0.13	0.95	0.23	0.05	0.48	3.47	4.77	96.97
1.0	850-900	73.60	15.66	0.19	0.73	0.28	0.04	0.59	3.98	4.93	96.66
1.5	950	74.65	15.56	0.27	0.35	0.38	0.01	0.62	3.89	4.19	96.68
2.1	940-960	71.98	16.31	0.26	0.25	0.28	0.07	0.98	4.56	5.27	96.26
3.0	1075[¶]	72.79	15.91	0.30	0.52	0.11	0.05	0.47	4.17	5.54	95.18
Metagreywacke (SBG)											
0.7	900-925	73.95	14.45	0.34	1.70	0.48	0.10	1.08	1.74	6.13	96.85
1.0	925–950	72.95	15.20	0.36	1.60	0.42	0.09	1.62	2.11	5.55	96.13
1.25	960	73.20	14.70	0.40	1.40	0.40	0.00	1.43	2.47	5.75	97.31
1.5	950-975	73.50	15.30	0.33	1.02	0.34	0.04	1.09	2.90	5.35	96.19
2.1	1025–1050	73.62	14.72	0.39	0.86	0.33	0.03	0.73	3.82	5.39	95.71
3.0	1075	73.73	14.24	0.35	0.73	0.29	0.03	0.41	3.08	7.03	93.15
Tonalite (CAT)											
0.4	950	73.82	12.96	0.52	2.22	0.29	0.05	1.22	3.19	5.31	96.83
0.8	950	74.58	14.64	0.26	0.83	0.34	0.04	1.60	3.18	5.04	94.67
1.5	950	73.44	15.07	0.30	0.89	0.33	0.02	1.11	3.93	4.86	96.56
2.1	940-960	72.46	15.56	0.30	0.83	0.27	0.00	1.17	4.03	5.25	96.01
3.0	1075	71.81	15.97	0.37	0.65	0.12	0.03	0.88	3.59	6.47	94.95

Melt compositions at the hydrous phase out boundary (see text), obtained by wavelength-dispersive electron probe microanalysis, and recalculated to volatile-free 100 wt%. MS at 0.6 and 1.0 GPa from Patiño Douce & Harris (1998); SBG at 0.7-1.5 GPa from Patiño Douce & Beard (1995); CAT at 0.4 and 0.8 GPa from Patiño Douce (1997). Where a range of temperatures is given, the range brackets the disappearance of the hydrous phase. In such cases the reported melt composition is the average of the two bracketing compositions. [†] Total Fe as FeO. [§] Microprobe totals before recalculation. [¶] Muscovite is present, but scarce. Patiño Douce & Harris (1998) have shown that the effect of this residual muscovite on melt composition is negligible.

Table 5. Representative melt compositions: experiments with added H_2O.

P GPa	T* °C	H_2O wt%	SiO_2	Al_2O_3	TiO_2	$FeO^{*†}$	MgO	MnO	CaO	Na_2O	K_2O	Total[§]
Muscovite schist (MS)												
0.6	750	1	74.95	15.73	0.14	1.01	0.34	0.04	0.56	4.53	2.70	93.77
1.0	700	4	73.20	16.01	0.04	0.70	0.19	0.02	1.46	6.67	1.71	93.93
1.0	750	1	74.47	16.00	0.04	0.71	0.19	0.02	0.87	5.74	1.91	91.45
2.1	700	2	72.56	18.57	0.09	0.41	0.13	0.04	1.14	4.07	3.34	91.18
2.7	950	1	74.60	16.75	0.14	0.47	0.05	0.00	0.75	3.88	3.36	92.10
Metagreywacke (SBG)												
0.7	925	1	74.10	14.30	0.28	1.63	0.36	0.05	1.20	2.17	5.74	97.10
1.0	925	1	72.30	14.70	0.31	1.99	0.44	0.03	2.12	3.32	4.65	97.10
1.5	925	1	72.00	15.00	0.28	1.11	0.28	0.02	0.98	5.83	4.56	95.20

Melt compositions obtained by wavelength-dispersive electron probe microanalysis, and recalculated to volatile-free 100 wt%. MS at 0.6 and 1.0 GPa from Patiño Douce & Harris (1998); SBG at 0.7-1.5 GPa from Patiño Douce (1996). [*] Water added to the experimental charges. [†] Total Fe as FeO. [§] Microprobe totals before recalculation.

pressure. As this happens, melts shift toward more felsic and alkali-rich (*i.e.*, "lower temperature") compositions.

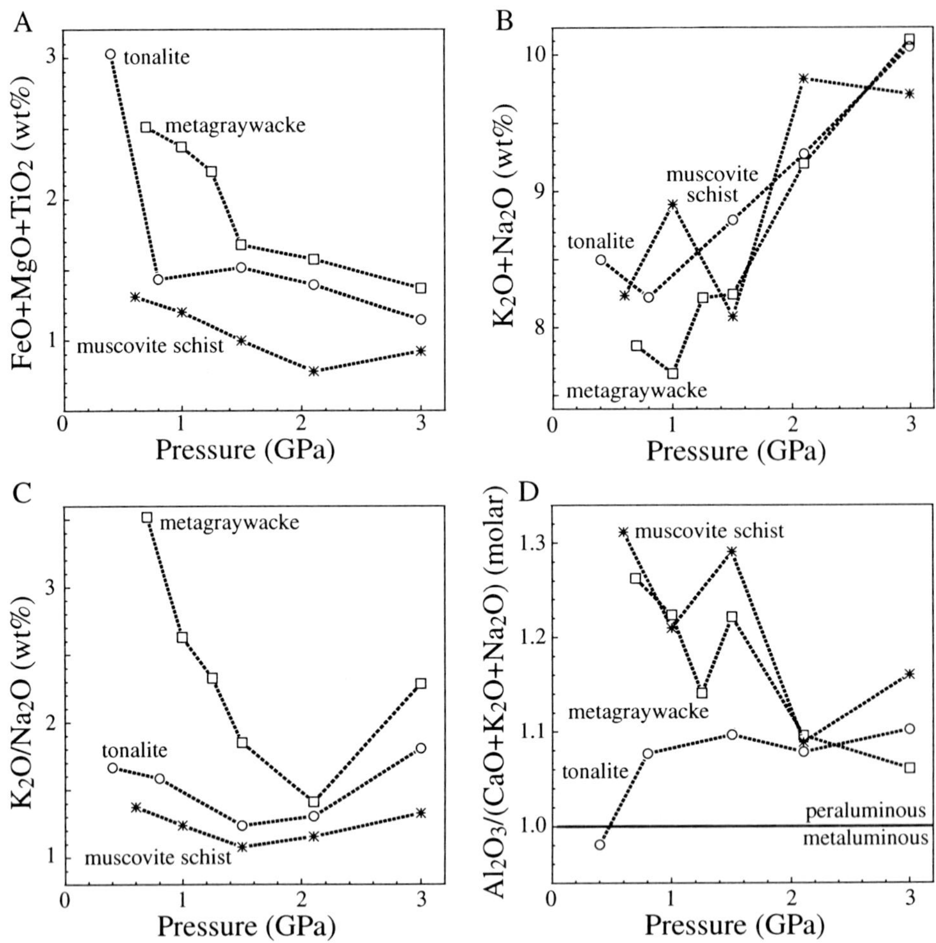

Figure 7. Compositions of melts produced by dehydration melting. Melt compositions are shown as a function of pressure, at the temperature at which the dominant hydrous phase disappears from each starting material (see text and Table 3). a) Total FeO + MgO + TiO_2. b) Total alkali content. c) Ratio of K_2O to Na_2O. d) Alumina saturation indices.

Melts formed by dehydration melting are granitic (Fig. 8), reflecting that dehydration melting of the three starting materials entails breakdown of micas. Variations in K_2O/Na_2O ratios are related to bulk composition and pressure. The more sodic melts are always from the muscovite schist, and the more potassic melts are from the metagreywacke (Figs. 7c, 8). Although the tonalite has a lower bulk K_2O/Na_2O ratio than the muscovite schist, it produces melts with higher K_2O/Na_2O ratios. This is a consequence of the different behaviors of micas and plagioclase (the dominant K and Na reservoirs, respectively) during melting reactions (Patiño Douce, 1996; Patiño Douce and Harris, 1998). Variations in K_2O/Na_2O ratios with pressure reflect the behavior of plagioclase and the stabilization of sodic pyroxenes at high pressure.

The trends are similar in the three starting materials, but are best displayed in the metagreywacke (Fig. 7c). At low pressure (P <1 GPa), plagioclase is a relatively minor contributor to the melting reaction, and in the tonalite is actually produced as a peritectic phase. In consequence, low-pressure melts are enriched in K_2O relative to Na_2O. As pressure increases, plagioclase breaks down incongruently, with the albite component partitioning into the melt phase. The greatest Na_2O enrichment in the melts is at pressures of 1.5–2.0 GPa, where plagioclase is no longer stable and clinopyroxene incorporates only modest jadeite component (or, in the muscovite schist, is not stable). The trend reverses at higher pressures (Fig. 7c), as Na-rich clinopyroxene (omphacite or jadeite, Fig. 4) becomes stable. High-pressure melts (3 GPa) are enriched in total alkalis and K_2O relative to Na_2O (Figs. 7b, 7c, Table 4).

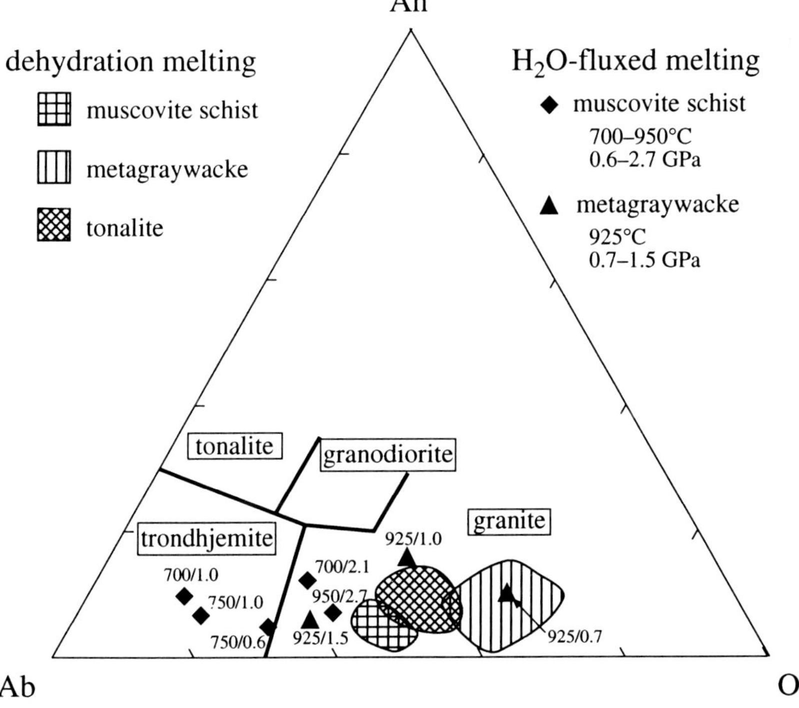

Figure 8. Normative albite-anorthite-orthoclase contents of melts formed by dehydration melting and melts formed by H_2O-fluxed melting. Fields for dehydration melting, symbols for H_2O-fluxed melting. Numbers next to symbols are temperature (°C) and pressure (GPa). Classification of silicic melts after Barker (1979).

Alumina saturation indices (ASI, molar $Al_2O_3/(CaO+Na_2O+K_2O)$) for melts generated by the three starting materials are shown in Fig. 7d. At low pressure (P <1 GPa), the melts produced from the tonalite are weakly metaluminous to weakly peraluminous (ASI = 0.98-1.08), whereas those produced from the metasediments are strongly peraluminous (ASI = 1.12–1.32). The difference is rooted in the presence of hornblende in the tonalite. At low pressure, the incongruent breakdown of hornblende + quartz produces Ca-rich plagioclase, which is an Al sink. The effect of peritectic plagioclase crystallization is particularly strong at 0.4 GPa, giving rise to Al-depleted and alkali-rich "A-type" granitic melts (Patiño Douce, 1997). In the metasediments, incongruent breakdown of muscovite or Al-rich biotite at low pressure produces peritectic assemblages dominated by alkali feldspar or orthopyroxene, respectively, that are less aluminous than the reactant assemblages. The melts are therefore strongly peraluminous. At high pressure (>2 GPa) all melts converge toward moderately peraluminous compositions (ASI ~1.1, *cf.* Fig. 7d). This convergence probably stems from the fact that high pressure melts from the three starting materials are saturated with similar eclogitic assemblages (garnet + Na-rich clinopyroxene).

Melts formed by the addition of H_2O at temperatures lower than the dehydration-melting solidus are enriched in Na_2O and depleted in K_2O compared to melts formed by dehydration melting (compare Tables 4 and 5). Results from the muscovite schist (Fig. 8, also Patiño Douce and Harris, 1998) show that the addition of H_2O at temperatures of 700-750°C generates trondhjemitic melts that coexist with muscovite-rich residues. Within the stability field of plagioclase, melts become more sodic and calcic with increasing pressure and decreasing temperature (Fig. 8). The H_2O-added experiment on the muscovite schist at 700°C and 2.1 GPa (Table 5, Fig. 8) suggests that the trend of Na enrichment with increasing pressure reverses when jadeite becomes stable, as in the case of dehydration melting.

As temperature approaches the dehydration-melting solidus, a greater proportion of micas enters the melting reaction, and melt compositions become more potassic. This is exemplified by the three experiments on the metagreywacke with 1 wt% added H_2O at 925°C (Table 5, Fig. 8). This temperature is ~50°C higher than the dehydration-melting solidus at 0.7 GPa, but just below the dehydration-melting solidus at 1.5 GPa (see Fig. 5). The melt composition at 0.7 GPa plots within the range of melts formed by dehydration melting, and melts become more sodic as pressure increases isothermally (Fig. 8).

7. MELTING OF CRUSTAL ROCKS DURING CONTINENTAL COLLISION AND SUBDUCTION

A synthesis of experimental results and comparison with possible thermal conditions in collisional orogens is shown in Fig. 9. All quartzofeldspathic rocks can melt in tectonically thickened crust if infiltration of aqueous fluids takes place. This is true even at the relatively low temperatures that prevail if the lithospheric mantle remains attached (field I). In general, the lower the temperature and the higher the pressure at which H_2O-fluxing induces melting in thickened continental crust, the more sodic are the melts. Melting of metapelites

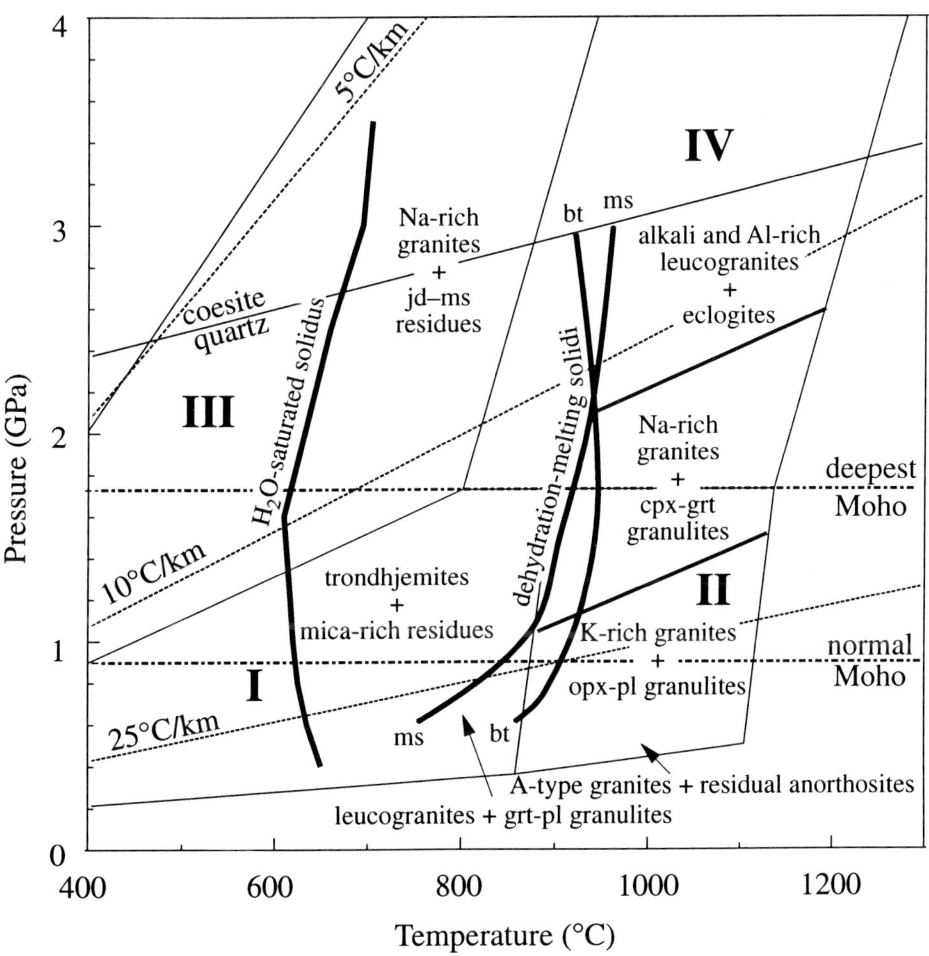

Figure 9. Melting behavior of common crustal protoliths, compared to thermal regimes in continental collision zones (from Fig. 1). H_2O-saturated solidus from Huang and Wyllie (1981), dehydration-melting solidi from Figs. 3 (Ms) and 5 (Bt).

and metagreywackes at temperatures of ~700°C in middle and deep levels of collisional orogens generates trondhjemitic melts and mica-rich residues. If H_2O-fluxed melting at these conditions takes place, it is likely to happen during, or even before, peak metamorphism, while contraction is still active. Synkinematic migmatites with trondhjemitic leucosomes and mica-rich melanosomes, which are relatively common in Barrovian metamorphic terranes, could thus be products of H_2O-fluxed melting in collisional orogens.

Dehydration melting in field I is only possible in muscovite schists. In this case, anatexis is restricted to pressures less than ~1 GPa, corresponding to the uppermost 35 km of the collisional belt. The products of crustal melting under these conditions are peraluminous leucogranites and garnet + plagioclase + K-feldspar granulitic residues. The conditions required to trigger anatexis are most likely to be generated by fast adiabatic decompression attending collapse of the orogen, such as in the case of the Himalayan leucogranites (path C in Fig. 1, Harris and Massey, 1994; Patiño Douce and Harris, 1998). Leucogranite generation in collisional orogens will therefore generally post-date peak metamorphic pressures and temperatures, as well as contraction. Decompression melting during orogenic collapse is likely to favor melt segregation, because low-pressure sites that "sponge up" viscous melt (Sawyer, 1994) should proliferate during extension. These sites could coalesce as extension progresses, eventually growing into km-scale intrusive bodies.

In the absence of H_2O infiltration, more-mafic metasediments such as metagreywackes, as well as igneous tonalites, should remain unmolten even during peak metamorphism in field I (Fig. 9). However, the large increase in temperature that results from detachment or erosion of the lithospheric mantle under a collisional orogen (field II) leads to extensive melting of most quartzofeldspathic lithologies. The most abundant products of melting at pressures less than 1.0-1.2 GPa are K-rich granites and plagioclase-rich residual granulites. There are variations in melt and residue composition, however, related to source lithology and pressure. Metasediments generate peraluminous melts, in equilibrium with either garnet + plagioclase + quartz residues (in the case of Al-rich metapelites) or orthopyroxene + plagioclase + quartz residues (when the sources are less aluminous, more mafic metagreywackes). A special, but perhaps common, case is that of rocks that have already undergone H_2O-fluxed melting in field I. Extraction of the relatively cold trondhjemitic melts formed by H_2O fluxing leaves mica-rich residues that are still fertile sources for granitic magmas. Remelting of these K-rich and Na-poor residual rocks requires temperatures higher than ~900°C (Patiño Douce and Johnston, 1991), as are attained in field II. The products of this second melting event are potassic (or perhaps ultrapotassic) granitic liquids, and residues composed almost exclusively of refractory ferromagnesian phases (garnet, cordierite or orthopyroxene, depending on pressure and bulk composition, cf. Patiño Douce and Johnston, 1991; Carrington and Harley, 1995).

Also in field II, melting of calc-alkaline tonalites at low pressure (P ≤0.4 GPa) yields Al-poor "A-type" granites (Patiño Douce, 1997) and residues composed chiefly of Ca-rich plagioclase ± orthopyroxene. These residues resemble anorthositic-noritic cumulate rocks that are often associated with A-type granites. The A-type signature of the granitic melts is lost with increasing pressure as clinopyroxene replaces plagioclase as the dominant residual phase. The shallow origin of A-type magmas explains why they are invariably found in "anorogenic" tectonic environments, where the continental crust is thin.

The deep crust of thickened orogenic belts is the critical region where residual assemblages formed during anatexis change from granulitic to eclogitic (Fig. 9). True eclogites are not formed as residues of dehydration melting at the maximum pressures that can be attained at the base of thickened continental crust (1.8 GPa, roughly corresponding to 70 km of crust). However, plagioclase + orthopyroxene are gradually replaced by garnet + clinopyroxene in the residue of dehydration melting at pressures of 1.2–1.8 GPa, corresponding to the lower third of the crust in thickened orogenic belts. Melts formed at these depths are more sodic and more felsic than shallower melts, and may be exemplified by the deep-seated plutonic rocks of the Idaho Batholith (Hyndman, 1983; Patiño Douce, 1995).

Melting in field III, the domain of UHP metamorphism, can only take place in the presence of free aqueous fluids (Fig. 9). Knowledge of the melting behavior of crustal rocks at these conditions is incomplete—the most exhaustive studies being those on granitic rocks by Huang and Wyllie (1981) and Stern and Wyllie (1981). Metapelites (Table 5, Fig. 3) yield Na-rich granites (but not trondhjemites, as in field I) and jadeite + quartz + garnet + muscovite residues. This residual assemblage is reminiscent of the jadeite quartzites of the Dora Maira massif (Western Alps), which have been interpreted as crystallized partial melts (Schreyer *et al.*, 1987; Sharp *et al.*, 1993; Philippot *et al.*, 1995). It is possible that partial melting was involved in the origin of the jadeite quartzites, but they appear more likely to represent solid residues from which melt has been efficiently extracted, rather than crystallized melts.

The melting behavior of crustal rocks in field IV (Fig. 9) is better understood, but whether crustal rocks undergo these "hot subduction" conditions in nature is largely speculative. The residual assemblages formed at these pressures are quartz eclogites in which garnet and sodic clinopyroxene are the chief constituents, accompanied by minor phases such as K-feldspar, phengite, zoisite and, perhaps, muscovite. The presence of hydrous phases in these high-temperature residual eclogites is noteworthy, because it shows that some H_2O and incompatible elements can be retained in rocks from which melt was extracted at very high pressure (>2.5 GPa). This H_2O can be transported deeper into the mantle if the rocks are not exhumed, or it can trigger additional melting if the hydrous phases break down during exhumation.

The melts formed in field IV are leucogranites rich in total alkalis, K_2O/Na_2O and alumina. There are natural occurrences of rocks consisting of the assemblage K-feldspar + quartz + garnet, with no primary plagioclase, in the Bohemian Massif (Vrána, 1989) and in lower crustal xenoliths from northern Australia (Rudnick and Taylor, 1987) and western North America (Hanchar *et al.*, 1994). The Bohemian "potassic granulites" were interpreted by Vrána to be partial melts of crustal protoliths generated at upper mantle pressures and relatively high temperatures (~1000°C). Experimental generation of potassic, alkali-rich and alumina-rich melts at high pressure (Fig. 7) supports this interpretation. Given that the Bohemian Massif also contains subtle evidence for UHP metamorphism at very high temperature (Becker and Altherr, 1992), the exciting possibility exists that crustal rocks are in some instances subjected to the extreme P-T conditions of field IV during continental subduction.

ACKNOWLEDGMENTS

Thoughtful reviews by T. Rushmer, J.G. Liou, W.G. Ernst, H.-J. Massonne and B.R. Hacker helped to improve the manuscript. Work supported by NSF grant EAR-9316304 to Patiño Douce.

REFERENCES

Ackermand, D., Herd, R.K.J., Reinhardt, M., and Windley, B.F. (1987) Sapphirine paragenesis from the Caraiba complex, Bahia, Brazil: stability of sapphirine in iron-bearing rocks, *Journal of Metamorphic Geology* **5**, 323–340.

Barker, F. (1979) Trondhjemite: definition, environment and hypotheses of origin, in F. Barker (ed.), *Trondhjemites, dacites and related rocks, Developments in Petrology*, 6, Elsevier, Amsterdam, pp. 1–12.

Becker, H. and Altherr, R. (1992) Evidence from ultra-high pressure marbles for recycling of sediments into the mantle, *Nature* **358**, 745–748.

Burnham, C.W. (1979a) The importance of volatile constituents, in H.S. Yoder (ed.), *The Evolution of the Igneous Rocks, Fiftieth Anniversary Perspectives*, Princeton University Press, Princeton, pp. 439–482.

Burnham, C.W. (1979b) Magmas and hydrothermal fluids, in H.L. Barnes (ed.), *Geochemistry of Hydrothermal Ore Deposits*, Wiley Interscience, New York, pp. 71–136.

Butler, R.W.H., Harris, N.B.W., and Whittington, A.G. (1997) Interactions between deformation, magmatism and hydrothermal activity during active crustal thickening: a field example from Nanga Parbat, Pakistan Himalayas, *Mineralogical Magazine* **61**, 37–52.

Carrington, D.P. and Harley, S.L. (1995) Partial melting and phase relations in high-grade metapelites: an experimental petrogenetic grid in the KFMASH system, *Contributions to Mineralogy and Petrology* **120**, 270–291.

Carroll, M.R. and Wyllie, P.J. (1990) The system tonalite-H_2O at 15 kbar and the genesis of calc-alkaline magmas, *American Mineralogist* **75**, 345–357.

Chopin, C., Henry, C., and Michard, A. (1991) Geology and petrology of the coesite-bearing terrain, Dora Maira massif, western Alps, *European Journal of Mineralogy* **3**.

Clemens, J.D. and Vielzeuf, D. (1987) Constraints on melting and magma production in the crust, *Earth and Planetary Science Letters* **86**, 287–306.

Conrad, W.K., Nicholls, I.A., and Wall, V.J. (1988) Water-saturated and -Undersaturated melting of metaluminous and peraluminous crustal compositions at 10 kb: evidence for the origin of silicic magmas in the Taupo Volcanic Zone, New Zealand, and other occurrences, *Journal of Petrology* **29**, 765–803.

England, P.C. and Thompson, A.B. (1986) Some thermal and tectonic models for crustal melting in continental collision zones, in M.P. Coward and A.C. Ries (eds.), *Collision Tectonics, Special Publication*, 19, Geological Society of London, London, pp. 83–94.

Green, T.H. (1981) Synthetic high-pressure micas compositionally intermediate between the dioctahedral and trioctahedral mica series, *Contributions to Mineralogy and Petrology* **78**, 452–458.

Hacker, B.R. and Peacock, S.M. (1994) Creation, preservation, and exhumation of coesite-bearing, ultrahigh-pressure metamorphic rocks, in R.G. Coleman and X. Wang (eds.), *Ultrahigh Pressure Metamorphism*, Cambridge University Press, Cambridge, United Kingdom

Hanchar, J.M., Miller, C.F., Wooden, J.L., Bennett, V.C., and Staude, J.-M.G. (1994) Evidence from Xenoliths for a dynamic lower crust, Eastern Mojave Desert, California, *Journal of Petrology* **35**, 1377–1415.

Harley, S.L. (1985) Garnet-orthopyroxene bearing granulites from Enderby Land, Antarctica: metamorphic pressure-temperature-time evolution of the Archaean Napier Complex, *Journal of Petrology* **26**, 819–856.

Harris, N. and Massey, J. (1994) Decompression and anatexis of the Himalayan metapelites, *Tectonics* **13**, 1537–1546.

Harris, N.B.W. and Holland, T.J.B. (1984) The significance of cordierite-hypersthene assemblages from the Beitbridge region of the central Limpopo belt: evidence for rapid decompression in the Archean?, *American Mineralogist* **69**, 1036–1049.

Houseman, G.A., McKenzie, D.P., and Molnar, P. (1981) Convective instability of a thickened boundary layer and its relevance for the thermal evolution of continental convergent belts, *Journal of Geophysical Research* **86**, 6115–6132.

Huang, W.L. and Wyllie, P.J. (1981) Phase relations of S-type granite with H_2O to 35 kbar: Muscovite granite from Harney Peak, South Dakota, *Journal of Geophysical Research* **86**, 10515–10529.

Hyndman, D.W. (1983) The Idaho Batholith and associated plutons, Idaho and western Montana, *Geological Society of America Memoir* **159**, 213–240.

Kretz, R. (1983) Symbols for rock-forming minerals, *American Mineralogist* **68**, 277–279.

Lambert, I.B. and Wyllie, P.J. (1972) Melting of gabbro (quartz eclogite) with excess water to 35 kilobars, with geological applications, *Journal of Geology* **80**, 693–708.

Lambert, I.B. and Wyllie, P.J. (1974) Melting of tonalite and crystallization of andesite liquid with excess water to 30 kilobars, *Journal of Geology* **82**, 88–97.

Le Breton, N. and Thompson, A.B. (1988) Fluid-absent (dehydration) melting of biotite in metapelites in the early stages of crustal anatexis, *Contributions to Mineralogy and Petrology* **99**, 226–237.

Liou, J.G., Zhang, R.Y., Eide, E.A., Maruyama, S., Wang, X., and Ernst, W.G. (1996) Metamorphism and tectonics of high-P and ultrahigh-P belts in Dabie-Sulu Regions, eastern central China, in A. Yin and T.M. Harrison (eds.), *The Tectonic Evolution of Asia*, Rubey IX, Cambridge University Press, Cambridge, United Kingdom, pp. 300–343.

Massonne, H.J. and Schreyer, W. (1987) Phengite barometry based on the limiting assemblage with K-feldspar, phlogopite, and quartz, *Contributions to Mineralogy and Petrology* **96**, 212–224.

Morimoto, N. (1988) Nomenclature of pyroxenes, *Mineralogical Magazine* **52**, 535–550.

Patiño Douce, A.E. (1995) Experimental generation of hybrid silicic melts by reaction of high-Al basalt with metamorphic rocks, *Journal of Geophysical Research* **15**, 623–639.

Patiño Douce, A.E. (1996) Effects of pressure and H_2O content on the compositions of primary crustal melts, *Transactions of the Royal Society of Edinburgh: Earth Sciences* **87**, 11–21.

Patiño Douce, A.E. (1997) Generation of metaluminous A-type granites by low-pressure melting of calc-alkaline granitoids, *Geology* **25**, 743–746.

Patiño Douce, A.E. and Beard, J.S. (1995) Dehydration-melting of biotite gneiss and quartz amphibolite from 3 to 15 kbar, *Journal of Petrology* **36**, 707–738.

Patiño Douce, A.E. and Beard, J.S. (1996) Effects of P, f(O2) and Mg/Fe ratio on dehydration-melting of model metagreywackes, *Journal of Petrology* **37**, 999–1024.

Patiño Douce, A.E. and Harris, N. (1998) Experimental constraints on Himalayan Anatexis, *Journal of Petrology*, in press.

Patiño Douce, A.E., Humphreys, E.D., and Johnston, A.D. (1990) Anatexis and metamorphism in tectonically thickened continental crust exemplified by the Sevier hinterland, western North America, *Earth and Planetary Science Letters* **97**, 290–315.

Patiño Douce, A.E. and Johnston, A.D. (1991) Phase equilibria and melt productivity in the pelitic system: implications for the origin of peraluminous granitoids and aluminous granulites, *Contributions to Mineralogy and Petrology* **107**, 202–218.

Patiño Douce, A.E., Johnston, A.D., and Rice, J.M. (1993) Octahedral excess mixing properties in biotite: a working model with applications to geobarometry and geothermometry, *American Mineralogist* **78**, 113–131.

Philippot, P., Chevallier, P., Chopin, C., and Dubessy, J. (1995) Fluid composition and evolution in cœsite-bearing rocks (Dora-Maira massif, Western Alps): Implications for element recycling during subduction, *Contributions to Mineralogy and Petrology* **121**, 29–44.

Puziewicz, J. and Johannes, W. (1990) Experimental study of a biotite-bearing granitic system under water-saturated and water-undersaturated conditions, *Contributions to Mineralogy and Petrology* **104**, 397–406.

Rudnick, R.L. and Taylor, S.R. (1987) The composition and petrogenesis of the lower crust: a xenolith study, *Journal of Geophysical Research* **92**, 13981–14005.

Sawyer, E.W. (1994) Melt segregation in the continental crust, *Geology* **22**, 1019–1022.

Schmidt, M.W. (1993) Phase relations and compositions in tonalite as a function of pressure: An experimental study at 650°C, *American Journal of Science* **293**, 1011–1060.

Schmidt, M.W. and Thompson, A.B. (1996) Epidote in calc-alkaline magmas: An experimental study of stability, phase relationships, and the role of epidote in magmatic evolution, *American Mineralogist* **81**, 462–474.

Schreyer, W., Massonne, H.J., and Chopin, C. (1987) Continental crust subducted to depths near 100 km: implications for magma and fluid genesis in collision zones, in B.O. Mysen (ed.), *Magmatic Processes: Physiochemical Principles*, 1, Geochemical Society, University Park, pp. 155–163.

Sengupta, P., Karmakar, S., Dasgupta, S., and Fukuoka, M. (1991) Petrology of spinel granulites from Araku, Eastern Ghats, India, and a petrogenetic grid for sapphirine-free rocks in the system FMAS, *Journal of Metamorphic Geology* **9**, 451–459.

Sharp, Z.D., Essene, E.J., and Hunziker, J.C. (1993) Stable isotope geochemistry and phase equilibria of coesite-bearing whiteschists, Dora Maira Massif, western Alps, *Contributions to Mineralogy and Petrology* **114**, 1–12.

Shatsky, V.S., Sobolev, N.V., and Vavilov, M.A. (1995) Diamond-bearing metamorphic rocks of the Kokchetav massif (Northern Kazakhstan), in R.G. Coleman and X. Wang (eds.), *Ultrahigh Pressure Metamorphism*, Cambridge University Press, Stanford, pp. 427–455.

Skjerlie, K.P. and Johnston, A.D. (1993) Fluid-absent melting behavior of an F-rich tonalitic gneiss at mid-crustal pressures: implications for the generation of anorogenic granites, *Journal of Petrology* **34**, 785–815.

Skjerlie, K.P. and Johnston, A.D. (1996) Vapour-absent melting from 10 to 20 kbar of crustal rocks that contain multiple hydrous phases: Implications for anatexis in the deep to very deep continental crust and active continental margins, *Journal of Petrology* **37**, 661–691.

Stern, C.R. and Wyllie, P.J. (1981) Phase relations of I-type granite with H_2O to 35 kilobars: The Dinkey Lakes biotite-granite from the Sierra Nevada batholith, *Journal of Geophysical Research* **86**, 10412–10422.

Thompson, A.B. (1982) Dehydration melting of pelitic rocks and the generation of H_2O - undersaturated granitic liquids, *American Journal of Science* **282**, 1567–1595.

Thompson, A.B. and Algor, J.R. (1977) Model system for anatexis of pelitic rocks. I. Theory of melting reactions in the system $KAlO_2$–$NaAlO_2$–Al_2O_3–SiO_2– H_2O, *Contributions to Mineralogy and Petrology* **63**, 247–269.

Tuttle, O.F. and Bowen, N.L. (1958) *Origin of granite in the light of experimental studies in the system $KAlSi_3O_8$–$NaAlSi_3O_8$–SiO_2– H_2O*, Geological Society of America Memoir 74, 153 pp.

Vielzeuf, D. and Holloway, J.R. (1988) Experimental determination of the fluid-absent melting relations in the pelitic system. Consequences for crustal differentiation, *Contributions to Mineralogy and Petrology* **98**, 257–76.

Vielzeuf, D. and Montel, J.M. (1994) Partial melting of metagreywackes. 1. Fluid-absent experiments and phase relationships, *Contributions to Mineralogy and Petrology* **117**, 375–393.

Vrána, S. (1989) Perpotassic granulites from southern Bohemia. A new rock-type derived from partial melting of crustal rocks under upper mantle conditions, *Contributions to Mineralogy and Petrology* **103**, 510–522.

Wain, A. (1997) New evidence for coesite in eclogite and gneisses; defining an ultrahigh-pressure province in the Western Gneiss region of Norway, *Geology* **25**, 927–930.

Wyllie, P.J. (1977) Crustal anatexis: an experimental review, *Tectonophysics* **43**, 41–71.

Wyllie, P.J. and Wolf, M.B. (1993) Amphibole dehydration-melting: sorting out the solidus, in H.M. Prichard, T. Alabaster, N.B.W. Harris, and C.R. Neary (eds.), *In Magmatic Processes and Plate Tectonics, Geological Society Special Publication No.*, 76, Geological Society of London, London, pp. 405–416.

Yardley, B.W.D. and Valley, J.W. (1997) The petrologic case for a dry lower crust, *Journal of Geophysical Research* **102**, 12173–12185.

Chapter 3

Rheology of Crustal Rocks at Ultrahigh Pressure

Bernhard Stöckhert and Jörg Renner
Research Group "High Pressure Metamorphism in Nature and Experiment", Faculty of Geosciences, Ruhr-University, D-44780 Bochum, Germany, Bernhard.Stoeckhert and Joerg.Renner@rz.ruhr-uni-bochum.de

Abstract: An improved understanding of tectonic processes in the deep levels of subduction zones and collisional belts requires information on the mechanical behavior of continental crust during ultrahigh-pressure (UHP) metamorphism. Predictions are based on the results of experimental deformation of minerals stable at ultrahigh pressure and on the anticipated effect of pressure on deformation mechanisms. Flow laws for dislocation creep of coesite and aragonite indicate that, at ultrahigh pressure, the strength of continental material remains well below a few MPa at natural strain rates. We question the high strength that has been inferred for eclogites, based on preliminary experimental data on jadeite and on natural microstructures of omphacite. The (micro)structural record of natural UHP rocks indicates that strain localization into weak shear zones, albeit not yet identified, must be common. We propose that the presence of dense fluids or hydrous melts at grain and solid phase boundaries could accomplish deformation analogous to liquid phase sintering in ceramics. The low strength of continental material at ultrahigh pressure precludes notable shear heating, thus cool geotherms and P-T paths are implied. The low strength also places upper bounds on the stress drop of seismic events in presently subducted continental crust and limits the size of coherent subducted continental slices.

1. INTRODUCTION

This chapter explores the rheological aspects of ultrahigh-pressure (UHP) metamorphism (UHPM). of continental crust in subduction and collision zones. We discuss theoretical concepts, experimental techniques and results, observed structures and microstructures of natural rocks, and combine these to draw tentative conclusions on the mechanical state of continental crust in the deepest levels of collisional belts. In view of the scarcity of observations and data

B. R. Hacker and J. G. Liou (eds.),
When Continents Collide: Geodynamics and Geochemistry of Ultrahigh-Pressure Rocks, 57–95.
© 1998 *Kluwer Academic Publishers. Printed in the Netherlands.*

available at present, a major purpose of this paper is to draw the attention of workers in UHP terrains to specific aspects of rheology.

The fundamental problem inherent to the study of Earth rheology is the contrast between the human and tectonic timescales. Consequently, deformation experiments on minerals and rocks have to be carried out at strain rates that are rapid compared to natural tectonic processes and the experimental results must be extrapolated over several orders of magnitude in time (Paterson, 1987). The only way to test the validity of extrapolated laboratory data is to compare the predictions to the structural and microstructural record of natural rocks (e.g., Schmid, 1982). When considering UHP, however, this approach is strongly hampered by the limited number of natural occurrences and by the high degree of modification and overprinting during exhumation, which is inherent to the exceptional tectonic history UHP rocks have undergone (e.g., Hacker and Peacock, 1994). A second problem is the complexity of natural rocks. Experimental rock deformation is generally restricted to simple and, with few exceptions, single phase systems. The applicability to natural polyphase materials has to be considered carefully. A third problem is that natural rock deformation proceeds by a variety of deformation mechanisms or deformation regimes that are not equally well addressable in the laboratory and whose microstructural record is subject to different potentials for preservation.

Lithospheric strength is commonly predicted following a concept proposed by Goetze and Evans (1979) and Brace and Kohlstedt (1980). Stresses at shallow depths are limited by a frictional strength envelope that, at some depth, intersects a flow law for dislocation creep of the mineral presumed to control the rheology of the crust. The concept relies on the assumption that dislocation creep is the relevant deformation regime, even though other deformation regimes—in particular diffusional creep—dominate in specific environments. Whereas frictional behavior is to a first order independent of the type of material (Byerlee, 1978), flow strength in the dislocation creep regime differs considerably over the range of rock-forming minerals.

2. SOME FUNDAMENTALS OF ROCK RHEOLOGY AT UHP

2.1 Effects of Confining Pressure

The deformation of rocks is realized by three principal processes: cataclasis, intracrystalline plasticity, or diffusive mass transfer, potentially accompanied by grain boundary sliding (e.g., Paterson, 1978; Frost and Ashby, 1982; Poirier, 1985; Evans and Kohlstedt, 1995; Kohlstedt *et al.*, 1995; Lockner, 1995; Poirier, 1995). The following account of deformation mechanisms concentrates on the effect of pressure because this is the constitutive geological variable which sets UHPM apart from normal crustal conditions.

2.1.1 Brittle Failure

In experiments at low temperatures, both the stress necessary to cause failure of an intact rock sample and the stress required for frictional sliding along a pre-existing fault increase with increasing confining pressure. The stress drop at failure diminishes with increasing pressure until continuous work hardening is observed at confining pressures higher than that for which the intersection of the failure and the friction criterion occurs (e.g., Byerlee, 1968; Renner and Rummel, 1996). On the sample scale, single faults are observed at low pressures, and increasing pressure suppresses macroscopic localization, giving rise to multiple conjugate faults or distributed cataclastic deformation. On the microscopic scale, this transition from brittle failure to cataclastic flow seems to be related to a transition from dominantly tensile (mode I) to shear (mode II) microcracking (e.g., Hirth and Tullis, 1994). In the laboratory this transition occurs at pressures above 1 GPa in dense crystalline rocks (Shimada, 1993; Hirth and Tullis, 1994). Such a transition at 1 GPa cannot be relevant for Earth's crust, because temperatures at pressures of 1 GPa are in most cases sufficient for plastic deformation. However, it has been argued that within Earth's crust the transition occurs at much lower pressures (some tens of MPa) due to the effect of sample size on the strength of brittle materials (Shimada, 1993).

The presence of a fluid phase counteracts the stabilizing effect of increased confining pressure. The effective pressure law (e.g., Jaeger and Cook, 1984) states that the stresses acting on a rock are reduced by the fluid pressure to yield effective stresses. This law is valid as long as strain rates are low, the viscosity of the fluid is low, and the permeability of the rock is sufficient to maintain a uniform fluid pressure (Brace and Martin, 1968). Brittle deformation is even possible at high temperatures where crystal plastic deformation of the solid components of the partially molten rock would normally occur if silicate melt lowers the effective pressure (e.g., Dell'Angelo and Tullis, 1988; Hacker and Christie, 1990).

2.1.2 Dissolution-Precipitation Creep

The presence of an aqueous fluid at moderate temperatures allows for dissolution-precipitation creep, often referred to as pressure solution (Elliott, 1973; Rutter, 1983), which is driven by stress-induced gradients in chemical potential. Grain contacts at elevated normal stress are the sites of dissolution. Transport of the dissolved material between these sources and the sinks, represented by strain shadows and veins in many metamorphic rocks, takes place through the pore fluid over variable distances. Unfortunately, experimental pressure solution is difficult (Rutter, 1983) and largely restricted to systems with only one solid phase, although experimental results show that the nature of the interface (grain boundary or interphase boundary, orientation) has a marked influence on the rate of dissolution (Hickman and Evans, 1991; 1995). In accord with this experimental evidence, high-pressure low-temperature metamorphic phyllites show evidence of dissolution-precipitation creep, whereas adjacent pure quartzites deformed by dislocation creep, with an effective viscosity about two orders of magnitude higher than that of the phyllites (Schwarz and Stöckhert, 1996; Stöckhert *et al.*, in press). Hence, dissolution-precipitation creep is controlled not only grain size but also by the distribution of interphase boundaries in polyphase materials. Dissolution-precipitation creep is likely facilitated by high pressure, as the solubilities of silicates in aqueous fluids tend to increase with pressure (Dove and Rimstidt, 1994; Manning, 1994). Furthermore, the dihedral angles between silicates and an aqueous pore fluid decrease with increasing pressure (Laporte and Watson, 1991) and increasing solute content (Brenan, 1991; Lee *et al.*, 1991), giving rise to enhanced intergranular permeability and eventually wetting of certain grain or interphase boundaries. These effects may be pronounced at UHP, though experimental evidence is not available. Diffusive mass-transfer processes along grain and interphase boundaries similar to dissolution-precipitation creep also occur in the presence of melt (e.g., Cooper and Kohlstedt, 1984; Cooper *et al.*, 1989).

2.1.3 Crystal Plasticity

At temperatures higher than roughly one half of the melting temperature, deformation of solid materials is accommodated by intracrystalline plasticity, the agents of deformation being point, line or planar defects. Increased pressure decreases the distances between the constituent atoms or molecules of a crystal and will therefore affect the ease of formation, multiplication and migration of defects. Usually, the effect of pressure on the behavior of a defect, and as such the effect on deformation, is described by an activation volume.

The formation of intrinsic point defects, *e.g.* vacancies, is thermally activated. A certain equilibrium concentration of intrinsic point defects is present at a given pressure and temperature. Impurities are point defects which can also give rise to additional extrinsic point defects in ionic crystals due to charge balance. For self diffusion to occur, atoms or constituents have to reach an activated state which has a larger volume than that occupied by the same constituent on a lattice site. The difference between the two volumes is the activation volume. Pressure opposes the expansion related to the activated state. Therefore, intrinsic diffusion is impeded by increasing pressure. In contrast, extrinsic diffusion can be enhanced or impeded by pressure. For example, the number of extrinsic vacancies has the same dependence on pressure as does the solubility of the respective impurity. Interstitials always have a negative activation volume (Nachtrieb and Coston, 1965), which implies that their mobility increases with increasing pressure. Obviously, changes of pressure can have various effects on point defect mobility and chemistry. A specific example of how pressure-dependent defects strongly affect the properties of a mineral is early partial melting, which occurs as a result of the pressure dependent solubility limit of point defects that depends strongly on mineral nonstoichiometry (Raterron *et al.*, 1995).

Activation volumes related to diffusion can be predicted by analyzing the elastic distortion caused by a point defect or by combining an empirical correlation between the activation energy for self diffusion and the melting temperature with the pressure dependence of the melting temperature (e.g., Sammis *et al.*, 1981). The change in volume of a crystal due to the presence of screw or edge dislocations, the endmember line defects, has been derived by Seeger and Haasen (1958), using elasticity theory. These calculations are based on the pressure dependence of the shear and bulk moduli as well as on reasonable approximations of the dislocation core radius. The apparent formation volumes tend to be smaller for screw dislocations than for edge dislocations because screw dislocations do not involve dilatation. As a specific example, α quartz is expected to have a negative apparent formation volume on the order of 1–2 molecular volumes for screw dislocations and a positive volume on the order of 3–4 molecular volumes for edge dislocations (Renner, 1996).

Multiplication and mobility of dislocations are controlled by the critical stress for the activation of a dislocation source, by the Peierls stress, and by the activation energy required to overcome discrete obstacles. These parameters depend on the shear modulus and the Burgers vector (Frost and Ashby, 1982). With increasing pressure, the shear modulus increases whereas the length of the Burgers vector decreases. Therefore, predictions of the effect of pressure on the multiplication and migration of dislocations are difficult. In general, dislocation multiplication and the overcoming of discrete obstacles are facilitated when the pressure dependence of the bulk modulus is larger than the pressure dependence of the shear modulus and hindered when the reverse is true. The pressure dependence of the lattice resistance (Peierls stress) seems to be dominated by changes in the dislocation core radius which cannot be treated on the basis of elasticity (Jesser and Kuhlmann-Wilsdorf, 1972). The extent of stacking faults shrinks with increasing pressure, facilitating cross slip of recombined dislocations and a reduction in strength (Poirier, 1985; de Bresser, 1996).

The mobility of high-angle grain boundaries, important for recrystallization, is generally determined by the diffusivity of the constituent species. In general, diffusion along or across grain boundaries or pipe diffusion along dislocations is much faster than volume diffusion (e.g., Hacker and Christie, 1991). Increased fluid pressure has been suggested to facilitate grain-boundary migration in quartz (e.g., Tullis and Yund, 1982).

2.1.4 Constitutive Laws

The relation between macroscopic rheological constitutive variables, such as strain rate $\dot{\varepsilon}$, differential stress σ, temperature T and pressure P, is expressed by constitutive equations whose particular forms depend on the active micromechanisms. Brittle behavior is most often described by yield and failure surfaces (e.g., Lockner, 1995). In flow laws (a term preferentially used for the constitutive equations of plastic deformation), parameters describing the lattice itself (unit cell, Burgers vector) or its response to stresses (shear and bulk modulus) appear rather independent of the particular micromechanism (Frost and Ashby, 1982). The pressure dependence of these quantities most likely causes a subordinate pressure dependence of flow stress or strain rate. A Boltzmann activation term is the consequence of the incorporation of a diffusion process (e.g., Evans and Kohlstedt, 1995). The power law flow law

$$\dot{\varepsilon} = A\sigma^n \exp(Q/RT)$$

has become popular as a general description of deformation controlled by dislocation climb. Strain rate and differential stress are related to each other by a pre-exponential factor A, a stress exponent n, and an apparent activation energy Q.

The pressure dependence of diffusion-controlled processes is most likely governed by the apparent activation energy (correctly enthalpy) $Q=H=E+PV$ where E denotes internal energy and V activation volume. For a power law material the effect of pressure on strain rate and stress is then $\partial \ln \dot{\varepsilon} / \partial P = -V / RT$ and $\partial \ln \sigma / \partial P = nV / RT$, respectively. For example, using an olivine activation volume of 10 cm^3/mol (Jaoul, 1990), the above relations yield a four-fold decrease in strain rate for each 1 GPa increase in hydrostatic pressure at constant applied stress. At constant strain rate the increase in flow stress with increasing pressure is enhanced by the stress exponent. However, the dominance of intrinsic diffusion is only expected at temperatures $T>0.6T_m$ which is perhaps not realized for UHPM. Note that partial melting of a polyphase rock occurs when the homologous temperatures of its constituents are still fairly low. Therefore, either extrinsic diffusion or in general a complex effect of pressure on defect chemistry must be considered. The point defect chemistry is sometimes empirically accounted for by introducing the fugacity of volatile constituents as an explicit parameter of the flow law (e.g., Bai *et al.*, 1991). Recently, the long-known weakening effect of free water on silicates (hydrolytic weakening, Griggs, 1967) has been correlated with the water fugacity f_{H_2O} that significantly increases with increasing pressure. Based on data from the literature, Kohlstedt *et al.* (1995) found a power law dependence of the strain rate of polycrystalline quartz on f_{H_2O}. The derived power law exponent close to 1 is in fair agreement with the dependence of the oxygen diffusion coefficient in quartz on f_{H_2O} determined by Farver and Yund (1991). In contrast, Post *et al.* (1996) derived an exponent greater than 2 from their experiments. Observations on natural quartzites favor the lower exponent (G. Hirth, pers. comm.).

2.2 Experimental Techniques

2.2.1 Starting Materials

Natural UHP rocks are in general unsuitable as starting materials, because they have undergone retrogression and deformation during exhumation. Consequently, experimental studies rely chiefly on synthetic materials. The types of defects and their concentration in these synthetic materials may deviate from those formed under natural conditions, however. The synthesis of polycrystalline aggregates of minerals stable at extreme P-T conditions with a microstructure appropriate for deformation experiments requires substantial preparatory work and elaborate techniques (e.g., Renner *et al.*, 1997). The restriction of experimental studies to single-phase materials is frequently an imperative consequence of the instability of phase assemblages at the high temperatures required to attain laboratory strain rates. Even if not so, the contrast in temperatures and strain rates between laboratory experiments and tectonic processes can lead to significant differences in the activated mechanisms and the relative strengths of minerals can be reversed. Hence, experimental deformation of monomineralic aggregates combined with theoretical modeling of the behavior of polyphase rocks (e.g. Ji and Zhao, 1993; 1994) represents a promising approach. Study of monomineralic rocks is not without application to nature, because deformation of rocks is frequently localized into monomineralic weak layers on all scales.

2.2.2 Apparatus

At present, only solid medium apparatus can operate at the P-T conditions characteristic of high- and ultrahigh-pressure metamorphism. In the range of 1–4 GPa the Griggs-Blacic apparatus (Tullis and Tullis, 1986) is used. It provides a reasonable stress resolution when the molten salt cell technology is applied (e.g., Tingle *et al.*, 1993; Rybacki *et al.*, in press). The multi-anvil apparatus allows higher pressures, however only rough estimates of stress, strain, and strain rate (Ando *et al.*, 1997; Karato and Rubie, 1997). In all these experimental setups the effect of pressure on the point defect chemistry can only be addressed by means of buffer techniques (e.g., Post *et al.*, 1996).

2.2.3 Use of Analogue Materials

Analogue materials provide a way to circumvent the technical problems of high pressure experimentation. Relying on a systematic relationship between rheological properties and crystal structure (Frost and Ashby, 1982) the investigation of low-P members of an "isostructural group" are used as proxies for phases stable at higher pressure (e.g., Karato *et al.*, 1995). Germanates and titanates have long been used as analogues for silicates (e.g., Beauchesne and Poirier, 1989; Burnley *et al.*, 1991). The structural equivalence between aluminum phosphate and quartz has been used to investigate the phenomenon of hydrolytic weakening at experimental P-T conditions that can be handled more easily (Boulogne *et al.*, 1988; Cordier *et al.*, 1988). The mechanical effect of a solid phase transformation has been investigated using sodium nitrate phases as analogues for the isostructural carbonates calcite and aragonite (Hams, 1993).

2.2.4 Perspectives

The perspectives for development in the future are twofold. On the one hand, improvements in high-pressure equipment and evaluation methods promise to enlarge the field of P-T conditions that can be realized (Ando *et al.*, 1997). On the other hand, the deduction of rheological parameters such as shear modulus and Burgers vector from molecular modeling may become feasible in the near future (e.g., Bukowinski, 1994).

2.3 Mineral Phases Relevant to UHPM

For subducted oceanic crust, which is compositionally and structurally uniform compared to continental crust, the mineralogical composition can be predicted as a function of depth and temperature (Peacock, 1993), subject to assumptions about the kinetics of phase transformations and the availability of water (Hacker, 1996). In fact, equilibrium may not be attained for large portions of subducted oceanic crust to considerable depth (Hacker, 1996; Kirby *et al.*, 1996). For continental crust, the variability of rock composition is large and any upper crustal rock type may be present in the deep levels of collision zones. Fortunately, the number of minerals that must be investigated can be reduced when first-order rheological information is required. First, major constituents must be identified because a certain amount of a mineral must be present to dominate the bulk flow behavior. Second, these major constituents have to be assigned to one of two rheological extremes, a stress-supporting framework or interconnected weak layers (Handy, 1994). Theoretical modeling of aggregates composed of two equant phases whose strength differs by a factor of 10 shows that if the weak phase makes up ≥30% of the rock, the aggregate strength is well approximated by the flow law for the weak phase (Ji and Zhao, 1994). The transition from strong to weak behavior is less pronounced when the grains of the strong phase have high aspect ratios or when the strengths of both compounds differ by less than a factor of 10. Less abundant mineral phases are not presumed to control bulk rheology (at least when plastic flow of purely solid phases is considered), because they form isolated inclusions within a stress-supporting framework, or isolated rigid inclusions in a flowing matrix.

Granites, gneisses and quartz-bearing schists dominate the continental crust. The microstructures of these rocks indicate that, under normal metamorphism, quartz controls the rheology, whereas feldspars are brittle or rigid at temperatures ≤500°C (e.g., Voll, 1976; Simpson, 1985; Vernon and Flood, 1987; Tullis *et al.*, 1990). Dislocation creep is effective in both minerals at higher temperatures. The fabrics of natural rocks, for example of granites transformed to augen gneiss under medium- to high-grade metamorphism, indicate that even then quartz constitutes the weak phase, although the contrast diminishes, in particular for fine-grained feldspar aggregates (Rutter and Brodie, 1992; Fitzgerald and Stünitz, 1993). Consequently, the mechanical behavior of continental crust is generally simulated using flow laws for quartz. Furthermore, the derivation of experimentally calibrated flow laws for feldspars and their application to natural rocks is subject to problems inherent to the low melting temperatures and the compositional and structural complexities of feldspars. Fortunately, these complexities are not relevant in the present context: plagioclase is not stable under conditions of high and UHP metamorphism and K-feldspar, which is a major constituent of granites and certain gneisses, rarely forms a stress-supporting framework.

The following minerals are expected to control the mechanical behavior of continental crust at UHP, with the phases presumably governing the bulk properties of the rocks italicized:

Mafic rocks: *Omphacite*, garnet
Granitoid rocks: *Coesite*, K-feldspar, phengite, jadeite
Metapelites: *Coesite*, phengite, kyanite, omphacite
Carbonate rocks: *Aragonite, dolomite*

The stability of coesite at pressures above 2.5–3 GPa defines the low-P boundary of UHPM (Schreyer, 1995). Carbonate rocks are included since these form massive and locally monomineralic layers of considerable extent in the continental crust and may thus be of potential significance for crustal scale rheology and strain localization.

3. MECHANICAL BEHAVIOR OF UHP MINERALS

3.1 Sodic Pyroxene (Omphacite, Jadeite)

Mechanical data for omphacite are not available. Jadeite has been studied only recently to use the exothermic disproportionation reaction albite → jadeite + quartz as an analogue for the key phase transformation accompanying eclogitization (Hacker and Kirby, 1990; Gleason and Green, 1996). Clinopyroxene aggregates with a composition between diopside and salite were deformed by Avé Lallement (1978), Kirby and Kronenberg (1984) and Boland and Tullis (1986). The experiments of Avé Lallement (1978) and Kirby and Kronenberg (1984) were carried out in solid-medium apparatus which have a limited stress resolution. In addition to these studies on polycrystalline material, diopside single crystals have been extensively investigated (e.g., Raterron *et al.*, 1994).

In eclogites a lattice preferred orientation of omphacite with [001] parallel to X and (010) parallel to the XY plane is common (Godard and van Roermund, 1995). The activated glide systems identified by Godard and van Roermund are (100)[001], (110)[001], and $(\bar{1}10)\frac{1}{2}$<110>. Irregularly sutured high-angle grain boundaries indicate migration recrystallization. Minor compositional differences across these grain boundaries and patchy compositional inhomogeneities in omphacite (D. Piepenbreier, pers. com., Van der Klauw *et al.*, 1997) suggest that chemical changes provided an additional driving force for grain-boundary migration. Exchange of the components with an external reservoir by diffusion along the high angle grain boundaries must have been fast compared to grain-boundary migration. This also suggests that diffusive mass transfer is likely to contribute to deformation of clinopyroxene under eclogite-facies conditions.

The mechanical data obtained in a gas apparatus for dislocation creep of "wet" clinopyroxenite (Boland and Tullis, 1986) may be taken as a tentative prediction of the flow strength of eclogites. A stress-temperature curve for constant strain rate is displayed in Fig. 1. It must be stressed, however, that the mechanical behavior of sodic pyroxene may be significantly different from diopside-hedenbergite. Godard and van Roermund (1995) have emphasized the lower melting temperature T_m of sodic pyroxene: at a confining pressure of 3 GPa, the melting temperature of jadeite is ~1410 °C (Bell and Roseboom, 1969), compared to ~1760 °C for diopside (Williams and Kennedy, 1969). For a given temperature, this difference causes the homologous temperature T/T_m of sodic pyroxene to be higher than diopside. At 600°C, the homologous temperature is 0.52 for jadeite and 0.43 for diopside. Such differences imply that jadeite or omphacite have significantly lower strength than predicted by a flow law for diopside. This consideration is consistent with the microstructural record of eclogites (e.g., D. Piepenbreier, pers. comm., van Roermund and Boland, 1981;

Philippot and van Roermund, 1992; Godard and van Roermund, 1995; Van der Klauw *et al.*, 1997), which indicates deformation of omphacite by dislocation creep at temperatures as low as 500-600°C and a low flow strength compared to garnet (Fig. 2). For these temperatures, the flow law of Boland and Tullis (1986) predicts that plastic flow of diopside is impossible.

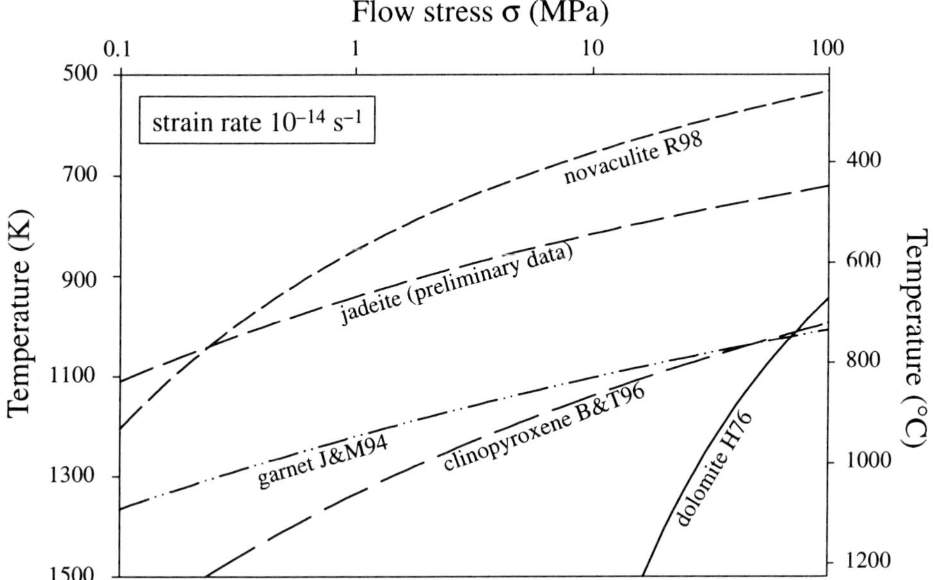

Figure 1. Power flow laws for UHP minerals (logarithmic stress scale); novaculite shown for comparison. dolomite: $ln(A[\text{MPa}^{-n}\text{s}^{-1}]) = -29.7 \pm 6.0$, $n = 9.1 \pm 0.9$, $Q = 349 \pm 25$ kJ/mol (Heard, 1976); clinopyroxenite: $ln(A[\text{MPa}^{-n}\text{s}^{-1}]) = 11.9$, $n = 3.3 \pm 0.9$, $Q = 490 \pm 159$ kJ/mol (Boland and Tullis, 1986); garnet: $ln(A[\text{MPa}^{-n}\text{s}^{-1}]) = 15.6$, $n = 2.2 \pm 0.1$, $Q = 485 \pm 30$ kJ/mol (Ji and Martignole, 1994); jadeite: $ln(A[\text{MPa}^{-n}\text{s}^{-1}]) = 22 \pm 2$, $n = 3.6 \pm 0.3$, $Q = 425 \pm 25$ kJ/mol; novaculite: $ln(A[\text{MPa}^{-n}\text{s}^{-1}]) = -8.1 \pm 1.1$, $n = 3.1 \pm 0.2$, $Q = 170 \pm 50$ kJ/mol (Rybacki *et al.*, in press).

We recently attempted to synthesize polycrystalline jadeite aggregates for deformation experiments at a pressure of 3.5 GPa and a temperature of 1100°C for up to 80 h, using either glass or gel as a precursor. However, crystallization was never complete due to sluggish kinetics and an amorphous phase remained in the central part of the cylindrical sample. SEM and cathodoluminescence microscopy revealed that no relics of the amorphous phase are preserved in the crystallized mantle, that makes up about 90% of the cross sectional area of the cylinder. Hence, we do not expect a drastic mechanical effect of the initially amorphous core on the strength of the sample. Furthermore, after deformation at temperatures as low as 800-900°C the samples were completely crystallized. Jadeite crystals are lath shaped, up to 0.04 mm long and show a characteristic aspect ratio of 3:5. Grain boundaries are unilaterally rational, as are the interfaces of pores. Porosity is inhomogeneously distributed and in places reaches a few percent. There is no marked microstructural contrast between the undeformed and the deformed sample. Five deformation experiments on these jadeite aggregates give preliminary power law parameters of $ln(A[\mathrm{MPa}^{-n}\mathrm{s}^{-1}]) = 22 \pm 2$, n = 3.6 ± 0.3, and $Q = 425 \pm 25$ kJ/mol. If these parameters are confirmed by further experimental results, a considerable strength difference between jadeite and diopside at geological strain rates is indicated (Fig. 1), which supports the inferred low strength of omphacite as compared to diopside under natural conditions. Whereas at low temperatures jadeite is considerably stronger than α quartz, at temperatures of about 600°C, the strength of jadeite approaches that of quartz, and the contrast may even be reversed at higher temperatures (Fig. 1). Thus, eclogites may not be as strong as expected (Mancktelow, 1993).

3.2 Garnet

Garnet is a mineral commonly assumed to be very strong. In fact, isolated garnet crystals characteristically form rigid clasts in a deformed matrix. However, a conspicuous example of elongate garnets has been described by Ji and Martignole (1994). They interpreted the grains to reflect crystal-plastic deformation at high temperatures, although others have suggested an origin by dissolution (see discussion by Ji and Martignole, 1994; Den Brok and Kruhl, 1996). For synthetic Gd-Ga garnets used in ceramics, the dominant glide system is $(110)\frac{1}{2}\langle 111 \rangle$ (Garem *et al.*, 1982). In natural garnets from peridotites and eclogites, dislocations with the Burgers vector $\langle 100 \rangle$ have also been found (Ando *et al.*, 1993). Based on experimental results obtained by Wang, Karato, Liu and Fujino (cf. Karato *et al.*, 1995), Ji and Martignole (1994) quoted the flow law parameters for almandine-pyrope in the caption to Fig. 1. Karato *et al.* (1995) studied the strength of garnets in a broad compositional range using the indenter technique. From these data they derived a universal flow law for garnets based on normalizing stress by the shear modulus μ and temperature by the melting temperature: $\dot{\varepsilon} = C(\sigma / \mu)^n \exp(-gT_m / T)$ with $C = 46 \pm 7\ \mathrm{s}^{-1}$, $n = 2.7 \pm 0.3$ and $g = 36 \pm 5$. According to these experimental results, the flow strength of garnet is similar to that of wet clinopyroxenite (Boland and Tullis, 1986) at natural strain rates (see Fig. 1). This prediction is at odds with microstructures in eclogites (Fig. 2). There, clinopyroxene aggregates are deflected around rigid garnet crystals. If the predictions of garnet strength are correct, this observation may support the above conclusion that sodic pyroxene is weaker than inferred from experimental studies of diopside-hedenbergite clinopyroxenes.

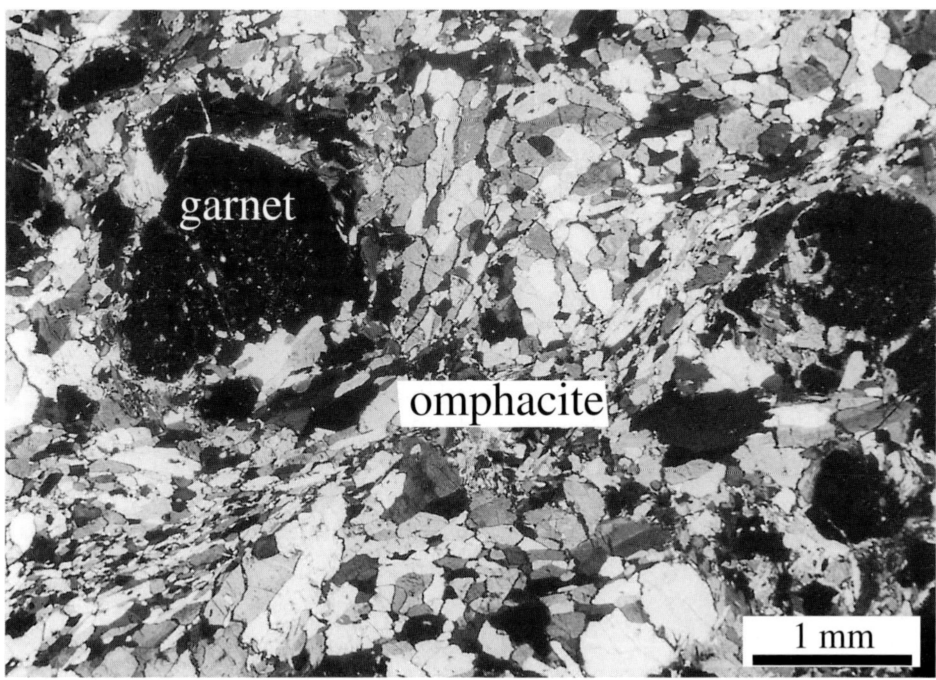

Figure 2. Microstructures of an UHP eclogite from Lago Cignana, Western Alps (Van der Klauw *et al.*, 1997). The foliation defined by omphacite flows around the garnet, revealing a contrast in strength. Stress concentration at the stiff garnet crystal causes variations in recrystallized grain size of omphacite.

3.3 Dolomite

In dolomite, the critical resolved shear stress for twinning $(\bar{1}012)\langle 10\bar{1}1\rangle$ and activation of the predominant glide system $(0001)\langle 2\bar{1}\bar{1}0\rangle$ is on the order of 100 MPa, about one order of magnitude higher than for calcite (Barber *et al.*, 1981). Barber *et al.* (1994) compared experimental observations and theoretical predictions of the microphysics of dislocation creep of polycrystalline dolomite. The marked plastic anisotropy results in a small number of activated slip systems per grain and deformation of polycrystalline aggregates requires activation of additional mechanisms, such as microcracking and grain-boundary sliding. The high flow strength and the exceptionally high stress exponent of 9.1 found by Heard (1976) may be a consequence of these properties. Heard's (1976) experimentally derived flow law is depicted in Fig. 1 and the parameters are given in the caption. It is expected that dolomite rocks remain rigid when associated with aragonite marbles at UHP.

3.4 Coesite

Due to its anticipated abundance as a replacement of quartz and as a breakdown product of plagioclase, coesite likely forms a continuous phase and controls the mechanical behavior of most continental crustal rocks at UHP. Triaxial compression tests were performed on synthetic polycrystalline coesite (Renner *et al.*, 1997) in a modified solid-medium apparatus (Rybacki *et al.*, in press) at confining pressures ≤3.7 GPa, at temperatures ≤1170°C and at strain rates of $6x10^{-7}$-$6x10^{-4}$ s^{-1} (Renner *et al.*, 1996). Microstructural homogeneity of the deformed samples precludes major gradients in the boundary conditions (in particular temperature) within the sample assembly. The stress resolution in a solid medium apparatus is limited by friction and we addressed this limitation by calculating lower and upper bounds to the flow stress (Rybacki *et al.*, in press). The differential stresses decrease with increasing temperature and decreasing strain rate regardless of the method of calculation. Flow laws have been separately fit to the stress limits. The lower stress limits are well fit by a law for deformation controlled by dislocation glide, with the lattice resistance represented by a Peierls stress σ_P:

$$\dot{\varepsilon} = B\sigma^n \exp\left\{-\frac{-\Delta G}{RT}\left[1-\left(\frac{\sigma}{\sigma_P}\right)^p\right]^q\right\}$$

with $ln(B[\mathrm{MPa}^{-m}\mathrm{s}^{-1}]) = 14.4 \pm 4.3$, $m = 1.0 \pm 0.1$, $\Delta G = 322 \pm 38$ kJ/mol, $\sigma_P = 42 \pm 19$ GPa, $p = 0.7 \pm 0.2$, and $q = 1.0 \pm 0.2$. For the upper limits we deduced power law parameters $ln(A[\mathrm{MPa}^{-n}\mathrm{s}^{-1}]) = -7.3 \pm 4.5$, $n = 3.1 \pm 0.5$ and $Q = 257 \pm 37$ kJ/mol.

The correlation between applied temperature, stress, strain rate and observed microstructures (deformation lamellae, undulatory extinction, grain-boundary migration, recrystallized grains, shape and crystallographic preferred orientation; see Fig. 3) indicates dislocation creep—in agreement with the mechanical data. The initial low water content of the synthetic coesite, 500-1000 x 10^{-6} H/Si, does not change during deformation but FTIR (Fourier Transform Infrared Spectroscopy) spectra indicate a modification of the type of incorporated hydrogen-related defects; several types of point defect indicated by distinct sharp absorption bands in the starting material are not present in deformed samples, leaving only broad band absorption (Renner, 1996). Comparing the above flow laws to a flow law for fine-grained novaculite (grain size ~10 μm) in the α quartz stability field using the same apparatus, but the molten salt technology, reveals that polycrystalline coesite aggregates are stronger than α quartz aggregates at laboratory strain rates (Fig. 4) and temperatures at least as high as 900°C. This

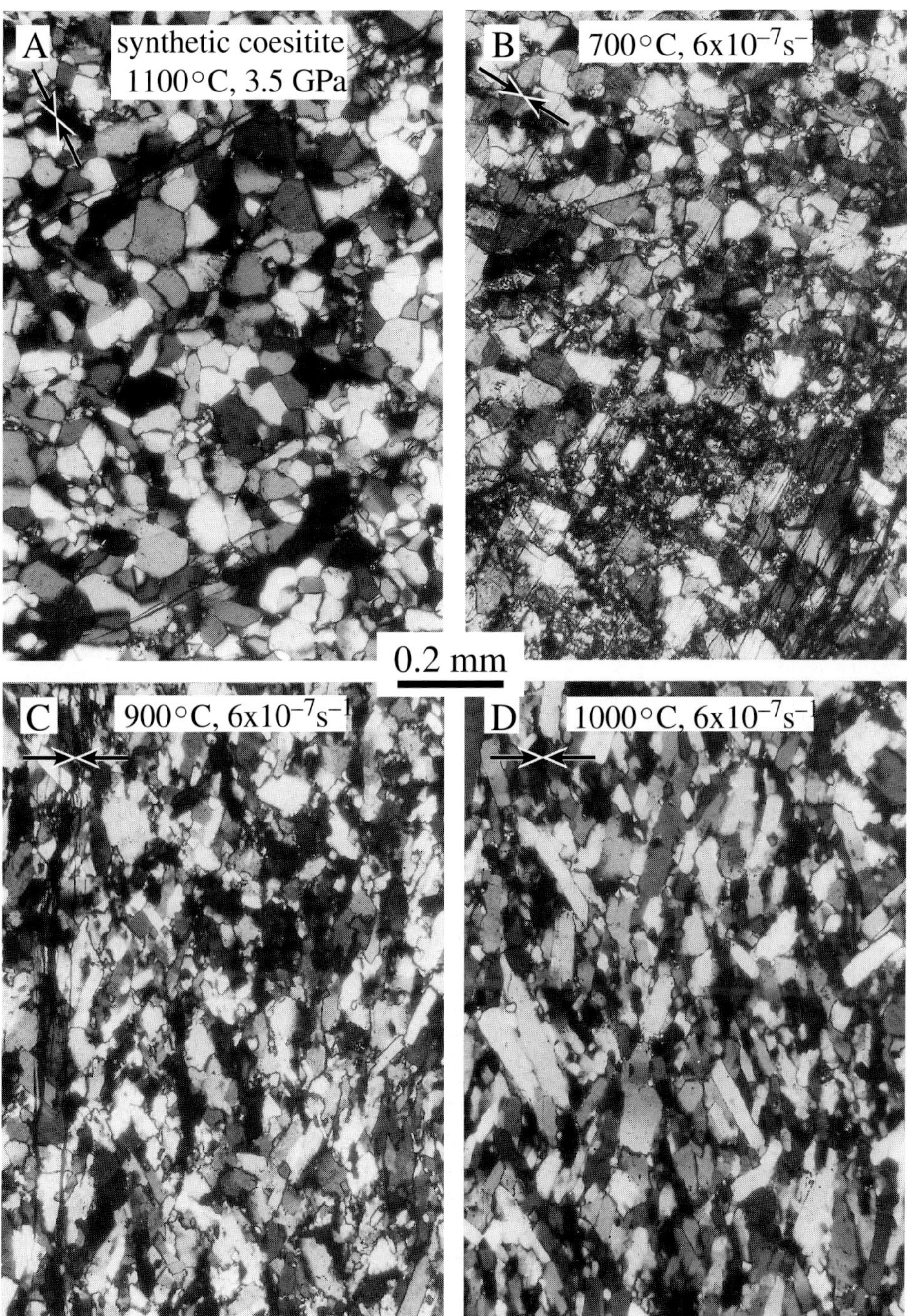

Figure 3. Microstructures of synthetic polycrystalline coesite synthesized at 1100°C, and deformed at a strain rate of $6x10^{-7}$ s^{-1}. The orientation of the cylinder axis (shortening direction in deformed samples) is indicated on each microphotograph.

Figure 3 continued. A) Undeformed starting material characterized by equant grains with straight or slightly curved high-angle grain boundaries. A tendency toward a foam structure is observed, though low-energy twin boundaries and unilaterally rational grain boundaries are still common. Up to ~1150°C, the interfacial free energy in coesite aggregates depends significantly on orientation (Renner *et al.*, 1997). B) Microstructure after deformation at high differential stress and 700°C. Coesite reveals undulatory extinction, deformation bands and lamellae, and recrystallization to a grain size of ~10 µm. C) Microstructure after deformation at 900°C. Grains have a shape preferred orientation with long axes normal to the shortening direction and are recrystallized to a grain size of ~20 µm. High-angle grain boundaries are strongly sutured. A few recrystallized grains have grown extensively to form blasts with a tabular shape controlled by unilaterally rational grain boundaries. D) Microstructures after deformation at 1000°C are similar to those at 900°C, but recrystallized grains are larger, and the proportion and size of tabular blasts have increased.

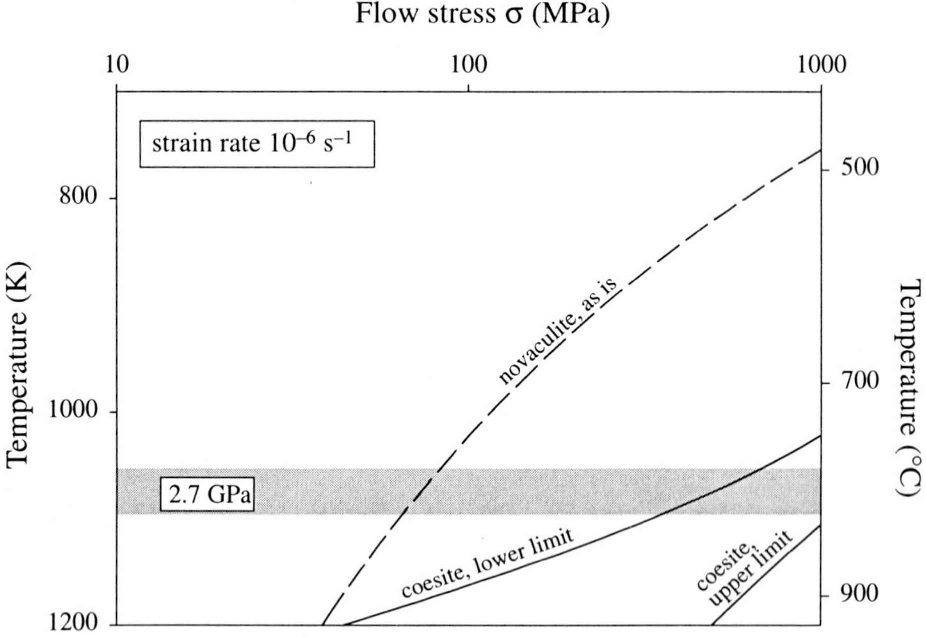

Figure 4. Experimentally determined flow laws of polycrystalline SiO_2 (novaculite, "as is", Rybacki *et al.*, in press; coesite, Renner, 1996) at a strain rate of 10^{-6} s^{-1}. This strain rate is assumed relevant for laboratory experiments on the coesite-quartz transformation at 800°C and 2.7 GPa (gray), where quartz has deforms by dislocation creep, whereas coesite relics are rigid clasts, suggesting a higher strength (Wirth and Stöckhert, 1995). This strength contrast of the two phases on laboratory time scales is consistent with observations on natural rocks (Ingrin and Gillet, 1986).

result agrees with independent microstructural observations in experimental products and natural rocks. In experiments on the transformation of coesite to quartz at 800°C and 2.7 GPa (Fig. 4), α quartz deforms readily by dislocation creep whereas coesite remains unaffected (Wirth and Stöckhert, 1995). The same observation is reported for natural rocks from the Dora Maira Massif (Ingrin and Gillet, 1986).

In Fig. 5 extrapolations of quartz flow laws from the literature are compared to the results for coesite. Rather than reviewing thc considerable number of flow laws available for quartzites, here we concentrate on the extreme limits of treatment with respect to water, namely vacuum drying of the sample prior to deformation and addition of free water. Additionally, two data sets on untreated "as is" samples are given since in both studies molten salt was used, which yields a better stress resolution compared to a classic solid medium. Though the respective fields of flow stresses for α and β quartz show considerable overlap, particularly at high stresses, α quartz aggregates tend to be weaker than β quartz aggregates at comparable water conditions. The validity of the stresses in Fig. 5 most likely ends within the interval between 100-1000 MPa where power law breakdown (Frost and Ashby, 1982) is cncountered and different deformation mechanisms become rate controlling. Hence, deformation experiments on synthetic polycrystalline coesite aggregates give an extrapolated stress range similar to that for α quartz. The assessment of changes in rheological behavior of the continental crust due to the replacement of quartz by coesite and the importance of related rheological discontinuities is not only limited by the wide stress range given for coesite, but also depends on whether flow laws for quartz can be extrapolated to pressures of the quartz-coesite transition. In particular, uncertainty remains with respect to the effect of water on the flow strength, for which an appropriate microphysical model is still missing. Post *et al.* (1996) proposed that strength is related to water fugacity, but the fugacity of water in aqueous solutions at ultrahigh pressures and temperatures is poorly constrained and presumably strongly affected by the presence of additional solids in impure natural systems.

3.5 Aragonite

Using the molten salt cell, Rybacki *et al.* (1995) and Konrad (1997) experimentally deformed synthetic polycrystalline aragonite in the aragonite stability field. In these studies the initial microstructure of the samples synthesized in a solid medium apparatus showed considerable variation, despite nominally uniform conditions of synthesis. In some samples even calcite was present. Hacker and Kirby (1993) had already noted that in solid medium apparatus both phases coexist in a broad P-T field at temperatures above 900°C, in conflict with extrapolation of the calcite-aragonite equilibrium determined in gas apparatus experiments. Stored strain energy may affect the respective stability fields. In the deformation experiments of Rybacki *et al.* (1995) and Konrad (1997), a rather homogeneous microstructure of equiaxed small aragonite grains developed after as little as 10% axial shortening. The size of the recrystallized grains in deformed samples varied systematically with the flow stress, as predicted (e.g., Twiss, 1977). The mechanical data are well described

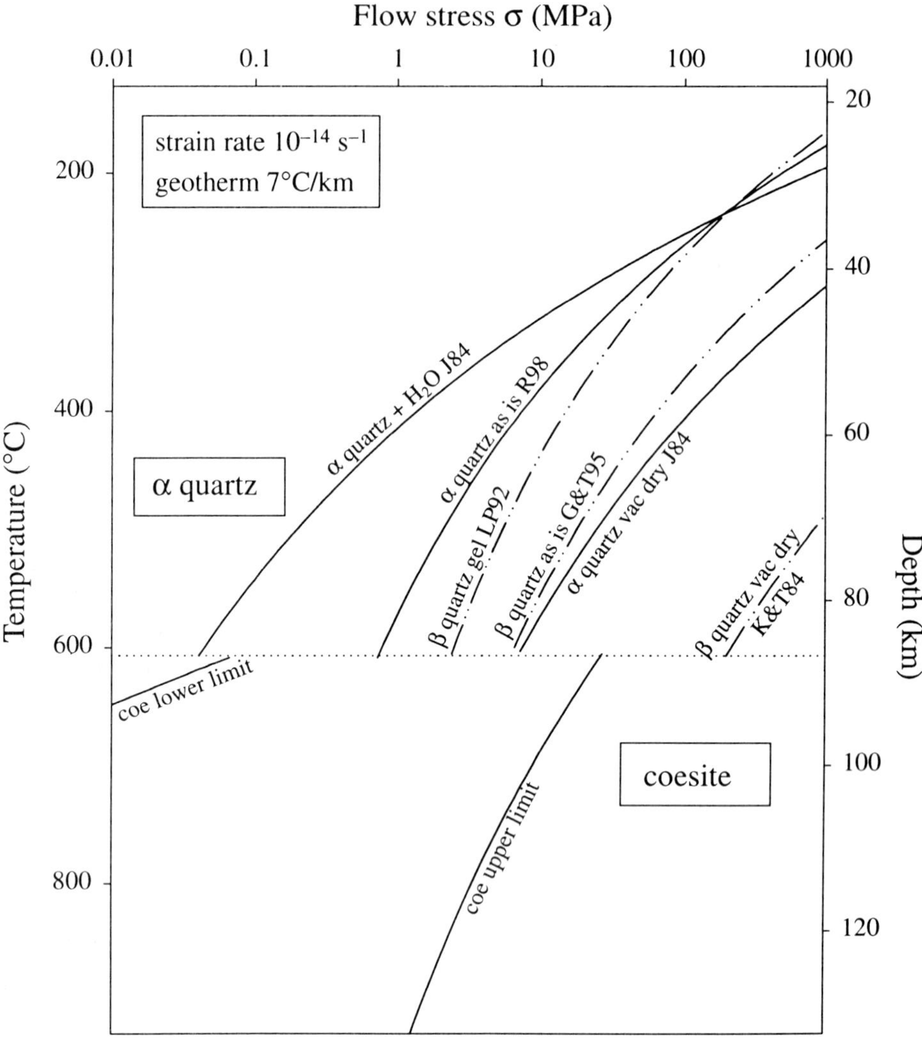

Figure 5. Experimentally determined flow laws of polycrystalline SiO_2 extrapolated to a strain rate of 10^{-14} s^{-1}, with the influence of water fugacity neglected. The transition from α quartz to coesite occurs at a depth of 80-90 km for the assumed geotherm of 7°C/km (Mirwald and Massonne, 1980). β quartz is not a relevant phase at UHP, however a substantial number of studies has been performed in the β field, generally with better stress resolution than in those in the α field. J84, vacuum-dried Heavitree quartzite with 0.4 wt% H_2O added (Jaoul *et al.*, 1984); R97, novaculite "as is" (Rybacki *et al.*, in press); L&P92, synthetic quartzite from gel precursor (Luan and Paterson, 1992) average given by Evans and Kohlstedt (1995); K&T84, vacuum-dried Heavitree quartzite (Kronenberg and Tullis, 1984); G&T95, "as is" Black Hills quartzite without melt (Gleason and Tullis, 1995); upper and lower limits of coesite (Renner, 1996).

by a power flow law with $A = 1 \pm 0.5$ MPa^{-n}s^{-1}, $n = 4.9 \pm 0.3$, and $Q = 255 \pm 40$ kJ/mol.

Compared to two natural rocks (Carrara marble and Solnhofen limestone), synthetic aragonite flows at lower stresses at identical strain rates (Fig. 6). However, synthetic calcite aggregates yielded the lowest flow stress at all conditions. Curiously, the synthetic aggregate is weaker than Carrara marble at temperatures above about 500°C, even for a strain rate 8 orders of magnitude higher. Most likely, some important parameter is missing when describing flow in calcite aggregates by a power law. In view of the small grain size of Solnhofen limestone, this parameter appears to be related to impurity content (Wang *et al.*, 1996) rather than to grain size. The general similarity of stresses at laboratory strain rates for Carrara marble and the synthetic aragonite aggregate is in agreement with the observation by Hacker and Kirby (1993) that Carrara marble transforming to aragonite docs not show systematic strength changes related to the transformation. In summary, a significant difference in flow strength of the two $CaCO_3$ polymorphs is unlikely for a strain rate of 10^{-14} s^{-1} representative for tectonic processes.

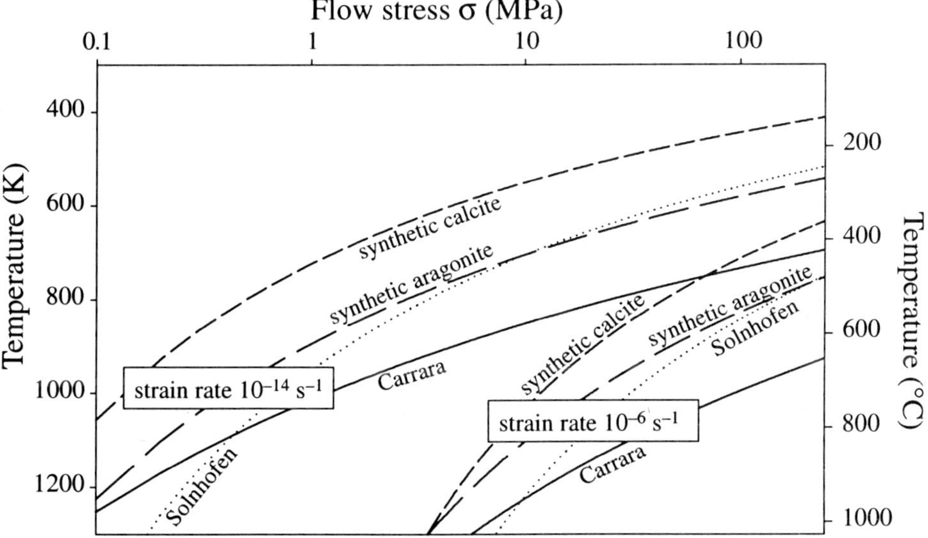

Figure 6. Experimentally determined flow laws of polycrystalline $CaCO_3$, for two strain rates, 10^{-6} and 10^{-14} s^{-1}, representing experiments of Hacker and Kirby (1993) and tectonic deformation, respectively. Carrara marble regime 3 (Schmid *et al.*, 1980); Solnhofen limestone regime 2 (Schmid *et al.*, 1977); synthetic aragonite (Konrad, 1997); synthetic calcite (Siddiqi *et al.*, 1997) ($ln(A[\text{MPa}^{-n}\text{s}^{-1}]) = -2.3$, $n = 4.1$, $Q = 180$)

Natural aragonite marbles are generally strongly retrogressed to calcite. The microstructures of well-preserved aragonite marble in a HP-LT metamorphic unit on Crete (Theye and Seidel, 1993) reveal deformation exclusively by dissolution-precipitation creep (Stöckhert et al., 1995 and in press: Wachmann, 1997). Thus, this occurrence cannot be used to test the predictions of a flow law for dislocation creep. At the same location, a systematic difference between the densities of fluid inclusions in aragonite and in quartz, which in both minerals had been stretched plastically due to internal overpressure during exhumation, has been taken by Küster and Stöckhert (1997) to hint at a greater flow strenght of aragonite as compared to quartz at temperatures near 300°C and strain rates on the order of 10^{-15} s^{-1}. As shown in Fig. 7, this tentative prediction of *relative* strength is consistent with the experimental results.

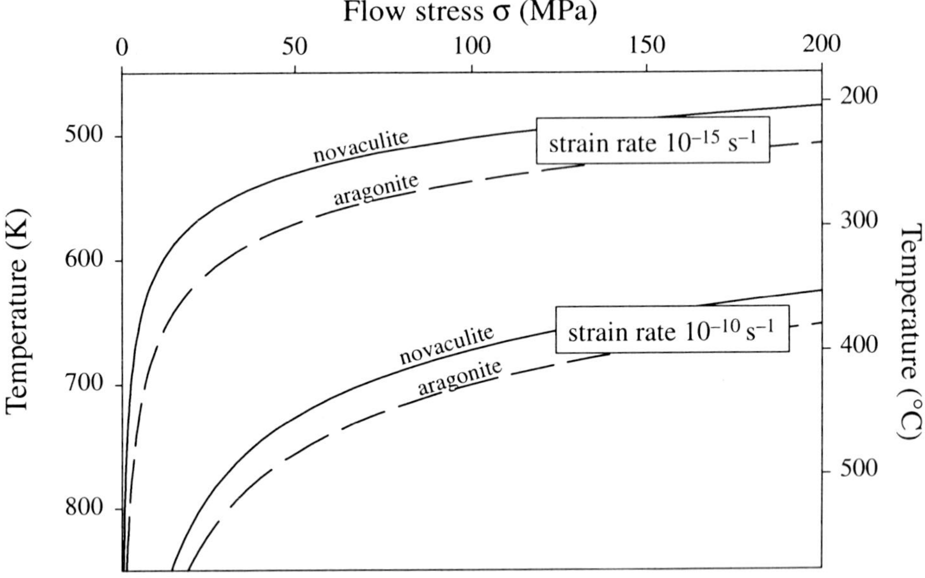

Figure 7. Extrapolated flow stresses for synthetic aragonite and natural novaculite. For a constant strain rate, aragonite has a strength comparable to that of quartz at about 30°C higher temperature. Such a difference was suspected by Küster and Stöckhert (1997) based on the mechanical behavior of fluid inclusions in both minerals.

3.6 Microstructural Record of Natural UHP Rocks

It is remarkable how little some UHP metamorphic rocks have been modified by deformation on their long return to the surface (e.g., Coleman and Wang, 1995). Apart from the transformation of coesite to quartz and of aragonite to calcite, the microstructures developed at UHP are well preserved in the Dora Maira Massif, Western Alps (Chopin *et al.*, 1991; Schertl *et al.*, 1991; Henry *et al.*, 1993; Michard *et al.*, 1995), despite the removal of 100 km of overlying crust or lithosphere subsequent to UHPM in an active tectonic environment. In other cases, UHP metamorphic rocks have undergone a subsequent stage of normal crustal metamorphism. In the Kokchetav Massif relics of minerals grown at UHP are preserved as inclusions in garnet or zircon (Shatsky *et al.*, 1995), whereas outside these strong containers the rocks are intensively retrogressed and microstructures related to UHP are completely destroyed (Fig. 8 B). This is also true for large parts of the Chinese UHP belt (B.R. Hacker, pers. comm., Coleman and Wang, 1995; Hacker *et al.*, 1995).

UHP crustal slices are commonly transected and bounded by shear zones (Schertl *et al.*, 1991; Henry *et al.*, 1993; Hacker *et al.*, 1995; Michard *et al.*, 1995) developed or reactivated under greenschist-facies conditions after the rocks had returned to the middle to upper crust. Within the coherent slices revealing UHP phase assemblages and microstructures, strain appears to be moderate. To our knowledge, high strain fabrics and shear zones unequivocally formed at UHP have not been identified. Henry *et al.* (1993) and Michard *et al.* (1995) described the structural and microstructural record of the Dora Maira UHP metamorphic rocks in some detail. This area deserves special attention because UHPM occurred in the Tertiary and exhumation into the upper crust was completed within a few million years (Gebauer *et al.*, 1997). The timing and the post-UHP overprint is better constrained than in the older UHP areas of Dabie Shan (Mesozoic), Kokchetav Massif, and Norwegian Caledonides (early Paleozoic).

In the Dora Maira Massif, slight deformation at UHP is unequivocally documented by strain shadows that surround pyrope crystals and are filled with UHP minerals, and by curved inclusion trails in the outer zones of some pyrope porphyroblasts. Most of the garnets have randomly oriented inclusions. Henry *et al.* (1993) and Michard *et al.* (1995) concluded that deformation at UHP was strongly partitioned and that high strain zones are possibly blurred by post-UHP reactivation and concomitant retrogression. In any case, strain in the bulk of the continental UHP metamorphic material in the Dora Maira Massif is conspicuously low. This is particularly well documented by the essentially undeformed Variscan Brossasco Granite, which survived UHP without loosing the macroscopic appearance of a granite (Biino and Compagnoni, 1992; Henry *et al.*, 1993; Michard *et al.*, 1995). In some parts the magmatic microstructure has remained undistorted (Fig. 8 C, D), whereas in others the granite has been transformed into a gneiss or mylonite. In these high strain regions, however, the quartz microfabrics show that the deformation occurred late under greenschist-facies conditions and not at UHP. In the undeformed granite, the most conspicuous evidence of UHP is the fine-grained quartz aggregates with random crystallographic orientation grown at the expense of coesite crystals that had replaced the undistorted original mm-sized magmatic quartz grains at UHP. The

Figure 8. Quartz microstructures of UHP rocks. A) Coarse foam structure in Dora Maira pyrope quartzite. Dislocation creep would have destroyed this grain shape; its absence implies low differential stress. B) Kokchetav diamond-bearing UHP gneiss reveals post-UHP dislocation creep at crustal conditions and exemplifies the erasure of UHP microstructures. C and D) Preserved original magmatic fabric in Brossasco Granite. Magmatic biotite surrounded by coronitic garnet survived UHP. Large magmatic quartz grains replaced by a fine-grained mosaic, attributed to rapid nucleation when quartz formed at the expense of coesite. Faint kink bands in biotite may be related to the volumetric strain of the coesite-quartz transformation.

small grain size is attributed to rapid nucleation during the coesite-quartz transformation. More clearly than any other UHP rock of the Dora Maira Massif, the Brossasco Granite shows that the differential stress remained too low to drive plastic flow throughout the history of burial to more than 100 km depth, at temperatures up to 800°C (Schertl *et al.*, 1991), and also on the way back to the surface.

It appears to be a general feature that the microstructural record of natural UHP rocks provides only limited evidence for mechanical behavior at UHP: The rocks appear to be at most slightly deformed. Furthermore, UHP rocks of the Dora Maira Massif, outside the greenschist facies shear zones, reveal a quartz microstructure of the foam type controlled by interfacial free energy (Fig. 8 A), although there is also a coarse palisade microstructure attributed to the coesite-quartz transformation (e.g., p. 974 of Henry *et al.*, 1993). Both types of microstructure indicate that the differential stress remained too low to drive deformation by dislocation creep along the exhumation path as well.

The only microstructural record of significant deformation by dislocation creep at UHP is found in eclogites. For example, in coesite-bearing eclogites from the Lago di Cignana area (Reinecke, 1991) in the Piemonte Zone of the Western Alps, van der Klauw *et al.* (1997) have described omphacite microstructures similar to those found in eclogites elsewhere (e.g., Godard and van Roermund, 1995). Omphacite with sutured grain boundaries forms interconnected weak layers flowing around rigid garnet (Fig. 2).

4. DISCUSSION

4.1 Low Strain in UHP Rocks—An Indication of Low Differential Stress

Although most UHP rocks lack evidence for high strain at UHP, considerable deformation and displacement along contacts between adjacent units is expected. Consequently, deformation must have been localized into shear zones (Michard *et al.*, 1995). These zones of weakness have either not been identified, or were not preserved due to intense retrogression or preferred reactivation during exhumation. At present, the available observations are insufficient for conclusive statements.

Localization of deformation into shear zones is considered to be a consequence of strain softening, which leads to an increased strain rate (lower effective viscosity) at constant or decreasing stress. For any simple geometry, the flow stress in shear zones cannot exceed the strength of the adjacent rigid blocks. Based on this argument, the undeformed Brossasco Granite in the Dora Maira Massif can be taken as a stress gauge that places upper bounds on the magnitude of differential stress transmitted through the bulk of the crustal slice along the P-T path established by Schertl *et al.* (1991). The absence of notable deformation is equivalent to a slow strain rate. Taking into account the presumably short durations and speeds of tectonic processes inherent in UHPM (e.g., Gebauer *et al.*, 1997), a strain rate on the order of 10^{-15} s^{-1} is taken as an upper bound for an apparently undeformed rock. Thus, the strength predicted by the above flow law for dislocation creep of coesite (Renner *et al.*, 1996) at 10^{-15} s^{-1} and at temperatures of UHPM can be taken as an upper bound for the stress transmitted through the rock at that stage. In the case of the Brossasco granite, the predicted differential stress during UHPM at 800°C remains below about 2 MPa, and below about 4 MPa at 700°C. The lack of notable syn-UHP deformation in other rock types where coesite has formed an interconnected weak layering suggests that this upper bound applies to the entire Dora Maira UHP crustal slice.

4.2 Potential Mechanisms of Strain Localization

There are many possibilities for strain softening in shear zones (e.g., White *et al.*, 1980; Tullis *et al.*, 1990). In the absence of well-preserved natural examples, the relative importance of the various mechanisms at UHP can hardly be assessed, but a short compilation of potential effects relevant to UHPM is given below.

Very fine-grained polycrystalline aggregates tend to deform at low stresses by diffusion-controlled granular flow accommodated by grain-boundary sliding. Such behavior has been experimentally observed for polycrystalline calcite and olivine (see review by Paterson, 1990), feldspar (Tullis and Yund, 1991), and anhydrite aggregates (Dell'Angelo and Olgaard, 1995), in which it may produce superplasticity. The small grain size develops by dynamic recrystallization during a preceding stage of dislocation creep (e.g., Schmid, 1982). The counteracting effect of normal grain growth can be inhibited by dispersed second-phase particles (Olgaard, 1990). Solids undergoing phase transformations have the potential of reaction-enhanced ductility (Rubie, 1983). A small grain size evolves during transformations when new phases nucleate and grain growth is either inhibited by insufficient temperature or interdispersion of different phases. Although theory suggests that transformational weakening is an important mechanism with the potential of effectively reducing stress or increasing strain rates in systems undergoing phase transitions (e.g., Poirier, 1982), experimental evidence or the identification of a respective microstructural record in UHP rocks is still missing. The metastable preservation and subsequent rapid transformation of mantle materials, offered as an explanation for deep earthquakes (Green and Burnley, 1989), seem to depend on thermodynamic prerequisites (e.g., Green and Zhou, 1996), that appear not to be met, for the calcite → aragonite transformation (Hacker and Kirby, 1993) or the albite → jadeite + quartz reaction (Gleason and Green, 1996).

In theory, solid state transformations can transiently affect the strength of a rock in several ways due to misfit stresses or changes in elastic properties (e.g., Poirier, 1982; Paterson, 1983). Kirby and Stern (1993) reviewed changes observed in the mechanical behavior of crystals associated with phase transformations. However, the question remains how these phenomena interact in a polycrystalline aggregate. Limited experimental evidence is available for materials of geological interest (Meike, 1993).

4.3 Mechanical Effect of Fluids and Melts at UHP

Where the upper crust is composed of metamorphic rocks, hydrostatic fluid pressures can be maintained to depths of ~10 km (Brace, 1980; Grawinkel and Stöckhert, 1997; Huenges *et al.*, 1997), but near-lithostatic fluid pressures are generally expected for deeper metamorphic conditions (Fyfe *et al.*, 1978). For instance, in geothermal wells fluid pressures were found to be nearly lithostatic at temperatures exceeding ~350°C (Fournier *et al.*, 1991). However, transient reductions related to major seismic events are possible (e.g., Sibson, 1990; Küster and Stöckhert, in press). A near-lithostatic fluid pressure implies a low brittle and frictional strength. The effect of temporal changes of fluid pressure due to dehydration or hydration reactions (e.g., Olgaard *et al.*, 1995; Hacker, 1997) depends on permeability and the rate of reaction. Triggering of seismic events may be possible if the state of stress is appropriate.

The presence of a fluid phase during UHPM has been inferred from dehydration reactions and mass-transfer considerations (Harley and Carswell, 1995; Schreyer, 1995). For the Dora Maira Massif, Philippot (1993) and Philippot *et al.* (1995) demonstrated the presence of fluids by analyzing fluid inclusions in UHP minerals. These fluid inclusions contain highly concentrated aqueous brines that precipitated a variety of daughter minerals. Experimental studies of the wetting angle between aqueous brines and silicates have revealed that, in general, the dihedral angle decreases with increasing salinity (e.g., Watson and Brenan, 1987; Lee *et al.*, 1991; Holness, 1992) and pressure (e.g., Laporte and Watson, 1991; Holness, 1993). Although experimental data for complex systems at extreme pressures are not available, it can be anticipated that at UHP the dihedral angles are small and the fluid phase may wet certain grain boundaries. Such a fluid phase distribution would reduce the tensile strength to low values and facilitate grain-boundary sliding and dissolution-precipitation creep, by analogy to liquid phase sintering in ceramics (e.g., Cawley and Lee, 1994; Cook and Pharr, 1994; Rabkin, 1994; Rabkin *et al.*, 1994). The dihedral angles and the wetting behavior are expected to depend strongly on microstructure and mineral assemblage because interfacial free energy depends on the kind of adjacent crystals, their orientations, and the orientation of the interface (e.g., Shaw and Duncombe, 1991; Laporte and Watson, 1995). Finally, interfaces that are not wetted at static conditions may be wetted by fluids or melts during deformation (Urai, 1983; Jin *et al.*, 1994; Tullis *et al.*, 1996).

The presence of hydrous melts may have an effect similar to that of dense aqueous solutions. Conditions of UHPM (Coleman and Wang, 1995; Harley and Carswell, 1995; Schreyer, 1995) are generally above the high-pressure "wet" solidus for granitic melts (Huang and Wyllie, 1975). For the Dora Maira Massif, the presence of partial melts during UHPM has been discussed by Schreyer *et al.* (1987), Schreyer (1995), Philippot (1993) and Sharp *et al.* (1993), though unequivocal evidence for melt segregation into veins is scarce (H.-P. Schertl, pers. comm.). Partial melts are expected to reduce mechanical strength (e.g., Brown *et al.*, 1995; Rutter and Neumann, 1995). The effect is expected to depend strongly on the geometry of melt distribution (e.g., Laporte, 1994), which is insufficiently known for the conditions and phase assemblages relevant to UHPM due to a lack of experiments. The absence of veins and migmatitic structures in UHP rocks of the Dora Maira Massif suggests that melt, wherever formed, was not easily segregated. Finally, experimental results on the deformation of partially molten rock (Hacker and Christie, 1990; Rushmer, 1995; Rutter and Neumann, 1995) are not quantitatively extrapolatable to natural strain rates and low stresses because constitutive laws are missing and the processes at low strain rates may differ significantly from those observed in the laboratory.

Thus, deformation by a process analogous to liquid phase sintering could allow for low crustal strength at UHP or, when confined to certain rock types and mineral assemblages, even localized flow. Microstructural evidence for this process should be sought in natural UHP rocks, although the microstructures of ceramics deformed by liquid phase sintering (e.g., German, 1996) leave little hope for an unequivocal identification of an analogous process in rocks.

4.4 Rheological Discontinuities Related to Phase Transformations

An assessment of the change in the rheological behavior of the continental crust due to the replacement of quartz by coesite and of the significance of potentially related rheological discontinuities, apart from aspects of transformational plasticity, depends on whether flow laws for quartz can be extrapolated to pressures of the quartz-coesite transition (Fig. 5). This extrapolation to high pressure is complicated by the insufficiently known pressure dependence and microphysical effects of the incorporation of water-related defects in quartz. An empirical correlation between flow strength and water fugacity has been suggested by Kohlstedt *et al.* (1995) and Post *et al.* (1996), based on experimental data. However, the activity of H_2O in the highly saline brines associated with UHPM (Philippot, 1993; Philippot *et al.*, 1995) is strongly reduced (see discussion in Harley and Carswell, 1995), offsetting the effect of high pressure. Likewise, a marked contrast in the flow strength of aragonite and calcite rocks is not predicted by the experimental data. It must be stressed that such rheological predictions, made using flow laws for dislocation creep, are only significant if the strength of the rocks is actually controlled by this deformation mechanism. As discussed above, this appears to be unlikely for known UHP continental crust, where deformation by diffusional creep in a broad sense, at differential stresses below those required to drive dislocation creep, are suggested by the microstructural record.

4.5 Low Strength and Buoyancy

Chemenda *et al.* (1995) proposed exhumation models for UHP metamorphic crust in collision zones based on the buoyant rise of steeply inclined coherent continental sections. For slabs that maintain mechanical integrity from the surface to depth, slab buoyancy produces a nonhydrostatic stress state. The order of magnitude for the resulting stress differences in a slab with a density difference of $\Delta\rho = 0.2$ g/cm^3 compared to the mantle at a depth of $z = 100$ km (e.g., Cloos, 1993) can be estimated roughly from $\Delta\sigma = \Delta\rho \cdot g \cdot z$ to be 200 MPa. These stresses drive the slab upward. Can continental crust sustain these stresses at UHP—independent of whether the deformation is homogeneous or localized? The mechanical data for coesite (Renner *et al.*, 1996) predict a strain rate of ~10^{-8} s^{-1} at a differential stress of 200 MPa at 800°C, which comes close to laboratory conditions. The strength of continental material at UHP is insufficient to sustain such a differential stress and the crust should become squeezed rather than return as a large, coherent slice. This simple consideration suggests that for a marked density contrast the size of UHP crustal slices at mantle depths must be comparatively small (cf. Molnar and Gray, 1979) and that buoyant rise of steeply inclined continental crust reaching from the surface far down into the mantle is probably not an appropriate scenario to explain the exhumation of UHP rocks.

4.6 Rheology and P-T Conditions in Subduction Zones

Thermomechanical simulations of subduction zones e.g. van den Beukel and Wortel, 1988; van den Beukel, 1992; Peacock, 1992; 1996) have shown that rheology has a pronounced effect on thermal structure. The contribution of shear heating—transformation of mechanical work into heat, and thus effective at high stress (e.g., Molnar and England, 1990)—is significant compared to the effects of the age of subducting lithosphere or the rate of convergence (Peacock, 1996). In an analysis of continental collision, van den Beukel (1992) concluded that the principles established for subduction of oceanic material also hold for continental crust. However, the low strength of continental crust at UHP predicted from the laboratory results, and the evidence for very low differential stress in UHP metamorphic rocks suggest that shear heating is insignificant. Microstructural evidence for decoupling and low shear stress has also been found in rocks exhumed from the shallower parts of subduction zones (Stöckhert *et al.*, 1997; Stöckhert *et al.*, in press). From this point of view, the thermal structures and P-T paths obtained by Peacock (1992; 1996) for a low shear stress may best reflect the actual conditions in convergent plate boundaries. However, the P-T conditions recorded by exhumed HP and UHP metamorphic rocks generally indicate somewhat higher temperatures for a given depth, compared to those predicted by the low-stress simulations. The discrepancy probably indicates that some other parameters or assumptions in the simulations need to be refined. Alternatively, exhumed HP and UHP metamorphic rocks may not necessarily represent steady state subduction, as assumed in most simulations.

4.7 Implications of UHP Rheology for Seismology

Active subduction of continental crust has been inferred from studies of seismicity and seismic properties at depth levels corresponding to UHPM in intracontinental shortening zones such as the Pamir-Hindu Kush (Roecker, 1982), and in continent-arc collision zones such as the Sunda-Banda arc (McCaffrey *et al.*, 1985). The properties identified in these studies on active convergence zones should be tested against the evidence from UHP rocks exposed at the surface. In particular, the low strength predicted from the extrapolation of the laboratory data and the absence of pervasive deformation in exhumed UHP rocks places constraints on the magnitude of stress drop of intracrustal subduction/collision zone earthquakes at 100-200 km depth. Even if the observed stress drop should be total (Scholz, 1990), as could be envisaged for systems whose mechanical strength is limited by the distribution of a fluid or melt phase, the magnitude of the stress drop is expected to remain below a few MPa.

5. SUMMARY AND CONCLUSIONS

In the present contribution we have examined aspects pertinent to the rheology of crustal materials at UHP. Despite the scarcity of experimental data and restricted information from natural UHP rocks, the following tentative conclusions can be drawn:

1) Some natural UHP rocks reveal evidence for weak or even essentially absent deformation at UHP. For the given geodynamic environment, this requires strain localization.

2) Shear zones developed at UHP have not been unequivocally identified. This could be due to the absence of clear indicators, a low potential for

preservation, or ease of reactivation. Shear zones that developed under late amphibolite to greenschist facies conditions after a significant portion of exhumation, are ubiquitous in UHP terrains, however.

3) A low strength of crustal rocks at UHP could be caused by the presence of a fluid phase liberated by metamorphic reactions. The high solute content of these fluids suggests small dihedral angles, though experimental evidence is not yet available. Wetted grain or interphase boundaries could allow for effective grain-boundary sliding with grain shape accommodation by dissolution precipitation, analogous to liquid phase sintering in ceramics. A similar effect may be produced by small melt fractions.

4) The experimentally derived flow law for dislocation creep of coesite places an upper bound of a few MPa on the differential stress supported by coesite-rich rocks that lack evidence for strong deformation at UHP.

5) The rheological contrast between quartz- and coesite-bearing materials is poorly known. In any case, a crustal-scale rheological discontinuity due to the quartz-coesite transition can only be expected if the strength is controlled by dislocation creep of these minerals. Based on the presently available evidence from UHP rocks this is not the case.

6) A low strength for crustal material at UHP implies the absence of effective shear heating and renders the "coolest" thermal structures and P-T paths obtained in thermomechanical simulations most realistic. To resolve discrepancies with the P-T record of exhumed metamorphic rocks, other boundary conditions chosen for the simulations should be re-examined. On the other hand, it must be emphasized that exhumed UHP rocks do not represent conditions of steady state subduction, as assumed in the simulations. The record of exhumed UHP rocks at the surface may be strongly biased.

7) The strength of continental crust at UHP is insufficient to support the forces originating from the density contrast when slabs of considerable vertical extent are subducted into the mantle.

8) The concept of low strength should be tested against the seismological and seismic record in subduction and collision zones, where continental crust is presently subducted to >100 km depth. The predicted low strength of continental crust at UHP restricts the stress drop of earthquakes, even if total, to values well below a few MPa.

ACKNOWLEDGMENTS

Most of the ideas and results presented in this paper were developed within the scope of the Research Group "High Pressure Metamorphism in Nature and Experiment", Faculty of Earth Sciences at Ruhr-University, generously funded by the Deutsche Forschungsgemeinschaft (DFG) from 1991–1996. This financial support and the trust the referees had in our proposals are gratefully acknowledged. First of all, we wish to thank our colleagues of the rock squeezer group, Fritz Rummel, Erik Rybacki, Andreas Zerbian and Karin Konrad for the perfect cooperation over the years and their permission to use unpublished results in this chapter. Experimental work would not have been possible without Wolfgang Harbott, the late Bernhard Kappernagel, and the machine shop crew. Thanks to Werner Schreyer, Thomas Reinecke, Hans-Joachim Massonne, Hans-Peter Schertl, Walter Maresch, Klaus Röller and Gunter Eggeler for numerous stimulating discussions, help and advice in various stages of our projects. Thanks also to Hans-Peter Schertl, Hans-Joachim Massonne and Nicolai Sobolev for the possibility to examine thin sections from their working areas, and to Dieter Dettmar, Thomas Baller, Michael Ress and Friedrich Eickhoff for technical assistance. Martina Küster is thanked for help and critical comments in various stages of the preparation of this manuscript. Constructive reviews by Gayle Gleason, Bradley Hacker, Shaocheng Ji and Jan Tullis significantly improved the final version and are gratefully acknowledged.

REFERENCES

Ando, J., Irifune, T., Takeshita, T., and Fujino, K. (1997) Evaluation of the non-hydrostatic stress produced in a multi-anvil high pressure apparatus, *Physics and Chemistry of Minerals* **24**, 139–148.

Ando, J.-i., Fujino, K., and Takeshita, T. (1993) Dislocation microstructures in naturally deformed silicate garnets, *Physics of the Earth and Planetary Interiors* **80**, 105–116.

Avé Lallement, H.G. (1978) Deformation of diopside and websterite, *Tectonophysics* **48**, 1–27.

Bai, Q., Mackwell, S.J., and Kohlstedt, D.L. (1991) High-temperature creep of olivine single crystals - 1. Mechanical results for buffered samples, *Journal of Geophysical Research* **96**, 2,441–2,463.

Barber, D.J., Heard, H.C., and Wenk, H.R. (1981) Deformation of dolomite single crystals from 20–800°C, *Physics and Chemistry of Minerals* **7**, 271–286.

Barber, D.J., Wenk, H.R., and Heard, H.C. (1994) The plastic deformation of polycrystalline dolomite: comparison of experimental results with theoretical predictions, *Materials Science and Engineering* **A175**, 83–104.

Beauchesne, S. and Poirier, J.-P. (1989) Creep of barium titanate perovskite: a contribution to a systematic approach to the viscosity of the lower mantle, *Physics of the Earth and Planetary Interiors* **55**, 187–199.

Bell, P.N. and Roseboom, E.H. (1969) Melting relations of jadeite and albite to 45 kilobars with comments on melting diagrams of binary systems at high pressures, *Mineralogical Society of America Special Paper* **2**, 151–161.

Biino, G. and Compagnoni, R. (1992) Very-high pressure metamorphism of the Brossasco coronite metagranite, southern Dora-Maira massif, Western Alps, *Schweizerische Mineralogische und Petrographische Mitteilungen* **72**, 347–363.

Boland, J.N. and Tullis, T.E. (1986) Deformation behavior of wet and dry clinopyroxenite in the brittle to ductile transition region, in B.E. Hobbs and H.C. Heard (eds.), *Mineral and Rock*

Deformation: Laboratory Studies - The Paterson Volume, Geophysical Monograph, 36, Washington, D.C., pp. 35–49.

Boulogne, B., Cordier, P., and Doukhan, J.-C. (1988) Defects and hydrolytic weakening in ?-berlinite AlPO4; A structural analog of quartz, *Physics and Chemistry of Minerals* **16**, 250–261.

Brace, W.F. (1980) Permeability of crystalline and argillaceous rocks, *International Journal of Rock Mechancs and Mining Sciences* **17**, 241–251.

Brace, W.F. and Kohlstedt, D.L. (1980) Limits on lithospheric stress imposed by laboratory experiments, *Journal of Geophysical Research* **85**, 6,248–6,252.

Brace, W.F. and Martin, R.J. (1968) A test of the law of effective stress for crystalline rocks of low porosity, *International Journal of Rock Mechanics and Mining Sciences* **5**, 415–426.

Brenan, J. (1991) Development and maintenance of metamorphic permeability: Implications for fluid transport, in D.M. Kerrick (ed.), *Contact Metamorphism, Reviews in Mineralogy*, **26**, Mineralogical Society of America, Washington, D.C., pp. 291–319.

Brown, M., Rushmer, T., and Sawyer, E.W. (1995) Introduction to special section: Mechanisms and consequences of melt segregation from crustal protliths, *Journal of Geophysical Research* **100**, 15,551–15,563.

Bukowinski, M.S.T. (1994) Quantum geophysics, *Annual Reviews of Earth and Planetary Science* **22**, 167–205.

Burnley, P.C., Green, H.W., and Prior, D.J. (1991) Faulting associated with the olivine to spinel transformation in Mg2GeO4 and its implications for deep-focus earthquakes, *Journal of Geophysical Research* **96**, 425–443.

Byerlee, J.D. (1968) Brittle-ductile transition in rocks, *Journal of Geophysical Research* **73**, 4,741–4,750.

Byerlee, J.D. (1978) Friction of rocks, *Pure and Applied Geophysics* **116**, 615–626.

Cawley, J.D. and Lee, W.E. (1994) Oxide ceramics, in M. Swain (ed.), *Structure and properties of ceramics*, Verlag Chemie, Weinheim, pp. 49–117.

Chemenda, A.I., Mattauer, M., Malavieille, J., and Bokun, A.N. (1995) A mechanism for syn-collisional rock exhumation and associated normal faulting: Results from physical modelling, *Earth and Planetary Science Letters* **132**, 225–232.

Chopin, C., Henry, C., and Michard, A. (1991) Geology and petrology of the coesite-bearing terrain, Dora Maira massif, western Alps, *European Journal of Mineralogy* **3**, 263–291.

Cloos, M. (1993) Lithospheric buoyancy and collisional orogenesis: subduction of oceanic plateaus, continental margins, island arcs, spreading ridges, and seamounts, *Geological Society of America Bulletin* **105**, 715–737.

Coleman, R.G. and Wang, X. (1995) Overview of the geology and tectonics of UHPM, in R.G. Coleman and X. Wang (eds.), *Ultrahigh Pressure Metamorphism*, Cambridge University Press, New York, pp. 1–31.

Cook, R.F. and Pharr, G.M. (1994) Mechanical properties of ceramics, in M. Swain (ed.), *Structure and properties of ceramics*, Verlag Chemie, Weinheim, pp. 341–407.

Cooper, R.F. and Kohlstedt, D.L. (1984) Solution-precipitation enhanced diffusional creep of partially molten olivine-basalt aggregates during hot-pressing, *Tectonophysics* **107**, 207–233.

Cooper, R.F., Kohlstedt, D.L., and Chyung, K. (1989) Solution-precipitation enhanced creep in solid-liquid aggregates which display a non-zero dihedral angle, *Acta Metallurgica* **37**, 1,759–1,771.

Cordier, P., Boulogne, B., and Doukhan, J.-C. (1988) Water precipitation and diffusion in wet quartz and wet berlinite AlPO4, *Bulletin de Mineralogie* **111**, 113–137.

de Bresser, J.H.P. (1996) Influence of pressure on the high temperature flow of Carrara Marble, *Transactions of the American Geophysical Union, Eos* **77**, F710.

Dell'Angelo, L.N. and Olgaard, D.L. (1995) Experimental deformation of fine-grained anhydrite: Evidence for dislocation and diffusion creep, *Journal of Geophysical Research* **100**, 15,425–15,440.

Dell'Angelo, L.N. and Tullis, J. (1988) Experimental deformation of partially melted granitic aggregates, *Journal of Metamorphic Geology* **6**, 495–515.

Den Brok, B. and Kruhl, J.H. (1996) Ductility of garnet as an indicator of extremely high temperature deformation: Discussion, *Journal of Structural Geology* **18**, 1,369–1,373.

Dove, P.M. and Rimstidt, J.D. (1994) Silica-water interactions, in P.J. Heaney, C.T. Prewitt, and G.V. Gibbs (eds.), *Silica: Physical Behavior, Geochemistry, and Materials Applications, Reviews in Mineralogy*, **29**, Mineralogical Society of America, Washington, D.C., pp. 259–308.

Elliott, D. (1973) Diffusion flow laws in metamorphic rocks, *Geological Society of America Bulletin* **84**, 2,645–2,664.

Evans, B. and Kohlstedt, D.L. (1995) Rheology of rocks, in T.J. Ahrens (ed.), *Rock Physics and Phase Relations, A Handbook of Physical Constants, AGU Reference Shelf*, 3, American Geophysical Union, Washington, D.C., pp. 148–165.

Farver, J.R. and Yund, R.A. (1991) Oxygen diffusion in quartz: Dependence on temperature and water fugacity, *Chemical Geology* **90**, 55–70.

Fitzgerald, J.D.F. and Stünitz, H. (1993) Deformation of granitoids at low metamorphic grade. I: Reactions and grain size reduction, *Tectonophysics* **221**, 269–297.

Fournier, M., Jolivet, L., Goffe, B., and Dubois, R. (1991) Alpine Corsica metamorphic core complex, *Tectonics* **10**, 1,173–1,186.

Frost, H.J. and Ashby, M.F. (1982) *Deformation-Mechanism Maps: The Plasticity and Creep of Metals and Ceramics*, Pergamon Press, Oxford.

Fyfe, W.S., Price, N.J., and Thompson, A.B. (1978) *Fluids in the Earth's crust*, Elsevier, Amsterdam.

Garem, H., Rabier, J., and Veyssiere, P. (1982) Slip systems in gadolinium, gallium garnet single crystals, *Journal of Material Science* **17**, 878–884.

Gebauer, D., Schertl, H.-P., Brix, M., and Schreyer, W. (1997) 35 Ma old ultrahigh-pressure metamorphism and evidence for very rapid exhumation in the Dora Maira massif, Western Alps, *Lithos* **41**, 5–24.

German, R.M. (1996) The microstructures of liquid phase sintered materials, in R.M. German, G.L. Messing, and R.G. Cornwall (eds.), *Sintering Technology*, Marcel Decker Inc, New York, pp. 213–220.

Gleason, G.C. and Green, H.W. (1996) Effect of differential stress on the albite to jadeite + coesite transition at confining pressures of > 3 GPa, *Transactions of the American Geophysical Union, Eos* **77**, F662.

Gleason, G.C. and Tullis, J. (1995) A flow law for dislocation creep of quartz aggregates determined with the molten salt cell, *Tectonophysics* **247**, 1–23.

Godard, G. and van Roermund, H. (1995) Deformation-induced clinopyroxene fabrics from eclogites, *Journal of Structural Geology* **17**, 1,425–1,443.

Goetze, C. and Evans, B. (1979) Stress and temperature in the bending lithosphere as constrained by experimental rock mechanics, *Geophysical Journal of the Royal Astronomic Society* **59**, 463–478.

Grawinkel, A. and Stöckhert, B. (1997) Hydrostatic pore fluid pressure to 9 km depth - Fluid inclusion evidence from the KTB deep drill hole, *Geophysical Research Letters* **24**, 3,273–3,276.

Green, H.W. and Burnley, P.C. (1989) The mechanism of failure responsible for deep-focus earthquakes, *Nature* **341**, 733–737.

Green, H.W. and Zhou, Y. (1996) Transformation-induced faulting requires an exothermic reaction and explains the cessation of earthquakes at the base of the mantle transition zone, *Tectonophysics* **256**, 39–56.

Griggs, D.T. (1967) Hydrolytic weakening of quartz and other silicates, *Geophysical Journal of the Royal Astronomical Society London* **14**, 19–31.

Hacker, B.R. (1996) Eclogite formation and the rheology, buoyancy, seismicity, and H2O content of oceanic crust, in G.E. Bebout, Scholl, D., Kirby, S.H., Platt, J.P. (ed.), *Dynamics of Subduction*, Monograph, American Geophysical Union, Washington, D.C., pp. 337–246.

Hacker, B.R. (1997) Diagenesis and the fault-valve seismicity of crustal faults, *Journal of Geophysical Research* **102**, 24,459–24,467.

Hacker, B.R. and Christie, J.M. (1990) Brittle/ductile and plastic/cataclastic transitions in experimentally deformed and metamorphosed amphibolite, *AGU Monograph* **56**, 127–147.

Hacker, B.R. and Christie, J.M. (1991) Observational evidence for a possible new diffusion path, *Science* **251**, 67–70.

Hacker, B.R. and Kirby, S.H. (1990) Effect of stress and deformation on albite breakdown, *Transactions of the American Geophysical Union, Eos* **71**, 639.

Hacker, B.R. and Kirby, S.H. (1993) High-pressure deformation of calcite marble and its transformation to aragonite under non-hydrostatic conditions, *Journal of Structural Geology* **15**, 1207–1222.

Hacker, B.R. and Peacock, S.M. (1994) Creation, preservation, and exhumation of coesite-bearing, ultrahigh-pressure metamorphic rocks, in R.G. Coleman and X. Wang (eds.), *Ultrahigh Pressure Metamorphism*, Cambridge University Press, Cambridge, United Kingdom, 159–181.

Hacker, B.R., Ratschbacher, L., Webb, L., and Dong, S. (1995) What brought them up? Exhumation of the Dabie Shan ultrahigh-pressure rocks, *Geology* **23**, 743–746.

Hams, S. (1993) Experimental study of transformation plasticity during the II-I-phase-transition of potassium nitrate as an analog to the aragonite-calcite transformation. Ph. D. Thesis. Bochum, p. 136. Ruhr University, Germany.

Handy, M.R. (1994) Flow laws for rocks containing two non-linear viscous phases: A phenomenological approach, *Journal of Structural Geology* **16**, 287–301.

Harley, S.L. and Carswell, D.A. (1995) Ultradeep crustal metamorphism: A prospective view, *Journal of Geophysical Research* **100**, 8367–8380.

Heard, H.C. (1976) Comparison of the flow properties of rocks at crustal conditions, *Philosophical Transactions of the Royal Society London* **A283**, 173–186.

Henry, C., Michard, A., and Chopin, C. (1993) Geometry and structural evolution of ultra-high-pressure and high-pressure rocks from the Dora-Maira Massif, Western Alps, Italy, *Journal of Structural Geology* **15**, 965–981.

Hickman, S.H. and Evans, B. (1991) Experimental pressure solution in halite, 1: The effect of grain/interphase boundary structure, *Journal of the Geological Society of London* **148**, 549–560.

Hickman, S.H. and Evans, B. (1995) Kinetics of pressure solution at halite-silica interfaces and intergranular films, *Journal of Geophysical Research* **100**, 13,113–13,132.

Hirth, G. and Tullis, J. (1994) The brittle-plastic transition in experimentally deformed quartz aggregates, *Journal of Geophysical Research* **99**, 11,731–11,747.

Holness, M.B. (1992) Equilibrium dihedral angles in the system quartz-H_2O-CO_2-NaCl at 800°C and 1–15 kbar: the effect of pressure and fluid composition on the permeability of quartzites, *Earth and Planetary Science Letters* **114**, 171–184.

Holness, M.B. (1993) Temperature and pressure dependence of quartz-aqueous fluid dihedral angles: the control of adsorbed H_2O on the permeability of quartzites, *Earth and Planetary Science Letters* **117**, 363–377.

Huang, W.-L. and Wyllie, P.J. (1975) Melting and subsolidus phase relationships for CaSiO3 to 35 kilobars pressure, *American Mineralogist* **60**, 213–217.

Huenges, E., Erzinger, J., Kück, J., Engeser, B., and Kessels, W. (1997) The permeable crust: Geohydraulic properties down to 9101 m depth, *Journal of Geophysical Research* **102**, 18,255 18,265.

Ingrin, J. and Gillet, P. (1986) TEM investigation of the crystal microstructures in a quartz-coesite assemblage of the Western Alps, *Physics and Chemistry of Minerals* **13**, 325–330.

Jaeger, J.C. and Cook, N.G.W. (1984) *Fundamentals of Rock Mechanics, 3rd. ed.*, Chapman and Hall, New York.

Jaoul, O. (1990) Multicomponent diffusion and creep in olivine, *Journal of Geophysical Research* **95**, 17,631–17,642.

Jaoul, O., Tullis, J., and Kronenberg, A.K. (1984) The effect of varying water contents on the creep behaviour of Heavitree quartzite, *Journal of Geophysical Research* **89**, 4,298–4,312.

Jesser, W.A. and Kuhlmann-Wilsdorf, D. (1972) The flow stress and dislocation structure of nickel deformed at very high pressure, *Material Science and Engineering* **9**, 111–117.

Ji, S. and Martignole, J. (1994) Ductility of garnet as an indicator of extremely high temperature deformation, *Journal of Structural Geology* **16**, 985–996.

Ji, S. and Zhao, P. (1993) Flow laws of multiphase rocks calculated from experimental data on the constituent phases, *Earth and Planetary Science Letters* **117**, 181–187.

Ji, S. and Zhao, P. (1994) Strength of two-phase rocks: a model based on fiber-loading theory, *Journal of Structural Geology* **16**, 253–262.

Jin, Z.-M., Green, H.W., and Zhou, Y. (1994) Melt topology during dynamic partial melting of peridotite, *Nature* **372**, 164–167.

Karato, S.-i. and Rubie, D.C. (1997) Towards an experimental study of deep mantle rheology: a new multi-anvil sample assembly for deformation studies under high pressures and temperatures, *Journal of Geophysical Research* **102**, 20,111–20,122.

Karato, S.-i., Wang, Z., Liu, B., and Fujino, K. (1995) Plastic deformation of garnets: systematics and implications for the rheology of the mantle transition zone, *Earth and Planetary Science Letters* **130**, 13–30.

Kirby, S.H. and Kronenberg, A.K. (1984) Deformation of clinopyroxenite: Evidence for a transition in flow mechanisms and semibrittle behavior, *Journal of Geophysical Research* **89**, 3,177–3,192.

Kirby, S.H., Stein, S., Okal, E.A., and Rubie, D.C. (1996) Metastable mantle phase transformations and deep earthquakes in subducting oceanic lithosphere, *Reviews of Geophysics* **34**, 261–306.

Kirby, S.H. and Stern, L.A. (1993) Experimental dynamic metamorphism of mineral single crystals, *Journal of Structural Geology* **15**, 1,223–1,240.

Kohlstedt, D.L., Evans, B., and Mackwell, S.J. (1995) Strength of the lithosphere: Constraints imposed by laboratory experiments, *Journal of Geophysical Research* **100**, 17,587–17,602.

Konrad, K. (1997) Experimental deformation of synthetic, polycrystalline aragonite aggregates. thesis. Bochum. Ruhr University, Germany.

Kronenberg, A.K. and Tullis, J. (1984) Flow strengths of quartz aggregates: grain size and pressure effects due to hydrolytic weakening, *J Geophys Res* **89**, 4281–4297.

Küster, M. and Stöckhert, B. (1997) Density changes of fluid inclusions in high-pressure low temperature metamorphic rocks from Crete - a thermobarometric approach based on the creep strength of the host minerals, *Lithos* **41**, 151–167.

Küster, M. and Stöckhert, B. (in press) High differential stress and sublithostatic pore fluid pressure in the ductile regime - microstructural evidence for short term postseismic creep in the Sesia Zone, Western Alps, *Tectonophysics*.

Laporte, D. (1994) Wetting behavior of partial melts during crustal anatexis: the distribution of hydrous silicic melts in polycrystalline aggregates of quartz, *Contributions to Mineralogy and Petrology* **116**, 486–499.

Laporte, D. and Watson, E.B. (1991) Direct observation of near-equilibrium pore geometry in synthetic quartzites at 600 °C-800 °C and 2–10.5 kbar, *Journal of Geology* **99**, 873–878.

Laporte, D. and Watson, E.B. (1995) Experimental and theoretical constraints on melt distribution in crustal sources: the effect of crystalline anisotropy on melt interconnectivity, *Chemical Geology* **124**, 161–184.

Lee, V.W., Mackwell, S.J., and Brantley, S.L. (1991) The effect of fluid chemistry on wetting textures in novaculite, *Journal of Geophysical Research* **96**, 10,023–10,037.

Lockner, D.A. (1995) Rock failure, in T.J. Ahrens (ed.), *Rock Physics and Phase Relations, A Handbook of Physical Constants, AGU Reference Shelf*, 3, American Geophysical Union, Washington, D.C., pp. 127–147.

Luan, F.C. and Paterson, M.S. (1992) Preparation and deformation of synthetic aggregates of quartz, *Journal of Geophysical Research* **97**, 301–320.

Mancktelow, N.S. (1993) Tectonic overpressure in competent mafic layers and the development of isolated eclogites, *Journal of Metamorphic Geology* **11**, 801–812.

Manning, C.E. (1994) The solubility of quartz in H_2O in the lower crust and upper mantle, *Geochimica et Cosmochimica Acta* **58**, 4,831–4,839.

McCaffrey, R., Molnar, P., and Roecker, S.W. (1985) Microearthquake seismicity and fault plane solutions related to arc-continent collision in the eastern Sunda Arc, Indonesia, *Journal of Geophysical Research* **90**, 4511–4528.

Meike, A. (1993) A critical review of investigations into transformation plasticity, in J.N. Boland and J.D. Fitzgerald (eds.), *Defects and Processes in the Solid State, The McLaren Volume*, Elsevier Science, New York, pp. 5–25.

Michard, A., Henry, C., and Chopin, C. (1995) Structures in UHPM rocks: A case study from the Alps, in R.G. Coleman and X. Wang (eds.), *Ultrahigh Pressure Metamorphism*, 132–158, Cambridge University Press, Stanford

Mirwald, P.W. and Massonne, H.-J. (1980) The low-high quartz and quartz-coesite transition to 40 kbar between 600 and 1600 °C and some reconnaissance data on the effect of NaAlO2component on the low quartz-coesite transition, *Journal of Geophysical Research* **85**, 6,983–6,990.

Molnar, P. and England, P. (1990) Temperatures, heat flux, and frictional stress near major thrust faults, *Journal of Geophysical Research* **95**, 4,833–4,856.

Molnar, P. and Gray, D. (1979) Subduction of continental lithosphere: some constraints and uncertainties, *Geology* **7**, 58–62.

Nachtrieb, N.H. and Coston, C. (1965) Self-diffusion in tin at high pressure, in C.T. Tomizuka and R.M. Emrick (eds.), *Physics of Solids at High Pressures*, Academic Press, New York, pp. 336–348.

Olgaard, D.L. (1990) The role of second phase in localizing deformation, in R.J. Knipe and E.H. Rutter (eds.), *Deformation, Mechanisms, Rheology and Tectonics, Geological Society Special Publication*, 54, Geological Society of London, London, pp. 175–181.

Olgaard, D.L., Ko, S., and Wong, T. (1995) Deformation and pore pressure in dehydrating gypsum under transiently drained conditions, *Tectonophysics* **245**, 237–248.

Paterson, M.S. (1978) *Experimental Rock Deformation - The Brittle Field*, Springer-Verlag, New York.

Paterson, M.S. (1983) Creep in transforming polycrystalline materials, *Mechanics of Materials* **2**, 103–109.

Paterson, M.S. (1987) Problems in the extrapolation of laboratory rheological data, *Tectonophysics* **133**, 33–43.

Paterson, M.S. (1990) Superplasticity in geological materials, in M.J. Mayo, M. Kobayashi, and J. Wadsworth (eds.), *Superplasticity in Metals, Ceramics, and Intermetallics, Materials Research Society Symposium Proceedings*, **196**, Materials Research Society, Pittsburgh, pp. 303–312.

Peacock, S.M. (1992) Blueschist-facies metamorphism, shear heating, and P-T-t paths in subduction shear zones, *Journal of Geophysical Research* **97**, 17,693–17,707.

Peacock, S.M. (1993) Large-scale hydration of the lithosphere above subducting slabs, *Chemical Geology* **108**, 49–59.

Peacock, S.M. (1996) Thermal and petrologic structure of subduction zones, in G.E. Bebout, D.W. Scholl, S.H. Kirby, and J.P. Platt (eds.), *Subduction Top to Bottom*, American Geophysical Union, Washington, D.C., pp. 119–134.

Philippot, P. (1993) Fluid-melt-rock interaction in mafic eclogites and coesite-bearing metasediments; constraints on volatile recycling during subduction, *Chemical Geology* **108**, 93–112.

Philippot, P., Chevallier, P., Chopin, C., and Dubessy, J. (1995) Fluid composition and evolution in cœsite-bearing rocks (Dora-Maira massif, Western Alps): Implications for element recycling during subduction, *Contributions to Mineralogy and Petrology* **121**, 29–44.

Philippot, P. and van Roermund, H.L.M. (1992) Deformation processes in eclogitic rocks: evidence for the rheological delamination of the oceanic crust in deeper levels of subduction zones, *Journal of Structural Geology* **14**, 1059–1077.

Poirier, J.P. (1982) On transformation plasticity, *Journal of Geophysical Research* **87**, 6,791–6,797.

Poirier, J.P. (1985) *Creep of crystals: high-temperature deformation processes in metals, ceramics, and minerals*, Cambridge University Press, Cambridge.

Poirier, J.P. (1995) Plastic rheology of crystals, in T.J. Ahrens (ed.), *Rock Physics and Phase Relations, A Handbook of Physical Constants, AGU Reference Shelf*, 2, American Geophysical Union, Washington, D.C., pp. 237–247.

Post, A.D., Tullis, J., and Yund, R.A. (1996) Effects of chemical environment on dislocation creep of quartzite, *Journal of Geophysical Research* **101**, 22,143–22,155.

Rabkin, E. (1994) Thin films on grain boundaries in metals and ceramics and their importance for the properties of the materials, in J. Nowotny (ed.), *Science of Ceramic Interfaces*, II, Elsevier Science, Amsterdam, pp. 371–398.

Rabkin, E., Ma, C.Y., and Gust, W. (1994) Diffusion induced grain boundary phenomena in metals and oxide ceramics, in J. Nowotny (ed.), *Science of Ceramic Interfaces*, II, Elsevier Science, Amsterdam, pp. 353–369.

Raterron, P., Doukhan, N., Jaoul, O., and Doukhan, J.C. (1994) High temperature deformation of diopside; IV, Predominance of <110> glide above 1000 degrees C, *Physics of the Earth and Planetary Interiors* **82**, 209–222.

Raterron, P., Ingrin, J., Jaoul, O., Doukhan, N., and Elie, F. (1995) Early partial melting of diopside under high pressure, *Physics of the Earth and Planetary Interiors* **89**, 77–88.

Reinecke, T. (1991) Very-high-pressure metamorphism and uplift of cœsite-bearing sediments from the Zermatt-Saas zone, Western Alps, *European Journal of Mineralogy* **3**, 7–17.

Renner, J. (1996) Experimental investigations on the rheology of coesite. Bochum, p. 178. Ruhr University, Germany.

Renner, J. and Rummel, F. (1996) The effect of experimental and microstructural parameters on the transition from brittle failure to cataclastic flow of carbonate rocks, *Tectonophysics* **258**, 151–169.

Renner, J., Rummel, F., and Stöckhert, B. (1996) Experimental deformation of synthetic coesite aggregates, *Transactions of the American Geophysical Union, Eos* **46**, F717.

Renner, J., Zerbian, A., and Stöckhert, B. (1997) Microstructures of synthetic coesite polycrystals - The effect of pressure, temperature, and time, *Lithos* **41**, 169–184.

Roecker, S.W. (1982) Velocity structure of the Pamir-Hindu Kush region: possible evidence of subducted crust, *Journal of Geophysical Research* **87**, 945–959.

Rubie, D.C. (1983) Reaction-enhanced ductility: the role of solid-solid univariant reactions in deformation of the crust and mantle, *Tectonophysics* **96**, 331–352.

Rushmer, T. (1995) An experimental deformation study of partially molten amphibolite: Application of low-melt fraction segregation, *Journal of Geophysical Research* **100**, 15,681–15,695.

Rutter, E.H. (1983) Pressure solution in nature, theory and experiment, *Journal of the Geological Society of London* **140**, 725–740.

Rutter, E.H. and Brodie, K.H. (1992) Rheology of the lower crust, in D.M. Fountain, R. Arculus, and R.W. Kay (eds.), *Continental lower crust, Developments in Geotectonics*, **23**, Elsevier, New York, pp. 201–267.

Rutter, E.H. and Neumann, D.H.K. (1995) Experimental deformation of partially molten Westerly granite under fluid absent conditions, with implications for the extraction of granitic magmas, *Journal of Geophysical Research* **100**, 15,697–15,715.

Rybacki, E., Renner, J., Konrad, K., Harbott, W., Stöckhert, B., and Rummel, F. (in press) A servohydraulically controlled deformation apparatus for rock deformation under conditions of ultra-high pressure metamorphism, *Pure and Applied Geophysics*.

Rybacki, E., Rummel, F., and Stöckhert, B. (1995) Synthesis and experimental deformation of polycrystalline aragonite, *Bochumer Geologische und Geotechnische Arbeiten* **44**, 189–194.

Sammis, C.G., Smith, J.C., and Schubert, G. (1981) A critical assessment of estimation methods for activation volumes, *Journal of Geophysical Research* **86**, 10,707–10,718.

Schertl, H.P., Schreyer, W., and Chopin, C. (1991) The pyrope-cœsite rocks and their country rocks at Parigi, Dora-Maira Massif, Western Alps: detailed petrography, mineral chemistry and PT-path, *Contributions to Mineralogy and Petrology* **108**, 1–21.

Schmid, S.M. (1982) Microfabric studies as indicators of deformation mechanisms and flow laws operative in mountain building, in K.J. Hsü (ed.), *Mountain Building Processes*, Academic Press, London, pp. 95–110.

Schmid, S.M., Boland, J.N., and Paterson, M.S. (1977) Superplastic flow in finegrained limestone, *Tectonophysics* **43**, 257–291.

Schmid, S.M., Paterson, M.S., and Boland, J.N. (1980) High temperature flow and dynamic recrystallization in Carrara marble, *Tectonophysics* **65**, 245–280.

Scholz, C. (1990) *The Mechanics of Earthquakes and Faulting*, Cambridge University Press, New York.

Schreyer, W. (1995) Ultradeep metamorphic rocks: The retrospective viewpoint, *Journal of Geophysical Research* **100**, 8353–8366.

Schreyer, W., Massonne, H.J., and Chopin, C. (1987) Continental crust subducted to depths near 100 km: implications for magma and fluid genesis in collision zones, in B.O. Mysen (ed.), *Magmatic Processes: Physiochemical Principles*, 1, Geochemical Society, University Park, pp. 155–163.

Schwarz, S. and Stöckhert, B. (1996) Pressure solution in siliciclastic HP-LT metamorphic rocks - constraints on the state of stress in deep levels of accretionary complexes, *Tectonophysics* **255**, 203–209.

Seeger, A. and Haasen, P. (1958) Density changes of crystals containing dislocations, *Philosophical Magazine* **3**, 470–475.

Sharp, Z.D., Essene, E.J., and Hunziker, J.C. (1993) Stable isotope geochemistry and phase equilibria of coesite-bearing whiteschists, Dora Maira Massif, western Alps, *Contributions to Mineralogy and Petrology* **114**, 1–12.

Shatsky, V.S., Sobolev, N.V., and Vavilov, M.A. (1995) Diamond-bearing metamorphic rocks of the Kokchetav massif (Northern Kazakhstan), in R.G. Coleman and X. Wang (eds.), *Ultrahigh Pressure Metamorphism*, Cambridge University Press, Stanford, pp. 427–455.

Shaw, T.M. and Duncombe, P.R. (1991) Forces between aluminum oxide grains in a silicate melt and their effect on grain boundary wetting, *Journal of the American Ceramic Society* **74**, 2,495–2,505.

Shimada, M. (1993) Lithosphere strength inferred from fracture strength of rocks at high confining pressures and temperatures, *Tectonophysics* **217**, 55–64.

Sibson, R.H. (1990) Conditions for fault-valve behavior, *Geological Society of London Special Publication* **54**, 15–28.

Siddiqi, G., Evans, B., Dresen, G., and Freund, D. (1997) Effect of semibrittle deformation on transport properties of calcite rocks, *Journal of Geophysical Research* **102**, 14,765–14,778.

Simpson, C. (1985) Deformation of granitic rocks across the brittle-ductile transition, *Journal of Structural Geology* **7**, 503–511.

Stöckhert, B., Massonne, H.-J., and Nowlan, E.U. (1997) Low differential stress during high-pressure metamorphism: The microstructural record of a metapelite from the eclogite zone, Tauern Window, Eastern Alps, *Lithos* **41**, 103–118.

Stöckhert, B., Wachmann, M., Küster, M., and Bimmermann, S. (in press) Low effective viscosity during high-pressure metamorphism due to dissolution precipitation creep: The record of HP-LT-metamorphic carbonates and siliciclastic rocks from Crete, *Tectonophysics.*

Stöckhert, B., Wachmann, M., and Schwarz, S. (1995) Rheology and structural evolution of high-pressure, low-temperature metamorphic rocks, *Bochumer Geologische und Geotechnische Arbeiten* **44**, 235–242.

Theye, T. and Seidel, E. (1993) Uplift-related retrogression history of aragonite marbles in Western Crete (Greece), *Contributions to Mineralogy and Petrology* **114**, 349–356.

Tingle, T.N., Green, H.W., Young, T.E., and Koczynski, T.A. (1993) Improvements to Griggs-type apparatus for mechanical testing at high pressures and temperatures, *Pure and Applied Geophysics* **141**, 523–543.

Tullis, J., Dell'Angelo, L., and Yund, R.A. (1990) Ductile shear zones from brittle precursors in feldspathic rocks: the role of dynamic recrystallization, in A.G. Duba, W.B. Durham, J.W. Handin, and H.F. Wang (eds.), *The Brittle-Ductile Transition in Rocks, Geophysical Monograph*, 56, American Geophysical Union, Washington, D.C., pp. 67–82.

Tullis, J. and Yund, R.A. (1982) Grain growth kinetics of quartz and calcite aggregates, *Journal of Geology* **90**, 301–318.

Tullis, J. and Yund, R.A. (1991) Diffusion creep in feldspar aggregates: experimental evidence, *Journal of Structural Geology* **13**, 987–1000.

Tullis, J., Yund, R.A., and J., F. (1996) Deformation-enhanced fluid distribution in feldspar aggregates and implications for ductile shear zones, *Geology* **24**, 63–66.

Tullis, T.E. and Tullis, J. (1986) Experimental Rock Deformation Techniques, in B.E. Hobbs and H.C. Heard (eds.), *Mineral and Rock Deformation: Laboratory Studies - The Paterson Volume, Geophysical Monograph*, **36**, American Geophysical Union, Washington, D.C., pp. 297–324.

Twiss, R.J. (1977) Theory and applicability of a recrystallized grain size paleopiezometer, *Pure and Applied Geophysics* **115**, 227–244.

Urai, J.L. (1983) Water assisted dynamic recrystallization and weakening in polycrystalline bischoffite, *Tectonophysics* **96**, 125–157.

van den Beukel, J. (1992) Some thermomechanical aspects of the subduction of continental lithosphere, *Tectonics* **11**, 316–329.

van den Beukel, J. and Wortel, R. (1988) Thermo-mechanical modelling of arc-trench regions, *Tectonophysics* **154**, 177–193.

Van der Klauw, S.N., Reinecke, T., and Stöckhert, B. (1997) Exhumation of ultrahigh-pressure metamorphic oceanic crust from Lago di Cignana, Piemontese zone, western Alps: the structural record in metabasites, *Lithos* **41**, 79–102.

van Roermund, H.L.M. and Boland, J.N. (1981) The dislocation substructures of naturally deformed omphacites, *Tectonophysics* **78**, 403–418.

Vernon, R.H. and Flood, R.H. (1987) Contrasting deformation and metamorphism of S- and I-type granitoids in the Lachlan Fold Belt, Eastern Australia, *Tectonophysics* **147**, 127–143.

Voll, G. (1976) Recrystallization of quartz, biotite and feldspars from Erstfeld to the Leventina Nappe, Swiss Alps, and its geological significance, *Schweizerische Mineralogische und Petrographische Mitteilungen* **56**, 641–647.

Wachmann, M. (1997) Structural evolution of high-pressure metamorphic carbonate rocks at Agios Theodori, SW-Crete, *Bochumer Geologische und Geotechnische Arbeiten* **48**, 1–122.

Wang, Z., Bai, Q., Dresen, G., Wirth, R., and Evans, B. (1996) High temperature deformation of calcite single crystals, *Journal of Geophysical Research* **101**, 20,377–20,390.

Watson, B. and Brenan, J.M. (1987) Fluids in the lithosphere, 1. Experimentally-determined wetting characteristics of CO_2–H_2O fluids and their implications for fluid transport, host-rock physical properties, and fluid inclusion formation, *Earth and Planetary Science Letters* **85**, 497–515.

White, S.H., Burrows, S.E., Carreras, J., Shaw, N.D., and Humphreys, F.J. (1980) On mylonites in ductile shear zones, *Journal of Structural Geology* **2**, 175–187.

Williams, D.W. and Kennedy, G.C. (1969) Melting curve of diopside to 50 kilobars, *Journal of Geophysical Research* **74**, 4,359–4,366.

Wirth, R. and Stöckhert, B. (1995) Transformation of coesite to quartz: microstructures in nature and experiment, *Bochumer Geologische und Geotechnische Arbeiten* **44**, 260–262.

Chapter 4

Thermal Controls on Slab Breakoff and the Rise of High-Pressure Rocks During Continental Collisions

J. Huw Davies
Department of Earth Sciences, University of Liverpool, UK, davies@liv.ac.uk

Friedhelm von Blanckenburg
Isotopengeologie, Universität Bern, Erlachstr. 9a, 3012 Bern, Switzerland, fredv@earth.ox.ac.uk

Abstract: We investigate the thermal controls on slab breakoff—the detachment of oceanic lithosphere from continental lithosphere during continental collision. We show that the depth of slab breakoff is sensitive to the thickness of the lithosphere and to the temperature of the convecting mantle, but comparatively insensitive to the distribution of radioactivity and shear heating. This is because lithospheric strength is largely controlled by the temperature of the uppermost mantle (well characterized by Moho temperature), and is largely insensitive to near-surface crustal temperatures (which is what foreland surface heat flow reflects). The insensitivity to subducting crustal temperatures allows us to confidently predict that slab breakoff will also be insensitive to erosion or accretion at the slab-wedge interface. It has been shown elsewhere that slab breakoff is also sensitive to the subduction velocity, and relatively insensitive to the angle of dip and critical strain rate. If the mantle was hotter in earlier Earth history, then we could expect shallower slab breakoff and hence less ultrahigh-pressure (UHP) metamorphism and greater amounts of collisional magmatism in the Archean.

If the subducted continental lithosphere is a craton (i.e., thick, cold lithosphere) then breakoff will be delayed and focused toward the oceanic side of the passive margin. In contrast, if the continental lithosphere is young and thin, breakoff will be accelerated and focused toward the continental side of the passive margin. Breakoff could also occur in oceanic lithosphere if the oceanic lithosphere is very young. We also investigate the case where the driving extensional body force is reduced when buoyant continental crust peels off from the downgoing plate and underplates the overriding plate or returns toward the surface. We find that even when all continental crust is removed, slab breakoff can still occur, but occurs deeper where the continental lithosphere is hot.

B. R. Hacker and J. G. Liou (eds.),
When Continents Collide: Geodynamics and Geochemistry of Ultrahigh-Pressure Rocks, 97–115.
© 1998 *Kluwer Academic Publishers. Printed in the Netherlands.*

We briefly discuss how slab breakoff relates to the exhumation of UHP rocks, and its predictions for their lower temperature bound P-T paths. If the rise of slivers along the interplate thrust has been the process that exhumed UHP rocks, the P-T paths observed in UHP metamorphic belts suggest that i) continental collisions proceed with convergence velocities much smaller than 50 mm/a, ii) the slivers rise sufficiently quickly that they do not attain thermal equilibrium with the interplate thrust, or iii) shear heating is significant.

1. SLAB BREAKOFF

Collisional orogens are associated with thermal metamorphism and uplift. These attributes can potentially be explained by collision models that view orogenesis as the result of thickening of colliding lithospheres (England and Houseman, 1988). Many continental collisions are associated with bimodal magmatism, i.e., a crustal rhyolitic component and a basaltic mantle component (Turner *et al.*, 1992). Ultrahigh-pressure metamorphic rocks are also associated with some collisional orogens. Simple models of lithospheric thickening, such as convective removal of a thickened thermal boundary layer (Houseman *et al.*, 1981), or lithospheric delamination (Bird, 1979), fail to explain the association of both these features in collisional belts (Davies and von Blanckenburg, 1995).

Slab breakoff is one model which has the potential for explaining all the features outlined above (Davies and von Blanckenburg, 1995). It has the additional advantage of starting from a reasonable initial condition; the subduction of the previously intervening oceanic basin. Wherever slab breakoff occurs it will provide deep controls on metamorphism in collisional orogens.

The consequences of slab detachment were first elaborated by von Blanckenburg and Davies (1992), while the influence of subduction velocity was discussed in Davies and von Blanckenburg (1995). Ton and Wortel (1997) further investigated the influence of rheology and dip on breakoff depth. Application to the European Tertiary Alps was discussed in von Blanckenburg and Davies (1995; 1996). Slab breakoff could have played a role in the development of many orogens, including the Norwegian Caledonides, Aegean, Dabie Shan (Davies and von Blanckenburg, 1995), the Betics, Carpathians, Appalachians, and western North America. Yoshioka and Wortel (1995) investigated the lateral propagation of slab detachment, once initiated, possibly by slab breakoff. Wortel and Spakman (1992) discussed how lateral propagation can be used to explain some features in the circum Mediterranean.

Figure 1a illustrates subducted oceanic lithosphere pulling a passive margin and its continental lithosphere into the mantle. The oceanic lithosphere is dense due to its cold thermal structure and exerts a downward force on the overriding passive margin. In contrast, the continental lithosphere is buoyant due to the thick

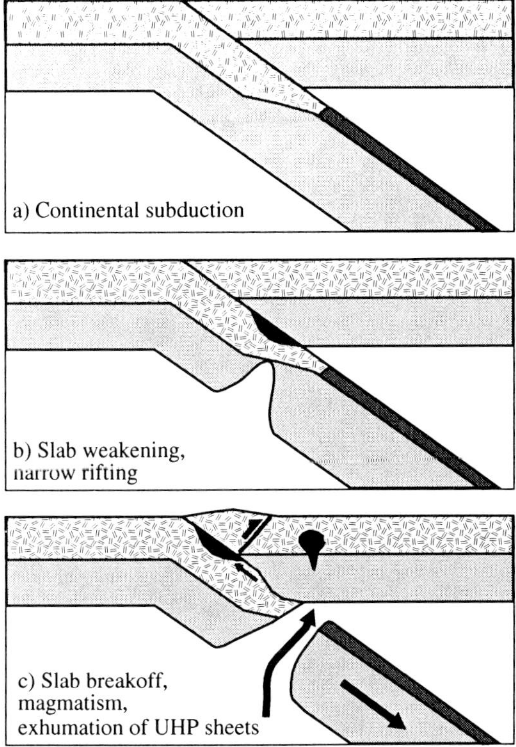

Figure 1. Schematic of slab breakoff. a) Initial subduction of continental crust leads to opposing tensional forces. b) Rapid, narrow rifting, illustrated here within the passive margin. c) Oceanic lithosphere has detached from the continental lithosphere and been replaced by hot inviscid asthenosphere from below. This generates a thermal anomaly at the base of the overriding lithosphere, and can lead to metamorphism. Removal of the dense load of the oceanic plate and the cessation of subduction with its dynamic downward pull, leads to uplift. Heating during subduction and breakoff can lead to release of a sliver of UHP rock that returns to the base of the crust.

layer of continental crust, and as it is subducted deeper, it exerts an ever increasing upward force on the passive margin. The net result of these body forces is a tensional force acting across the passive margin.

As lithosphere is subducted into the mantle it is heated, becoming weaker; and the tensional force increases. Ultimately the force acting across the passive margin exceeds the strength, and the passive margin can no longer hold the oceanic and continental lithosphere together. One possibility is that the oceanic lithosphere detaches from the continental lithosphere (slab breakoff) by means of a narrow rifting process (Fig. 1b).

A narrow rifting process results from a lithospheric-scale strain localization. One means of developing this feedback is if the locally thinning lithosphere is replaced by hot and weak asthenosphere from below. This replacement is very likely because the underlying asthenosphere is the least viscous material near the deforming lithosphere. The next additional increment of strain will then be localized in this region, because it is weakest, leading to further thinning and weakening. Kusznir and Park (1987) and Houseman and England (1986) demonstrated that the process of strain localization requires a critical high strain rate. At that critical strain rate the velocity of the upwelling asthenosphere is sufficiently high such that its advection of heat is greater than conduction of heat out through the top, hence the asthenosphere cannot cool to become rigid lithosphere. Once such a high strain rate is achieved, strain localization and slab breakoff is inevitable. We assume a strain rate of $5x10^{-15}$ s^{-1}, similar to or slightly higher than suggested in the references above. The exact value of the critical strain rate has limited influence on estimates of lithosphere strength (Davies and von Blanckenburg, 1995).

We evaluate the strength (defined as the maximum vertically integrated force that the lithosphere can sustain at a certain strain rate) of lithosphere at the critical strain rate and compare it to an estimate of the driving body force. Breakoff would occur at the depth where the extensional force is greater than the lithospheric strength at the critical strain rate.

As the oceanic lithosphere breaks off from the continental lithosphere, the gap in the descending lithosphere is filled by the least viscous material—which is the underlying asthenosphere. If the asthenosphere rises quickly to depths shallower than ~50 km it will melt by adiabatic decompression. Breakoff is unlikely to occur at such shallow depths, so the upwelling asthenosphere is unlikely to melt, but the thermal anomaly that the asthenosphere will introduce into the base of the orogen can lead to melting of metasomatized or hydrated overriding lithosphere (Fig. 1c). The resulting melts of the mantle migrate into the crust, producing crustal melting, and leading to bimodal magmatism. The thermal anomaly can also lead to metamorphism. The removal of the deep load of the oceanic lithosphere and the termination of subduction and its dynamic downward pull will lead to uplift and exhumation.

The heating of the downgoing plate, and the removal of the down-dip force on the leading edge of the subducting continental lithosphere following detachment of oceanic lithosphere, will make it easier for buoyant slivers of subducted crust to return to the surface. Slab breakoff is not required for UHP rocks to return to the surface (van den Beukel, 1992), but the decrease in the strength of the crust as it is heated and the disappearance of the subducting force arising from the negative buoyancy of the oceanic lithosphere, would both increase the probability of the development of buoyant crustal slivers and ease their return to the surface (Davies and von Blanckenburg, 1995; von Blanckenburg and Davies, 1995).

2. MODELING

In this paper we investigate the roles of thermal controls, e.g., radioactivity, mantle temperature and lithosphere thickness, on slab breakoff. Following Davies and von Blanckenburg (1995), where the modeling method is described in detail, we have used i) the linear prism adaptation of the analytic steady state thermal model of Royden (1993a; 1993b), ii) a brittle-ductile rheology following van den Beukel (1992), Sibson (1974) and Goetze (1978), and iii) a simple density model. The continental lithosphere is assumed to have a crust that is thicker (36 km) than the oceanic lithosphere (7 km) and a mantle that is less dense than either depleted or undepleted oceanic mantle lithosphere. The model assumes that the overriding lithosphere is rigid and that the angle of underthrusting is low. Parameters for the thermal model are listed in Table 1.

Table 1. Values of parameters for Royden's (1993b) thermal model.

Variable	Definition	Value
T_M	Moho temperature at base of foreland lithosphere	1370°C†
l	Lithosphere thickness	100 km
d	Angle of dip of subduction zone	20°
v	Convergence velocity	10 mm/a
A_l	Radioactivity in lower plate	2.5 μWm^{-3}
A_u	Radioactivity in upper plate	0 μWm^{-3}
τ	Shear stress	0 MPa
K	Thermal conductivity	2.5 W/mK
α_D	Thermal diffusivity	$10^{-6}m^2/s$
a	Accretion rate	0 mm/a
e	Erosion rate	0 mm/a
z_r	Thickness of radioactive layer in lower plate	10 km

† Average of Stein and Stein (1992), McKenzie and Bickle (1988)

The power law creep parameters assumed have been derived from laboratory experiments on olivine (Goetze and Evans, 1979) for the mantle, wet diorite for the oceanic crust (Hansen and Carter, 1982), wet quartzite for the upper crust (Koch *et al.*, 1989) and diabase for the lower crust (Caristan, 1982). Using the power law creep parameters from more recent experiments with improved methodologies (e.g., Chopra and Paterson, 1984; Gleason and Tullis, 1995) would not affect the conclusions regarding the thermal controls on slab breakoff, since the relative strengths of the minerals remain unchanged. It must also be remembered that there is large uncertainty in the extrapolation of experimental results to geologic conditions (Carter and Tsenn, 1987). For example, Gleason and Tullis (1995) commented that though their experiments indicate that quartzite is stronger than suggested by Koch *et al.* (1989), they argued that values lower than predicted by their own experiments (i.e., closer to Koch *et al.*, 1989) probably apply in the crust. Davies and von Blanckenburg (1995) showed that the difference between using the parameters of Chopra and Paterson, (1984) and

Goetze and Evans (1979) for the critical mantle (olivine) rheology is small at lithospheric conditions. The density estimates of the mantle section of the oceanic and continental lithosphere are based on Jordan (1979).

Focal mechanisms of intermediate-depth earthquakes suggest that subducting oceanic plates are generally in a tensional or near-tensional state (Apperson and Frohlich, 1987). The estimated change in tension that results from the change from subduction of oceanic lithosphere to buoyant continental lithosphere is therefore probably a lower bound on the actual tensional force. In addition, we have considered the effects of reductions in the tensional force caused by underplating of i) the lower crust or ii) all the crust to the base of the overriding lithosphere.

To simplify comparisons of the lithosphere strength to the tensional force we have calculated the vertically integrated strength, i.e., the summation of the strength of the lithosphere from its base to its surface, perpendicular to its surface. This is a fair comparison if the lithosphere deforms by pure shear and is not affected by large shear stresses at its upper and lower boundary. The integrated strength in Fig. 2 is plotted as a function of the depth of the surface of the subducted lithosphere.

3. RESULTS

The integrated strength for lithospheres with different thermal parameters are compared to estimates of the extensional force in Fig. 2. In all the panels, the thermal parameters are as in Table 1, except for the one parameter under consideration. The oceanic lithosphere is always indicated by triangles, while solid lines indicate the standard conditions listed in Table 1. Continental lithosphere with its thicker crust is always weaker than oceanic lithosphere with its thinner crust, illustrating the strong control of crustal thickness on integrated strength. The straight gray lines are the tensional force estimated from the change in buoyancy force as continental crust is subducted, and is estimated from the simple density model (Davies and von Blanckenburg, 1995). The breakoff depth is indicated by the point at which the tensional force line crosses the integrated strength curve. From the estimated breakoff depth and the known dip angle, the down-dip breakoff distance can be calculated, and hence the time to breakoff, because collision commencement can be calculated using the convergence velocity. Table 2 shows how the Moho temperature and surface heat flow of the foreland lithosphere change with parameter values for the following investigations.

Lithosphere thickness was varied by changing the thickness of its mantle part. All other parameters are unchanged from Table 1. Small changes in lithosphere

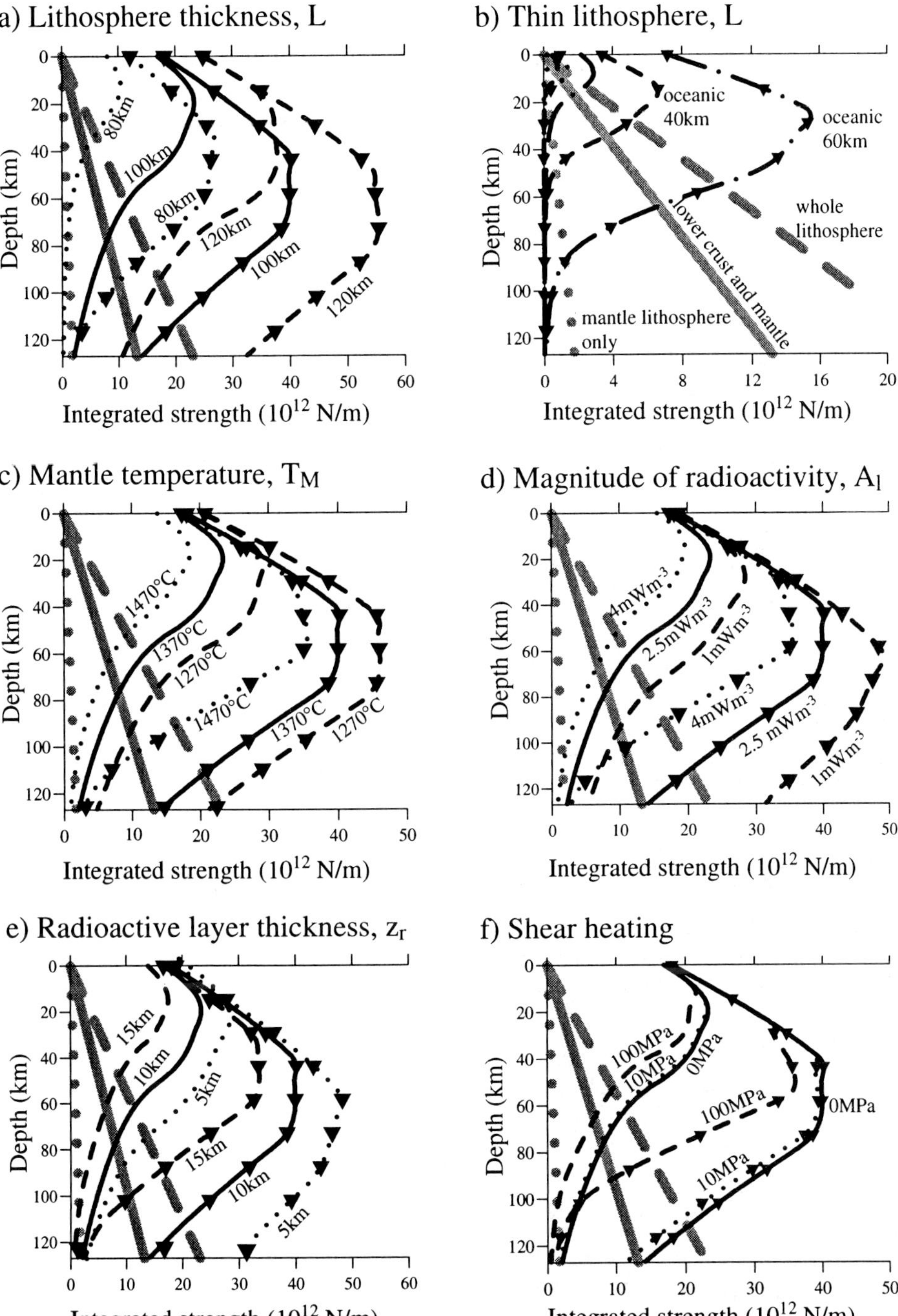

Figure 2. Integrated extensional lithospheric strength versus depth. Solid curves in all panels (except b) represent standard conditions of Table 1. Lines with (without) triangles represent oceanic

(continental) lithosphere. Three straight gray lines represent the tensional force arising from the density changes due to the subduction of the different continental lithosphere structure to increasing depths. The dashed line represents subduction of the whole lithosphere, the dotted line represents subduction of only the mantle part, while the solid line represents subduction of the lower crust and mantle are subducted and delamination of the upper crust. The depth at which the gray straight lines cross the strength curves indicates the depth of slab breakoff. In b), the dotted curve represents a lithosphere 20 km thick with no radioactivity, the dashed line 40 km thick with radioactivity and the dashed-dotted line a lithosphere 60 km thick with radioactivity. Integrated strength scale on this figure is expanded. In f), changes in shear heating have little effect on strength, except oceanic lithosphere at large shear stresses that are unlikely to be reasonable.

thickness produces large changes in strength and Moho temperature (Fig. 2a); e.g., ~20% change in lithosphere thickness leads to ~50% change in integrated strength. Since the strength drops dramatically at depth, it is likely that even with a low extensional force, slab breakoff would ultimately occur, though perhaps at relatively great depth. Equally, slab breakoff is unlikely to occur at depths shallower than 40 km because the strength is generally high at these depths. This is because the lithosphere is largely failing brittlely and hence the strength depends only on pressure (depth) and is independent of temperature. Hence slab breakoff is likely to occur from 50-120 km depth for a wide range of conditions.

Table 2. Variation of Moho temperature and surface heat flow of foreland lithosphere with change in value of single parameter.

Parameter	**Surface Heat Flow**	**Moho Temperature**
mWm^{-2}	**at depth (Z_M) = 36km**	**°C**
Lithosphere thickness (km)		
L=120	52.5	446
L=100 standard	57.5	525
L=80	66.3	644
Mantle temperature (°C)		
T_M=1270	55.5	489
T_M=1370 standard	57.5	525
T_M=1470	60.5	561
Volumetric radioactive heat generation rate (μWm^{-3})		
A=1	43.8	506
A_l =2.5 standard	57.5	525
A_l =4	72.2	544
Thickness of radioactive layer (km)		
z_r =5	46.4	501
z_r =10 standard	57.5	525
z_r =15	68.9	565
Shear stress (MPa)		
0 standard	57.5	525
10 (at 10 mm/a)	60.7	525
100 (at 10 mm/a)	89.2	525
Very thin lithosphere (km)		
60	80.0	842
40	107.5	1238
20 with z_M=7 km and A_l=0	171.1	480 (at $z=z_M$=7 km)

Thinner lithosphere is weaker because there is less thickness to integrate, and because the temperatures are slightly hotter.

Figure 2b extends the variation to consider very thin lithosphere. We have considered one very thin lithosphere of 20 km with the radioactivity set to zero, so as to better mimic conditions in very small and young oceanic basins. The strength decreases dramatically with these very thin lithospheres. Hence, breakoff may occur in oceanic lithosphere if it is very young and the continental lithosphere is cold. Furthermore, because subducting oceanic slabs are generally under tension at relevant depths, thin oceanic lithosphere can experience breakoff without subduction of continental crust.

Similar to lithosphere thickness, small variations in mantle temperature have large effects on the integrated strength (Fig. 2c). The heat flow does not change greatly, but the Moho temperature does. Hence, Moho temperature is a better indicator of lithosphere strength than surface heat flow (see also Sonder and England, 1986).

Figure 2d illustrates the large variations in the concentration of radioactivity required to make a change in the integrated strength. This could be predicted from the small resultant changes in the Moho temperature, even though the surface heat flow changes dramatically (Table 2). Similarly, the thickness of the radioactive layer has a weak influence on the integrated strength unless the thickness approaches the thickness of the crust (Fig. 2e).

In lithospheres with different levels of shear heating, the surface heat flow changes dramatically (Table 2), but the integrated strength requires unlikely levels of shear stress (100 MPa) before there is significant weakening (Fig. 2f). It can be seen to be even more unlikely when one remembers that this shear stress is modeled to generate heating along the full length of the interface.

4. DISCUSSION OF CONDITIONS FOR SLAB BREAKOFF

Shallow slab breakoff has been shown to be favored by high mantle temperature, thin lithosphere, high levels of radioactive heat production per unit volume, thick layers of radioactivity, and high levels of shear heating—all factors that lead to hotter or thinner lithospheres.

All the examples illustrate how factors that affect the deep temperature are more effective in reducing the lithosphere strength (Fig. 3). Lithosphere thickness and mantle temperature introduce strong changes in strength, whereas the amount and distribution of radioactivity have only small effects. The integrated strength is better indicated by the Moho temperature than by heat flow, which reflects only the near surface gradient, and is therefore poorly correlated with strength (Fig. 4a and 4b).

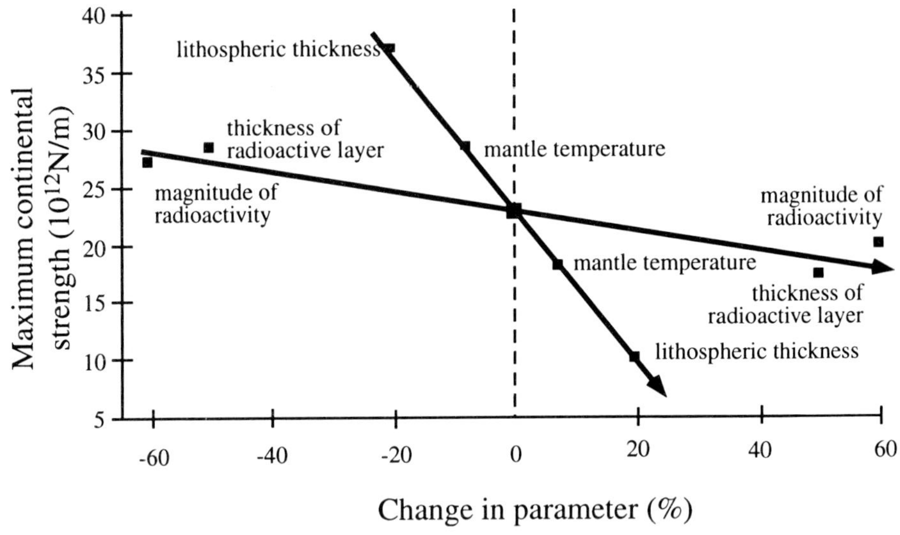

Figure 3. Influence of parameter value on strength of continental lithosphere. Parameters that directly affect mantle temperature—e.g., temperature at the base of lithosphere and lithosphere thickness—have a bigger effect than parameters such as the thickness and magnitude of crustal radioactivity, which mainly affect crustal temperatures.

Another control on the thermal structure and hence the strength of subducting lithosphere is the subduction velocity, which was investigated in Davies and von Blanckenburg (1995). Fast subduction velocities lead to colder and hence stronger lithosphere. It was shown that a subduction velocity as high as 50 mm/a is unlikely to lead to shallow breakoff, while a subduction velocity as low as 2 mm/a will lead to very shallow breakoff. A subduction velocity of 10 mm/a, similar to the velocity of many continental collisions, has been shown to lead to estimates of slab breakoff at intermediate depths. This depth of breakoff leads to predictions of little or no asthenospheric melting, but some lithospheric melting and consequent crustal melting (Davies and von Blanckenburg, 1995). The dip of underthrusting and the value of the critical strain rate also affect slab breakoff, but only weakly (Davies and von Blanckenburg, 1995). Ton and Wortel (1997) also found that the thickness of the continental crust, the thermal structure parameterized by surface heat flow, and the convergence velocity have a strong influence on slab breakoff.

5. APPLICATIONS TO COLLISIONAL OROGENS

The collision of India with Asia is the result of closure of the Tethyan ocean at ~50-60 Ma (Searle, 1996). After that, convergence proceeded with the attempted subduction of Indian shield lithosphere at velocities of ~50 mm/a

(DeMets *et al.*, 1990). If such shield lithosphere is thicker than 100 km, slab breakoff could be expected to be delayed. Similarly, the high convergence velocities would favor delayed breakoff. If Indian lithosphere subducted horizontally beneath Tibet, this delay could mean that slab breakoff would occur at a long horizontal distance from the suture. If subduction was not horizontal

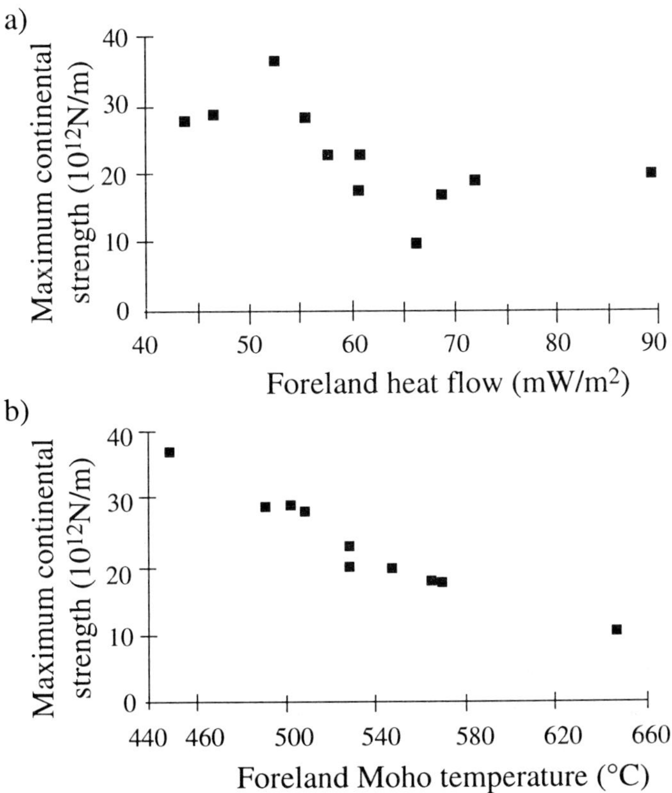

Figure 4. a) Maximum continental strength for all cases investigated *vs.* foreland heat flow. The striking result is that the correlation is weak and the data scattered. b) Maximum continental strength for all the cases investigated *vs.* foreland Moho temperature. The relationship is strong and nearly linear.

then this delay implies breakoff at greater depth. The explanation for the presence of deep subduction of supposed continental lithosphere in the Hindu Kush (Roecker, 1982) and beneath the Pamir (Burtman and Molnar, 1993) might be that the subducted lithosphere was thick and cold, and hence sufficiently strong to resist detachment at shallow depth. Given the rapid convergence, we might expect slab breakoff ~2–15 m.y. after closure of the basin, i.e., at ~40-50 Ma, provided subduction was not horizontal. Eclogites of ~50 Ma age have been reported for metasediments of the Indian Tethyan margin now exposed in the

NW Himalaya, but none have been reported from the eastern Himalaya (Tonarini *et al.*, 1993; de Sigoyer *et al.*, 1997). The mafic volcanics and granitoids of 40-60 Ma age in the eastern Himalayan Gangdese belt (Coulon *et al.*, 1986) might be related to slab breakoff, but are more likely to be related to magmatic activity resulting from the preceding oceanic subduction. It will be very difficult to differentiate between oceanic subduction magmatism and slab breakoff magmatism, especially given that there might only be a brief interval between closure of the Tethys and arrival of subducted continental lithosphere at depth critical for slab breakoff. If on the other hand, the subduction of India beneath Tibet has been horizontal, then slab breakoff could be delayed and some of the Miocene to Recent volcanic, metamorphic and tectonic activity within the Tibetan Plateau could be assigned to slab breakoff.

Gaps in seismicity beneath the Banda arc and the rapid uplift of Timor have led some to suggest that slab breakoff is proceeding (Milsom and Audley-Charles, 1986). The continental lithosphere being subducted beneath the Banda Arc should be strong because much of the Australian lithosphere is cratonic or shieldlike. In contrast, subduction of young lithosphere, as with Europe under Adria forming the European Tertiary Alps, might lead to slightly shallower slab breakoff (von Blanckenburg and Davies, 1995). Since it is difficult to predict the thermal structure of subducted continental lithosphere, and even harder to predict its rheology, it is clear that *a priori* prediction of breakoff depth will be difficult, as will definitive tests. If the conjugate margin still survives and can be confidently identified then its nature can be used to provide some estimate of the properties of the subducted margin.

The models for the strength of very young, thin oceanic lithosphere (Fig. 2b) could be applied to the case of ridge subduction (e.g., Hacker, 1991)—especially subduction of a young small basin; e.g., a back-arc basin. In this case, breakoff could readily occur at the subducted ridge, rather than at the passive margin or in the continental lithosphere, due to the weakness of the ridge. Whether breakoff occurs at the ridge or in the continental lithosphere also depends on the tensional force, which could be much less at the ridge than at the continental passive margin. The Alps, with its possibly small ocean basins, could have suffered breakoff at a ridge. Breakoff at subducting ridges (Gorring *et al.*, 1997) or oceanic plate boundaries (van der Lee and Nolet, 1997) has been suggested to be taking place along the east Pacific margin.

During collisional orogenesis, some of the crust of the downgoing plate is peeled off and not subducted (Molnar and Gray, 1979; Cloos, 1993). As shown by the lower bounds on the change in buoyancy force, peeling off of the whole crust dramatically affects the expected depth of breakoff. Most of the lower crust in the Alps has been subducted, while much of the upper crust has been peeled off and folded into nappes. The Himalaya also largely consist of upper crust of the Indian plate. Sometimes the crust is peeled off at great depth in orogens and returns to the surface, as evident from UHP rocks. This return to the surface is driven initially by the buoyancy contrast between the crust and the surrounding mantle. Given the lack of lower crust exposed in collisional orogens, it is probably rare for the entire crustal section of the downgoing plate to be detached from its underpinning mantle. Hence we feel that the appropriate tensional force in most orogens is rarely close to the lower bound (the straight gray dotted line of Fig. 2).

5.1 Secular Changes in the Style of Metamorphic Terrains

It has been suggested that Earth's mantle in Archean times was hotter than today by hundreds of degrees (Wasserburg *et al.*, 1964; Abbott *et al.*, 1994). As a consequence, oceanic crust was young and warm (Martin, 1986), and there is a possibility that continental crust was thinner, hotter, and softer (Choukroune *et al.*, 1995). In a plate-tectonic environment, all these features favor shallow breakoff. This prediction is supported by the absence of blueschist- and eclogite facies metamorphic rocks in Archean orogenic belts, their paucity in Proterozoic belts (Grambling, 1981), and their much higher abundance in Phanerozoic orogenies. Similarly the much higher abundance of granulite-facies metamorphic terrains in the Archean is generally explained by higher magmatic activity. This might be due to more vigorous convective mantle removal events (Sandiford, 1989), but an important contribution might be igneous activity triggered by shallow breakoff.

5.2 Implications of Slab Breakoff for UHP Rocks

Deep subduction of continental crust subjects it to UHP. The upper surface of the crust is a weak, interplate interface. Weak intracrust planes can also be present, as pre-existing planes of weakness, or they can form as temperatures increase and specific layers weaken relative to their surroundings. As shown by van den Beukel (1992), an interplate fault and an intraplate fault could create a sliver of crust that is free from the underlying subducting slab. This sliver can then squeeze back up the weak inter-plate thrust due to its own buoyancy until it reaches the base of the overriding plate (Fig. 5) (von Blanckenburg and Davies, 1995). This process is similar to the one modeled as "low compression mode of continental subduction" by Chemenda *et al.* (1996), and that invoked for exhumation of UHP rocks in the Dabie Shan (Ernst and Liou, 1995).

Slab breakoff is not required for the above process, but is consistent with it. Slab breakoff frees the leading edge of the continental crust and hence can ease the development of a sliver. If the sliver is thin and migrating sufficiently slowly, it will be cooled from both above and below and will be at the local temperature of the interplate thrust (Ernst and Peacock, 1996). Figure 6a shows the thermal paths for such cases for convergence velocities of 50, 10, and 2 mm/a—using otherwise standard conditions (Table 1). These are lower bound temperatures because we exclude shear heating on the thrust. The actual temperatures would be higher if the sliver is thick or squeezed up quickly such that thermal equilibrium with the thrust is not achieved. Following the sliver's arrival in the crust, the density contrast wanes and the remaining pressure decrease will be by tectonic or erosional unroofing, leading to medium-P thermal metamorphism in the sliver.

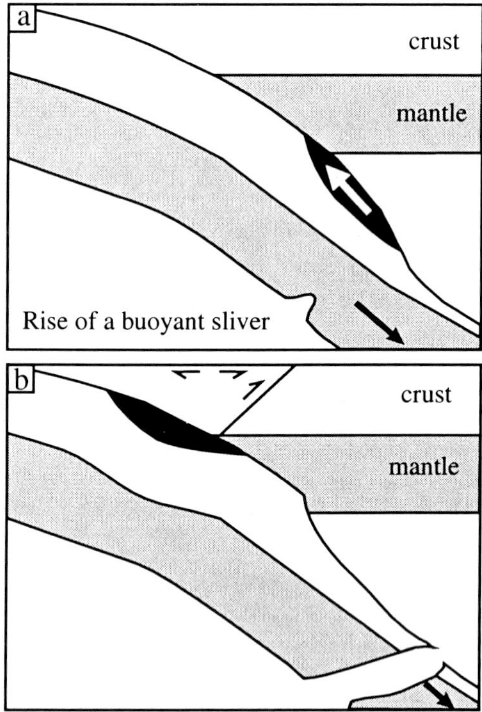

Figure 5. Release and rise of a buoyant sliver of continental crust. a) Before breakoff. b) After breakoff and arrival of the sliver at the base of the crust. From hereon a different exhumation mechanism must be active.

A comparison with PT paths observed in UHP terranes (Fig. 6b) shows that no paths have been observed as cold as those predicted for convergence velocities of 50 mm/a. This either means that collisions are always slower than 50 mm/a, that shear heating is significant, that slivers move quickly back toward the surface, or that such cold paths are simply not preserved in the geological record.

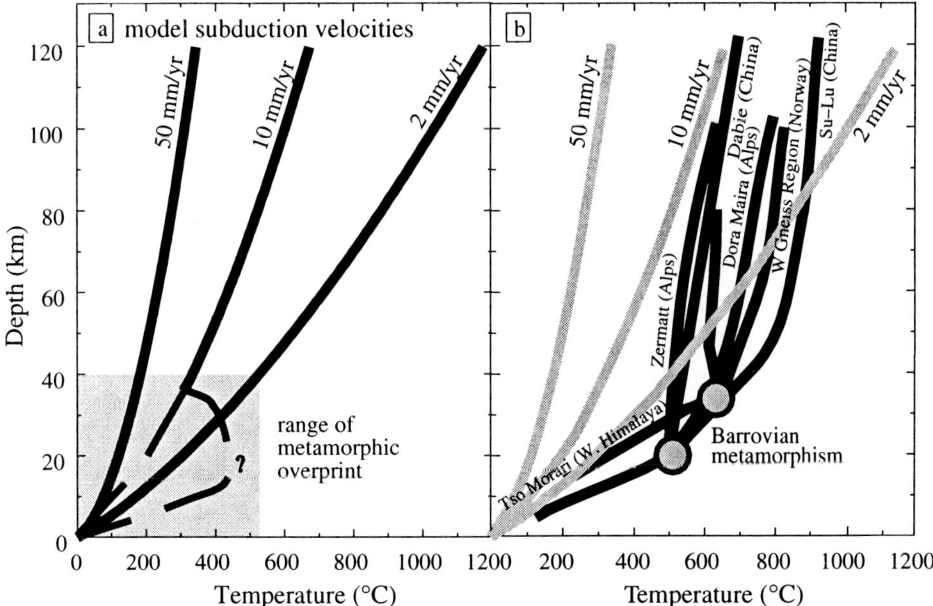

Figure 6. Calculated (gray) and reported (black, Alice Wain, personal communication 1997, Monié and Chopin, 1991; Hacker and Peacock, 1994; de Sigoyer *et al.*, 1997; Van der Klauw *et al.*, 1997) P-T paths for exhumation of UHP rocks. Calculated paths are for a sliver returning back up the interplate thrust at a rate such that it attains thermal equilibrium with the thrust. Three convergence velocities are modeled: 50, 10, and 2 mm/a. Temperatures could be higher due to shear heating or if the sliver does not reach thermal equilibrium, but should not be colder. b) PT paths of UHP metamorphic rocks.

6. CONCLUSIONS

The depth of breakoff controls the magnitude of the resulting consequences in the developing orogen. If breakoff occurs at a shallow depth, a large thermal anomaly is introduced that can lead to large magmatic and metamorphic consequences. If breakoff occurs at great depth, the thermal anomaly is weaker and more dissipated by the time it affects the enriched regions of the overriding lithosphere and the crust, leading to minor or no observed consequences. UHP rocks cannot return to the surface from deeper than the depth of slab breakoff, so their presence requires relatively deep or no slab breakoff.

The depth of slab breakoff is controlled by the strength of the mantle lithosphere. The Moho temperature is a good indicator of the total lithospheric strength. The foreland surface heat flow is a poor indicator of the deep thermal structure and hence the strength of the lithosphere unless we have a good understanding of shear heating and the distribution and magnitude of crustal radioactivity. Given the likely hotter mantle temperatures and possibly thinner lithospheres at earlier times in Earth history, it is likely that breakoff could have occurred at shallower depths with greater effects at the surface.

Given current best estimates of lithosphere thickness and mantle temperatures, most collisions will lead to slab breakoff at some depth. The identity of the continental lithosphere next to the passive margin can play a significant role in determining the location of the strain localization. Where the continental lithosphere is young and hot, strain localization is likely to be at the continental side of the passive margin, while if the continental lithosphere is old and cold (e.g., cratonal) the strain localization is likely to be deeper and at the oceanic side of the passive margin. In the seemingly rare situation where the bulk of the downgoing continental crust detaches from its underlying mantle, slab breakoff would be pushed to deeper depth.

UHP rocks are suggested to be slivers of subducted crust that peel off from their underpinnings and return by their own buoyancy to the base of the crust of the overriding lithosphere. Computations of the temperature of the uprising path show that collisions are either always slower than 50 mm/a, or that shear heating is significant, or that slivers move quickly back toward the surface.

7. ACKNOWLEDGMENTS

We would like to acknowledge detailed reviews from B. Hacker and A. Yin that helped improved the manuscript. JHD would like to acknowledge support from the University of Liverpool Research Development Fund.

REFERENCES

Abbott, D.H., Burgess, L., Longhi, J., and Smith, W.H. (1994) An empirical thermal history of the earth's upper mantle, *Journal of Geophysical Research* **99**, 13835–13850.

Apperson, K.D. and Frohlich, C. (1987) The relationship between Wadati-Benioff zone geometry and P, T and B axes of intermediate and deep focus earthquakes, *Journal of Geophysical Research* **92**, 13,821–13,831.

Bird, P. (1979) Continental delamination and the Colorado Plateau, *Journal of Geophysical Research* **84**, 7561–7571.

Burtman, V.S. and Molnar, P. (1993) *Geological and geophysical evidence for deep subduction of continental crust beneath the Pamir*, Geological Society of America Special Paper **281**, Geological Society of America, Boulder.

Caristan, Y. (1982) The transition from high-temperature creep to fracture in Maryland diabase, *Journal of Geophysical Research* **87**, 6781–6798.

Carter, N.L. and Tsenn, M.C. (1987) Flow properties of continental lithosphere, *Tectonophysics* **136**, 27–63.

Chemenda, A.I., Mattauer, M., and Bokun, A.N. (1996) Continental subduction and a mechanism for exhumation of high-pressure metamorphic rocks: new modeling and field data from Oman, *Earth and Planetary Science Letters* **143**, 173–182.

Chopra, P.N. and Paterson, M.S. (1984) The role of water in the deformation of dunite, *Journal of Geophysical Research* **89**, 7861–7876.

Choukroune, P., Bouhallier, H., and Arndt, N.T. (1995) Soft Archean lithosphere during periods of crustal growth or reworking, *Geol. Soc. Spec. Publ.* **95**, 67–86.

Cloos, M. (1993) Lithospheric buoyancy and collisional orogenesis: subduction of oceanic plateaus, continental margins, island arcs, spreading ridges, and seamounts, *Geological Society of America Bulletin* **105**, 715–737.

Coulon, C., Maluski, H., Bollinger, C., and Wang, S. (1986) Mesozoic and Cenozoic volcanic rocks from central and southern Tibet: ^{39}Ar-^{40}Ar dating, petrological characteristics and geodynamical significance, *Earth and Planetary Science Letters* **79**, 281–302.

Davies, J.H. and von Blanckenburg, F. (1995) Slab breakoff: a model of lithosphere detachment and its test in the magmatism and deformation of collisional orogens, *Earth and Planetary Science Letters* **129**, 85–102.

de Sigoyer, J., Guillot, S., Lardeaux, J.M., and Mascle, G. (1997) Glaucophane-bearing eclogites in the Tso Morari dome (eastern Ladakh, NW Himalaya), *European Journal of Mineralogy* **9**, 1073–1083.

DeMets, C., Gordon, R.G., Argus, D.F., and Stein, S. (1990) Current plate motions, *Geophysics Journal International* **101**, 425–478.

England, P.C. and Houseman, G.A. (1988) The mechanics of the Tibetan plateau, *Philosophical Transactions of the Royal Society of London* **A326**, 301–320.

Ernst, W.G. and Liou, J.G. (1995) Contrasting plate-tectonic styles of the Qinling-Dabie-Sulu and Franciscan metamorphic belts, *Geology* **23**, 353–356.

Ernst, W.G. and Peacock, S.M. (1996) A thermotectonic model for preservation of ultrahigh-pressure phases in metamorphosed continental crust, in G.E. Bebout, D.W. Scholl, S.H. Kirby, and J.P. Platt (eds.), *Subduction Top to Bottom*, American Geophysical Union, Washington, D.C., pp. 171–178.

Gleason, G.C. and Tullis, J. (1995) A flow law for dislocation creep of quartz aggregates determined with the molten salt cell, *Tectonophysics* **247**, 1–23.

Goetze, C. (1978) The mechanisms of creep in olivine, *Philosophical Transactions of the Royal Society of London* **A288**, 99–119.

Goetze, C. and Evans, B. (1979) Stress and temperature in the bending lithosphere as constrained by experimental rock mechanics, *Geophysical Journal of the Royal Astronomic Society* **59**, 463–478.

Gorring, M.L., Kay, S.M., Zeitler, P.K., Ramos, V.A., Rubiolo, D., Fernandez, M.I., and Panza, J.L. (1997) Neogene Patagonian plateau lavas: continental magmas associated with ridge collision at the Chile triple junction, *Tectonics* **16**, 1–17.

Grambling, J.A. (1981) Pressures and temperatures in Precambrian metamorphic rocks, *Earth and Planetary Science Letters* **53**, 63–68.

Hacker, B.R. (1991) The role of deformation in the formation of metamorphic field gradients: Ridge subduction beneath the Oman ophiolite, *Tectonics* **10**, 455–473.

Hacker, B.R. and Peacock, S.M. (1994) Creation, preservation, and exhumation of coesite-bearing, ultrahigh-pressure metamorphic rocks, in R.G. Coleman and X. Wang (eds.), *Ultrahigh Pressure Metamorphism*, Cambridge University Press, Cambridge, United Kingdom, 159–181.

Hansen, F.D. and Carter, N.L. (1982) Creep of selected crustal rocks at 1000 MPa, *Transactions of the American Geophysical Union, Eos* **63**, 437.

Houseman, G. and England, P. (1986) A dynamical model of lithosphere extension and sedimentary basin formation, *Journal of Geophysical Research* **91**, 719–729.

Houseman, G.A., McKenzie, D.P., and Molnar, P. (1981) Convective instability of a thickened boundary layer and its relevance for the thermal evolution of continental convergent belts, *Journal of Geophysical Research* **86**, 6115–6132.

Jordan, T.H. (1979) Mineralogies, densities and seismic velocities of garnet lherzolites and their geophysical implications, in F.R. Boyd and H.O.A. Meyer (eds.), *The Mantle Sample: Inclusions in Kimberlites and Other Volcanics*, American Geophysical Union, Washington, D. C., pp. 1–14.

Koch, P.S., Christie, J.M., Ord, A., and George, R.P., Jr. (1989) Effect of water on the rheology of experimentally deformed quartzite, *Journal of Geophysical Research* **94**, 13,975–13,996.

Kusznir, N.J. and Park, R.G. (1987) The extensional strength of the continental lithosphere: its dependence on geothermal gradient, and crustal composition and thickness, in M.P. Coward, J.F. Dewey, and P.L. Hancock (eds.), *Continental Extensional Tectonics, Geological Society of London Special Publication* **28**, Blackwell, Oxford, pp. 35–52.

Martin, H. (1986) Effect of steeper Archean geothermal gradient on geochemistry of subduction-zone magmas, *Geology* **14**, 753–756.

McKenzie, D.P. and Bickle, M.J. (1988) The volume and composition of melt generated by extension of the lithosphere, *Journal of Petrology* **29**, 625–680.

Milsom, J. and Audley-Charles, M.G. (1986) Post-collision isostatic readjustment in the Southern Banda Arc, in M.P. Coward and A.C. Ries (eds.), *Collision Tectonics, Geological Society Special Publication* **19**, pp. 353–364.

Molnar, P. and Gray, D. (1979) Subduction of continental lithosphere: some constraints and uncertainties, *Geology* **7**, 58–62.

Monié, P. and Chopin, C. (1991) $^{40}Ar/^{39}Ar$ dating in coesite-bearing and associated units of the Dora Maira massif, western Alps, *European Journal of Mineralogy* **3**, 239–262.

Roecker, S.W. (1982) Velocity structure of the Pamir-Hindu Kush region: possible evidence of subducted crust, *Journal of Geophysical Research* **87**, 945–959.

Royden. (1993a) Correction to "The steady state thermal structure of eroding orogenic belts and accretionary prisms, *Journal of Geophysical Research* **98**, 20039.

Royden, L.H. (1993b) The steady state thermal structure of eroding orogenic belts and accretionary prisms, *Journal of Geophysical Research* **98**, 4487–4507.

Sandiford, M. (1989) Secular trends in the thermal evolution of metamorphic terrains, *Earth and Planetary Science Letters* **95**, 85–96.

Searle, M.P. (1996) Cooling history, erosion, exhumation and kinematics of the Himalaya-Karakorum-Tibet orogenic belt, in A. Yin and T.M. Harrison (eds.), *Tectonics of Asia*, Cambridge University Press, Cambridge, pp. 110–137.

Sibson, R.H. (1974) Frictional constraints on thrust, wrench and normal faults, *Nature* **249**, 542–544.

Sonder, L.J. and England, P. (1986) Vertical averages of the continental lithosphere: relation to thin sheet parameters, *Earth and Planetary Science Letters* **77**, 81–90.

Stein, C.A. and Stein, S. (1992) A model for the global variation in oceanic depth and heat flow with lithospheric age, *Nature* **359**, 123–129.

Ton, S.Y.M.W.A. and Wortel, M.J.R. (1997) Slab detachment in continental collision zones : An analysis of controlling parameters, *Geophysical Research Letters* **24**, 2095–2098.

Tonarini, S., Villa, I.M., Oberli , F., Meier, M., Spencer, D.A., Pognante, U., and Ramsay, J.G. (1993) Eocene age of eclogite metamorphism in Pakistan Himalaya: implications for India-Eurasia collision, *Terra Nova* **5**, 13–20.

Turner, S., Sandiford, M., and Foden, J. (1992) Some geodynamic and compositional constraints on "postorogenic" magmatism, *Geology* **20**, 931–934.

van den Beukel, J. (1992) Some thermomechanical aspects of the subduction of continental lithosphere, *Tectonics* **11**, 316–329.

Van der Klauw, S.N., Reinecke, T., and Stöckhert, B. (1997) Exhumation of ultrahigh-pressure metamorphic oceanic crust from Lago di Cignana, Piemontese zone, western Alps: the structural record in metabasites, *Lithos* **41**, 79–102.

van der Lee, S. and Nolet, G. (1997) Seismic image of the subducted trailing fragments of the Farallon plate, *Nature* **386**, 266–269.

von Blanckenburg, F. and Davies, J.H. (1992) Slab break-off. Explanation of mantle magmatism in the Alps?, *Transactions of the American Geophysical Union, Eos* **73**, 546.

von Blanckenburg, F. and Davies, J.H. (1995) Slab breakoff: a model for syncollisional magmatism and tectonics in the Alps, *Tectonics* **14**, 120–131.

von Blanckenburg, F. and Davies, J.H. (1996) Feasibility of double slab breakoff (Cretaceous and Tertiary) during the Alpine convergence, *Eclogae Geologica Helvetica* **89**, 111–127.

Wasserburg, G.J., MacDonald, G., Hoyle, J.F., and Fowler, W.A. (1964) Relative contributions of uranium, thorium, and potassium to heat production in the earth, *Science* **143**, 465–467.

Wortel, M.J.R. and Spakman, W. (1992) Structure and dynamics of subducted lithosphere in the Mediterranean region, *Proceedings of the Koninklijke Nederlandse Akademie van Wetenschappen* **95**, 325–347.

Yoshioka, S. and Wortel, M.J.R. (1995) Three-dimensional numerical modeling of detachment of subducted lithosphere, *Journal of Geophysical Research* **100**, 20223–20244.

Chapter 5

Exhumation of Ultrahigh-Pressure Rocks: Thermal Boundary Conditions and Cooling History

Bernhard Grasemann
Institut für Geologie, Universität Wien, Austria, Bernhard.Grasemann@univie.ac.at

Lothar Ratschbacher
Institut für Geologie, Universität Würzburg, Germany, lothar@geologie.uni-wuerzburg.de

Bradley R. Hacker
Department of Geological Sciences, University of California, Santa Barbara, CA, 93106–9630, USA, hacker@geology.ucsb.edu

Abstract: We investigate the exhumation of ultrahigh pressure (UHP) and high-pressure (HP) rocks in the framework of a dynamic simulation that considers heat advection, heat conduction, heat production, and consequent time-dependent changes in the geothermal gradient. In the absence of lateral heating, rocks exhuming from great depth cool or decompress isothermally and the main cooling period follows the main period of exhumation. Even for a constant exhumation rate, UHP rocks undergo a two-stage cooling history at the end of which the pressure-temperature (P-T) paths of all rocks approach a steady state or "final" geotherm at crustal levels; the shape of the steady-state or final geotherm is mainly a function of exhumation rate. Reconstruction of pressure-temperature-time (P-T-t) paths permits a qualitative distinction between "fast" and "slow" UHP exhumation: fast exhumation is characterized by extremely rapid crustal cooling following small temperature increases or isothermal decompression, whereas slow exhumation is characterized by steady cooling following more modest heating. Rocks exhuming from different depths (e.g., crustal and mantle levels) follow substantially different PT paths (e.g., heating and cooling during decompression), even if all rocks in an orogen are exhumed by the same orogen-scale process.

P-T paths of UHP rocks of the Qinling-Dabie-Hong'an area of central China are consonant with our modeling in that the rocks exhumed from the greatest depth show nearly isothermal decompression, whereas more modestly buried rocks underwent heating during exhumation. The shapes of the P-T-t paths suggest an

B. R. Hacker and J. G. Liou (eds.),
When Continents Collide: Geodynamics and Geochemistry of Ultrahigh-Pressure Rocks, 117–139.
© 1998 *Kluwer Academic Publishers. Printed in the Netherlands.*

exhumation rate closer to 5 mm/a rather than 1 mm/a, and the apparent two-stage cooling history often interpreted for the Dabie area could have been produced by a single stage, constant exhumation rate.

1. INTRODUCTION

Deeply exhumed rocks that were metamorphosed under diamond-eclogite or coesite-eclogite facies—ultrahigh-pressure (UHP) rocks—are known from various places around the world (e.g., Liou *et al.*, 1996). Subduction is deemed responsible for the process by which crustal rocks reach depths in excess of 100 km (Hacker and Peacock, 1994), although it remains unclear which boundary conditions allow buoyant continental crust to reach UHP depths (e.g., Molnar and Gray, 1979). Exhumation of UHP rocks from great depth requires either the removal of the overburden by tectonic processes and/or erosion or the transport of the UHP rocks through the overburden (Platt, 1987). This second process is probably important for the emplacement of high-pressure tectonic blocks in melanges of subduction zones (Cloos, 1982) however, we focus on the exhumation of regional UHP terranes. Exhumation via tectonic processes can be either by thrusts and coeval, synthetic normal faults or by conjugate normal faults. In realistic tectonic situations, neither process exhumes rocks unless an additional process such as erosion or special deformation histories like hinge migration or fault rotation are involved.

Although UHP rocks occur in highly deformed zones (e.g., Rubie, 1984; Hacker *et al.*, 1995), structural methods commonly constrain only the final stages of the exhumation process and structures leading to and related to deformation during the UHP event are commonly not observed (Henry *et al.*, 1993). Therefore, most of the information about the UHP event and subsequent cooling has been derived by petrological and thermochronological methods.

Pressure-temperature (P-T) paths of rocks with a high P-T ratio differ from regional metamorphic rocks typical of collision orogens (Ernst, 1988). A pressure maximum at an early stage of orogeny and a subsequent increase in temperature characterize the exhumation histories of most metamorphic rocks from collisional orogens (Thompson and England, 1984). Petrological constraints, however, demand that rocks with a high P-T ratio underwent decompression with little or no heating (Fig. 1, Liou *et al.*, 1996). Some exhumation histories even require a retrograde P-T path subparallel to the prograde path (Cloos, 1982; Ernst, 1988). The establishment of a pressure-temperature-time (P-T-t) path relies on being able to tie P-T conditions to specific times. Thermochronological methods are, in general, restricted to certain closure temperatures, and in particular, the early, hotter stages of the cooling history are often incompletely constrained.

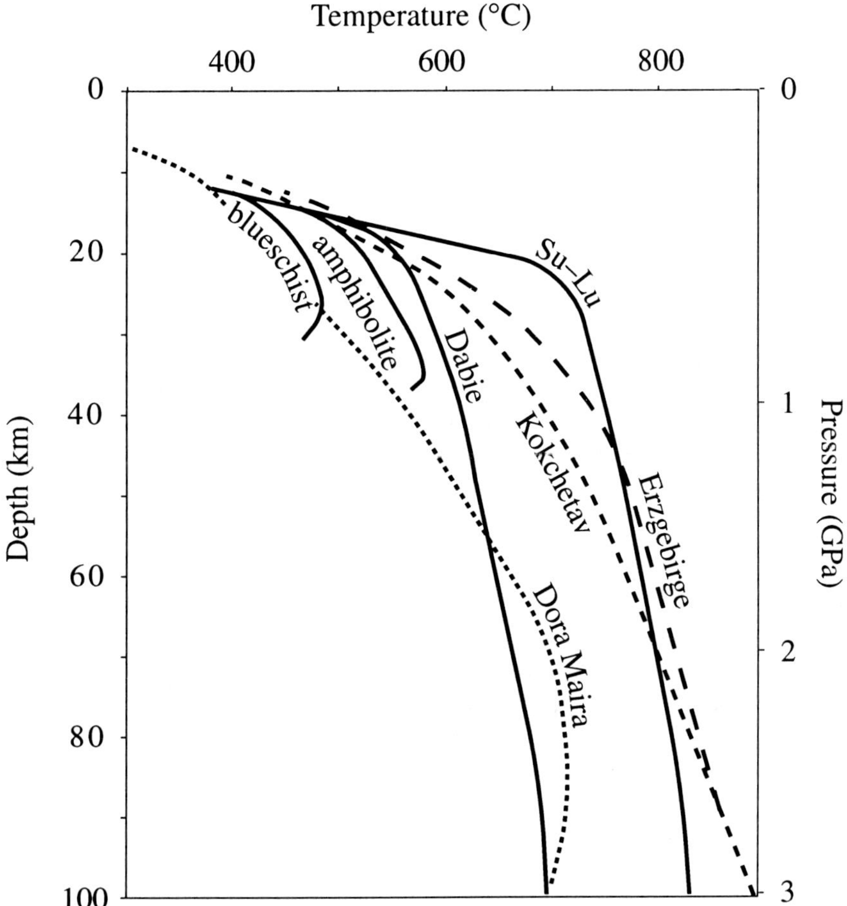

Fig. 1. Decompression paths for UHP rocks from Eurasia and China (Chopin *et al.*, 1991; Schmaedicke *et al.*, 1992; Okay, 1993; Zhang *et al.*, 1994; Liou *et al.*, 1996; Zhang *et al.*, 1997).

The most difficult task is to translate cooling rates into exhumation rates, as this requires needs information about the initial geothermal gradient and the change of the geothermal gradient during exhumation. Here, we investigate exhumation of UHP rocks in the framework of a dynamic simulation that considers heat advection, heat conduction, heat production, and the consequent time-dependent progressive change of the geothermal gradient. The interplay between heat conduction, heat production, heat advection and movement of rocks is complex and dynamic models simulate exhumation more realistically than models with a static thermal gradient (Mancktelow and Grasemann, 1997). An important result of this transient behavior is that heating and cooling of rocks is not only a function of burial and exhumation rates but also of the rate of relaxation of the isotherms. This result underlies why it is invalid to correlate rapid cooling with rapid exhumation.

Based on these physical considerations we aim to clarify by means of one-dimensional thermal calculations and by introducing the instantaneous temperature change (ITC) and the local geothermal slope (LGS) the conditions under which deeply buried rocks heat or cool during exhumation. We show that i) it is unlikely that rocks from great depth heat significantly during exhumation even if exhumation rates are slow, and ii) most of the exhumation occurs well before the rapid cooling recorded by thermochronology. The goal of the presented models is to highlight the influence of some important first-order effects on the exhumation of UHP rocks.

2. Temperature History During Exhumation of UHP Rocks

High exhumation rates are typically invoked to explain the preservation of UHP rocks (e.g., Michard *et al.*, 1993; Eide *et al.*, 1994; Nie *et al.*, 1994). However, the temperature history of a metamorphic rock not only depends on the rate of exhumation but also on the initial thermal state of the lithosphere and on the boundary conditions before and during exhumation (Ruppel and Hodges, 1994).

We define the *local geothermal slope,* LGS, as the slope of the temperature gradient at a given depth (Fig. 2):

$$LGS = \lim_{\Delta z_g \to 0} \frac{\Delta T_x}{\Delta z_x} = \lim_{\Delta z_g \to 0} \frac{T_0 - T_1}{z_0 - z_1} \qquad (1)$$

where T_g is a temperature range and z_g is a depth interval. If the LGS is positive, temperatures increase downward at that depth; if it is negative, temperatures decrease. Whether a rock cools or heats during movement from depth z_0 to z_1 depends on the transient state of the geotherm. The difference in temperature ΔT over a time increment is defined as positive if the temperature is increasing and negative if the temperature is decreasing. The term *temperature change rate* is preferred to *cooling rate* because the rocks may heat rather than cool during their movement through the lithosphere. The difference in depth Δz is defined to be positive when a rock moves toward the surface.

The rate of change in temperature with respect to change in depth at a given time (ITC; Fig. 2) is defined as:

$$ITC = \lim_{\Delta t \to 0} \frac{\dot{T}}{|\dot{z}|} = \lim_{\Delta t \to 0} \frac{dT/dt}{dz/dt} \qquad (2)$$

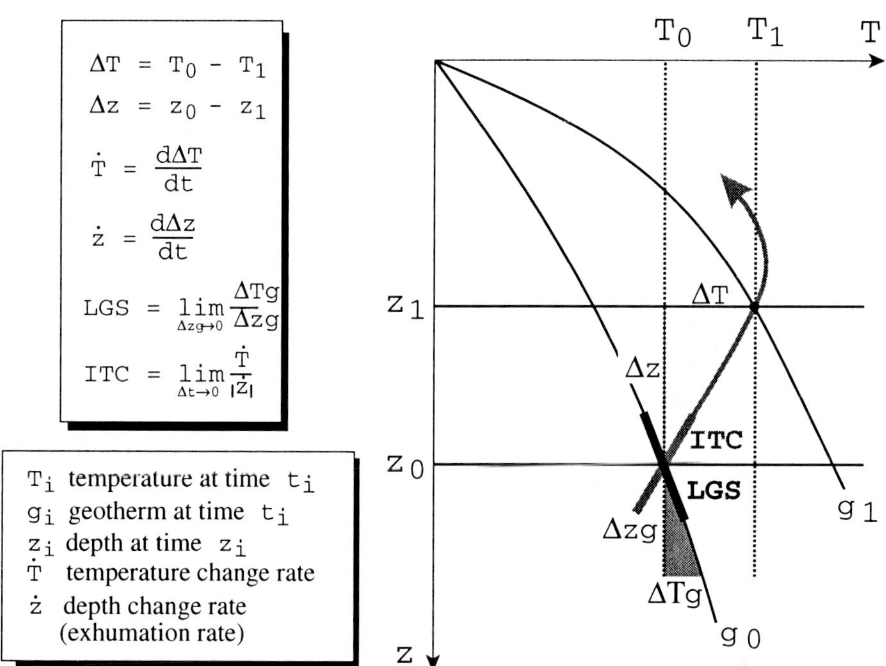

Fig. 2. The instantaneous temperature change (ITC) is the limit of the rate of temperature change divided by the exhumation rate as the time approaches zero. It characterizes the relationship between the change of the thermal regime and the exhumation rate and can be used to indicate if rock at a given depth will heat or cool during exhumation. If the ITC is positive the rocks will heat, if it is negative the rocks will cool. A zero ITC indicates a constant temperature. The relationship between the ITC and the local geothermal slope (LGS) defines whether the rock exhumes above, below, or at steady state.

where $\dot{T}$ is the temperature change rate, $\dot{z}$ is the depth change rate, and Δt is the time increment. By definition, a negative ITC defines heating during exhumation, positive ITC means cooling during exhumation, and an ITC of zero indicates isothermal decompression. When the ITC is equal to the LGS, the rock changes temperature along the steady state geotherm. Equation (2) can also be applied to burial, in which case $\dot{z}$ is negative and represents the burial rate. Therefore, in the general definition of the ITC the term *depth change rate* is preferred. However, the following discussion focuses on exhumation (sensu England and Molnar, 1990), for which $\dot{z}$ is positive. The *steady state geothermal gradient* is the temperature as a function of depth, which does not change within a reasonable limit under the applied boundary conditions. This definition includes quasi-steady state, where mathematically a true steady state cannot be reached (e.g., erosion of a heat-producing layer) but the temperatures do not change significantly over geologic time. For example, at an exhumation rate of 1 mm/a, a long time (>40 m.y.) and corresponding deep exhumation (>40 km) are required to approach steady state (for a more comprehensive discussion of boundary conditions see Mancktelow and Grasemann, 1997). Given that UHP rocks exhume from depth >100 km it is likely that steady state will be reached during exhumation. Considering the possible changes in a geothermal gradient during exhumation the following three regimes are distinguished (Fig. 3), defined by the values of the ITC and the LGS.

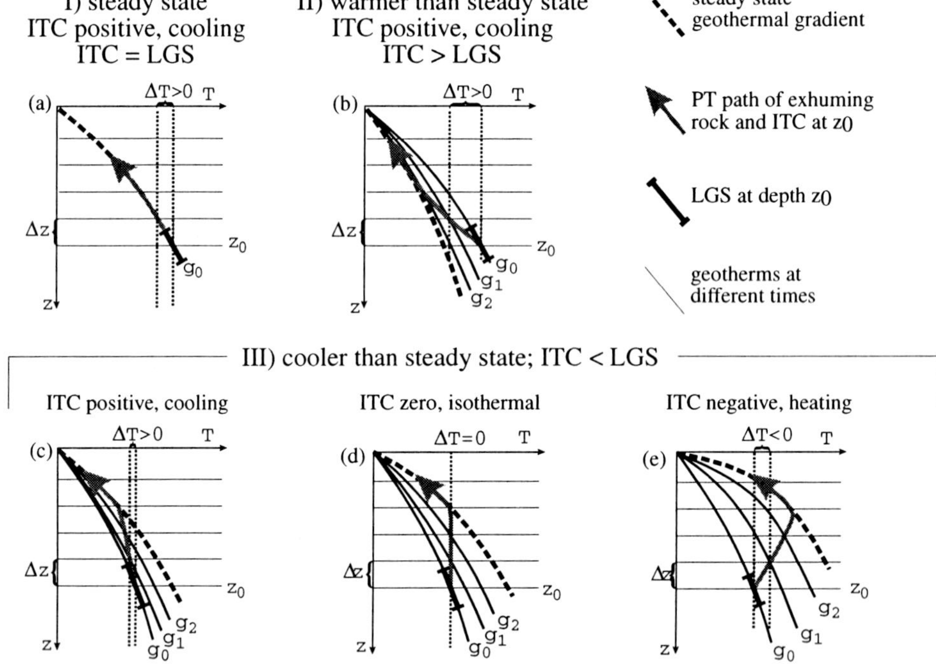

Fig. 3. During exhumation the geothermal gradient can become steeper or gentler. (I) If the ITC and the LGS are equal, rocks exhume at steady state. (II) If the initial geothermal gradient is warmer than the steady state, rocks cool and the ITC is always positive and greater than the LGS. III) If the initial geothermal gradient is cooler than steady state (i.e., ITC < LGS), three different conditions obtain: either the ITC is positive and rocks cool, the ITC is zero and rocks decompress isothermally, or the ITC is negative and rocks cool.

1) **Regime I, ITC = LGS: Steady state exhumation**: The geothermal gradient is at steady state for the given boundary conditions (exhumation rate, radioactive heat production, thickness of lithosphere, etc.). The ITC is equal to the LGS. During exhumation rocks *cool* along the path of the geotherm (Fig. 3a). In special tectonic environments (e.g., long-term subduction refrigeration) the ITC can be locally zero or negative, a situation that is not considered here.

2) **Regime II, ITC > LGS: Warmer than steady state exhumation**: The geothermal gradient is hotter than the steady state condition and the geothermal

gradient relaxes toward lower temperatures. The ITC is always positive and greater than the LGS and rocks at a certain depth must cool during exhumation (Fig. 3b). This situation applies to exhumation after magmatism or where fast exhumation is followed by slower.

3) **Regime III, ITC < LGS: Cooler than steady state exhumation**: The geothermal gradient is cooler than steady state and relaxes during exhumation toward higher temperatures. The ITC is always smaller than the LGS. This is probably the most common situation during the exhumation of rocks because exhumation advects heat to shallow levels, resulting in a steeper steady state geothermal gradient; and crustal thickening or subduction prior to exhumation typically produces an unusually cold geothermal gradient. The latter is considered responsible for the clockwise P-T paths often recorded by amphibolite-facies rocks, where the temperature maximum is reached after the pressure maximum (e.g., England and Richardson, 1977; Spear *et al.*, 1984). The relative rates of advection and conduction determine whether the rocks actually undergo heating or cooling during exhumation:

i) $\Delta T>0$: The ITC is positive and rocks *cool* during exhumation (Fig. 3c).

ii) $\Delta T=0$: The ITC is zero and rocks decompress *isothermally* (Fig 3d).

iii) $\Delta T<0$: The ITC is negative and rocks *heat* during exhumation (Fig 3e).

These regimes are not exclusive and a single exhumation history may be characterized by one or more regimes. Particularly for exhumation from mantle depths, the exhumation history may have two stages. An early stage is characterized by Regime III with continuous relaxation of the geothermal gradient toward steady state where the ITC can have any value. The ITC can even switch between negative and positive during exhumation. A late stage has a stable geothermal gradient for the given boundary conditions and the rocks move toward the surface along the steady state geothermal gradient with a positive ITC (Regime I). A complex change of the exhumation regimes and the ITC values is most likely if the exhumation history is not characterized by a single tectonic event.

The later stages of exhumation, particularly long-term exhumation or exhumation from deep levels, are characterized by thermal steady state (e.g., the paths in Fig. 3, where the arrows follow the dashed steady state curve). In particular, the exhumation of UHP rocks through crustal depths should be along the steady state geotherm. This is only true if the exhumation rate is more or less constant. However, if it is possible to determine this steady state geotherm by means of petrology and/or thermochronology, important parameters like the exhumation rate can be derived (for a more detailed discussion of this idea see below).

An important implication of Equation (2) is that $\dot{z}$ appears as the denominator of a limit function. This implies that even very rapid exhumation cannot change a rock from heating conditions (negative ITC) to cooling conditions (positive ITC) or vice versa. In fact, rapid exhumation rates force the ITC to converge toward zero (isothermal decompression) regardless of whether the initial ITC is positive or negative. The limit of this process is the instantaneous exhumation from z_0 to z_1, resulting in perfectly isothermal decompression with ITC equal to zero. Note that high exhumation rates tend to hold the temperature of an exhuming rock constant. Thus, steady cooling during the whole exhumation period indicates slower exhumation while faster exhumation keeps temperatures relatively constant for a long period of the exhumation history. Thus it is incorrect to equate rapid exhumation with rapid cooling, although rapid exhumation and isothermal decompression are likely to be followed by rapid cooling. This cooling at the end of or after rapid exhumation reflects the continued, slower movement of rocks through closely spaced isotherms or the relaxation of isotherms after exhumation has stopped.

In summary, cooling of rocks during exhumation requires that the transient geothermal gradient be either at steady state (Regime I), warmer than steady state (Regime II), or cooler than steady state (Regime III) with a positive ITC. The latter is most likely to characterize exhumation of UHP metamorphic rocks generated at convergent margins where relatively cold lithosphere is subducted into the mantle. Thus regardless of which exhumation model is favored, exhumation of UHP rocks starts with a geothermal gradient that is cooler than steady state.

3. MODELING OF EXHUMATION

Based on simple boundary conditions we aim to study by means of one-dimensional thermal calculations the first-order effects that influence the exhumation P-T paths of deeply buried rocks. We emphasize that the transient thermal frame of subduction zones in which rocks are brought to great depth can only be described by two-dimensional models (e.g., Minear and Toksöz, 1970; Andrews and Sleep, 1974; Anderson *et al.*, 1978; Hsui and Toksöz, 1979; van den Beukel and Wortel, 1988; Peacock, 1990; Peacock *et al.*, 1994; Ernst and Peacock, 1996). Although the subduction of cool oceanic lithosphere exerts a primary control on the thermal structure of subduction zones by depressing the isotherms, other parameters (e.g., shear heating, induced flow in the mantle wedge, hydrothermal circulations, metamorphic dehydration reaction) influence the transient temperature distribution (Peacock, 1996). The magnitude of these effects is often poorly constrained and we do not consider them in our models.

In the present model we assume that exhumation occurs within a steady state subduction zone. Lateral heat conduction is not considered and therefore the calculated P-T paths are too warm compared with more realistic two-dimensional models where the rising rocks undergo additional heat loss (Ernst and Peacock,

1996). However, the general characteristic of our modeled P-T paths is similar to the more complex models and for high exhumation rates the difference between one- and two-dimensional models is less significant (Ruppel and Hodges, 1994). The advantage of the presented models is that the influence of some important first-order effects can be clearly demonstrated.

To explore exhumation regimes in which the ITC is either positive, zero, or negative, we discuss two models with different boundary conditions at the bottom of the lithosphere. Model I uses a constant temperature-constant depth boundary condition allowing the geothermal gradient to reach steady state. Model II uses a variable temperature-constant depth or constant temperature-variable depth boundary condition, which precludes the establishment of steady state. The commonly used constant heat flux-constant depth or constant heat flux-variable depth boundary condition at the base of the crust (e.g. see discussion in England and Thompson, 1984; Sandiford and Dymoke, 1991) is an inappropriate boundary condition for the base of the lithosphere because unrealistically high steady state temperatures develop at great depth, leading to melting of large parts of the lithosphere. The rheological consequences and the importance of specifying the lower thermal boundary condition are discussed by Stüwe and Sandiford (1995). Temperatures as a function of depth are obtained from the heat transfer differential equation, which is solved numerically by an approach discussed in the Appendix.

3.1 Model I (constant temperature–constant depth)

Model I investigates the thermal evolution of rocks exhumed within the hanging wall of a subduction zone. This tectonic setting has been widely invoked for the exhumation of UHP rocks (e.g., Platt, 1987; Avigad and Garfunkel, 1991; Chopin *et al.*, 1991; Cloos, 1993; Chemenda *et al.*, 1996). The temperature at the surface is 0°C and the lower boundary condition is a constant temperature of 800°C at the base of the hanging wall at 100 km depth. This constant temperature-constant depth boundary condition (Stüwe and Sandiford, 1995) is chosen to model subduction refrigeration. The geothermal gradient chosen for the initial condition is non-steady state and is a low, near-surface gradient of approximately 10°C/km that decreases to 8°C/km at the base of the slab. Other thermal parameters are listed in Table 1.

Results of the model for two different exhumation rates of 1 and 5 mm/a are plotted in temperature-depth plots (Fig. 4). The initial geotherm represents the initial temperature profile before exhumation begins. The final geotherm is the gradient after 100 m.y. for an exhumation rate of 1 mm/a and after 20 m.y. for a rate of 5 mm/a—in both cases, rocks initially buried 100 km are exhumed at the end of the model. The constant temperature at the boundaries (i.e., surface and base of the slab), the long time interval, and the corresponding deep exhumation allow the final temperatures to reach steady state. Mainly due to the advection of heat the steady state geotherms have a pronounced convex-upward shape. The difference in shape between the steady state geotherm of the model with 1 and 5 mm/a is only a result of the different exhumation rates; all other thermal parameters are the same.

Table 1. Thermal modeling parameters.

Parameter	**Model I**	**Model II**
starting depth for exhumation history (km)	10-100	10-100
erosion rate(mm a^{-1})	1, 5	1, 5
temperature of the exposed upper surface (°C)	0	0
depth of lower boundary (km)	100	100 or 140
temperature at lower boundary (°C)	800	600-800°C or 800°C
thermal diffusivity (m^2 s^{-1})	1E-6	1E-6
surface volumetric heat production (W m^{-3})	2.5E-6	2.5E-6
depth at which heat production drops to 1/e (km)	30	30
heat capacity (kJ kg^{-1} K^{-1})	1100.0	1100.0
density (kg m^{-3})	2800	2800

P-T paths for rocks exhuming from 10 to 100 km are shown as loops between the initial and steady state geotherms. The time needed for the exhumation from different depths is different due to the constant exhumation rates. In order to visualize the depths where rocks are heating, cooling, or under isothermal conditions, the diagrams are contoured for negative, zero, and positive ITC. In other words the temperature evolution of a rock for the given boundary conditions at a certain depth and temperature (i.e., the location in the diagrams of Fig. 4) behaves according to the contoured area. Comparing the contoured fields of the ITC in Fig. 4a and b, we stress the following implications:

1) ITC < 0: In a broad range of depths from 5 and 80 km, rocks heat during the initial stages of exhumation. These well-known loop-shaped P-T paths are typical for Barrovian-type metamorphic rocks in collisional orogens (e.g., Thompson and England, 1984). The most extreme ITC values—which correspond to the greatest heating during exhumation—are reached by rocks exhuming from a depth of ~20 km. Higher exhumation rates (Fig. 4b) increase the depth range over which rocks heat, but simultaneously decrease the magnitude of the heating effect. A rock exhuming from 20 km depth at 1 mm/a heats until it reaches a depth of 16 km; the temperature difference is ~40°C. For an exhumation rate of 5 mm/a, heating lasts until the rock reaches a depth of 12.5 km but the temperature difference is only ~20°C.

2) ITC = 0: For reasons of numerical resolution the field of isothermal decompression in Fig. 4 was defined as that part of the exhumation path where temperature changes <2°C. A black bar indicates these quasi-isothermal conditions. This temperature interval is an order of magnitude smaller than the resolution of any geological method. There is only a small depth range around

80 km where exhumation begins isothermally, and this is an effect of the imposed boundary conditions. The duration or the depth range through which rocks exhume isothermally is increased by more rapid exhumation rates: at very high exhumation rates the depth range where rocks decompress isothermally expands toward shallower and deeper levels, consuming the fields of ITC < 0 and > 0. This process is limited by the magnitude of possible exhumation rates (Craw *et al.*, 1994; Burg *et al.*, 1997; Hovius *et al.*, 1997; Blythe, this volume). Another possibility for increasing the size of this isothermal decompression field, not considered here, is variable exhumation rate (Draper and Bone, 1981; Thompson *et al.*, 1997).

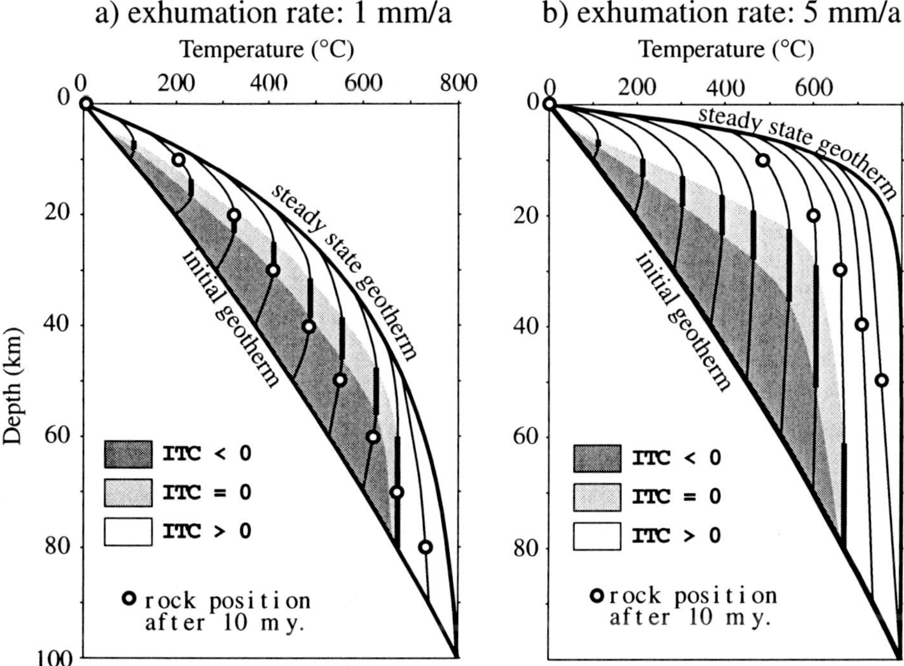

Fig. 4. Results of Model I, where the lower boundary condition at 100 km depth is a constant temperature of 800°C. Temperature/depth plot with an initial geothermal gradient and the steady state condition. P-T-t paths for rocks exhuming from 10-100 km are indicated, and positions of rocks after 10 m.y. of exhumation are indicated with open circles. To visualize where heating, cooling or isothermal conditions prevail, the diagram is contoured for the ITC. Comparison of Fig. 4a and 4b indicates that rocks exhuming from great depths above a constant temperature lower boundary condition always cool during exhumation, regardless of denudation rate. Rocks exhuming from crustal levels heat during their first stage of exhumation.

3) ITC > 0: Any rock that exhumes to the surface must reach the field where the ITC is positive (i.e., cooling). There are two fields where the ITC is greater than 0 at the onset of exhumation. At shallow crustal levels (<5 km) the rocks are

too close to the fixed 0°C surface boundary condition to allow heating during exhumation. At depths >80 km the temperature difference between the initial and the steady state conditions is too small (~100°C) to allow any heating during relaxation. In contrast, the greatest difference in temperature between the initial and steady state thermal gradients occurs at intermediate depths and is ~300°C for the 1 mm/a scenario and 600°C for the 5 mm/a exhumation rate. This is because the chosen initial geotherm is nearly linear (e.g., due to homogeneous crustal thickening) whereas the steady state geotherm has a pronounced convex upward shape produced mainly by advection. Because the heat difference between surface and basal boundary condition within our model lithosphere must remain constant (i.e., 800°C), a steep near-surface thermal gradient (e.g., 60°C/km, Fig. 4b) must be balanced by a gentler gradient at greater depth. This important behavior does not hold true for models with a constant heat flux basal boundary condition at a constant or variable depth (e.g., England and Thompson, 1984), because an increasing geothermal gradient results in increasing temperatures at the base of the slab and a significantly different thermal evolution especially at long time scales (cf. Benfield, 1949; Mancktelow and Grasemann, 1997).

The most important conclusion is that rocks exhuming from mantle depths above a constant-temperature lower boundary condition cool throughout their exhumation regardless of slow or fast exhumation rates. On the other hand rocks exhuming from crustal levels undergo heating followed by cooling. This observation has important implications for the temperature-time histories of exhuming UHP rocks. For example, in an orogen including a suite of rocks exhumed from a range of depths, blueschist-facies rocks should show heating followed by cooling, whereas coesite-containing UHP rocks should show only cooling. Rocks exhumed from depths of 20 km should show the most extreme heating (Fig. 4). Thus, one should question the assumption that all rocks exhumed in an orogen from different depths should follow the same style of P-T path during exhumation.

Another important result of this model is that rocks exhuming under a constant rate from great depth should initially show slow cooling or—within the uncertainty of geological methods—isothermal decompression, followed by much faster cooling. This is the direct consequence of the advection of heat to shallow levels. During the first stage, a rock moves from great depth toward the surface together with the isotherms. However, in the upper part of the crust the rock must pass through narrowly spaced isotherms and cool rapidly. The geologically important implication of this process is that at shallow levels rocks cool along nearly the steady state path (Regime I steady state exhumation). Thus if one can determine the rapidly cooling part of the P-T path of rocks exhuming continuously from great depths, it is possible to approximate the steady state geotherm for this exhumation. As the shape of the steady state geotherm mainly depends on the exhumation rate (compare Fig. 4a and 4b) it is possible, by the thermal modeling techniques outlined here, to estimate the exhumation rate if it is assumed to be constant.

3.2 Model II (variable temperature–constant depth or constant temperature–variable depth boundary condition)

Model II is identical to Model I except that the lower boundary condition is not a constant temperature of 800°C but increases steadily from 600°C to 800°C. Such an increase in temperature is typical for orogenesis where subduction of oceanic lithosphere is followed by subduction of continental lithosphere at slower rates (e.g., Davies and von Blanckenburg, 1995). The mathematical effect of this modified lower boundary condition is that the steadily increasing temperatures at the base of the slab preclude the establishment of steady state during exhumation and the temperatures at all depths (except the surface 0°C boundary condition) increase with time (Fig. 5). Due to the lower temperatures of 600°C at the beginning of exhumation the initial geothermal gradient is slightly less than in Model I.

Comparison of Fig. 4 and 5 shows that an increasing temperature at the base of the slab influences the P-T-t paths of rocks only slightly. For instance, the amount of heating for rocks at depths >90 km is ~10°C. Another result of this lower boundary condition and the fact that steady state can never be established is that rocks, especially those at initial depths >50 km, reach the final geotherm distinctly later than those in Model I. For example, in Model I a rock exhuming from 90 km depth at a rate of 1 mm/a approaches the steady state geotherm at a depth of ~50 km at a temperature >650°C (Fig. 4a). The corresponding rock in Model II approaches the final geotherm at a depth of 30 km at a temperature of only 450°C (Fig. 5a).

The higher exhumation rate dramatically increases the field where the ITC is zero (Fig. 5b). Samples from 60 km down to the base of the slab decompress isothermally for several tens of kilometers. Even at crustal levels where the ITC at the start of the exhumation is negative, rocks heat <10°C. All samples exhuming from 100-140 km depth will steadily cool.

Thus, comparison of Model I and II reveals that whether the boundary condition is constant temperature-constant depth, variable temperature-constant depth or constant temperature-variable depth, rocks exhuming from great depth either decompress isothermally or steadily cool during exhumation. Although there is a temperature difference of as much as 100°C between the steady state geotherms in Fig. 4 and the final geotherms in Fig. 5, the overall shape is very similar and is mainly controlled by the exhumation rate. Even though it might be difficult to estimate these geotherms by petrological methods, a determination of the shape of the P-T path of UHP rocks should allow differentiation between "fast" and "slow" exhumation.

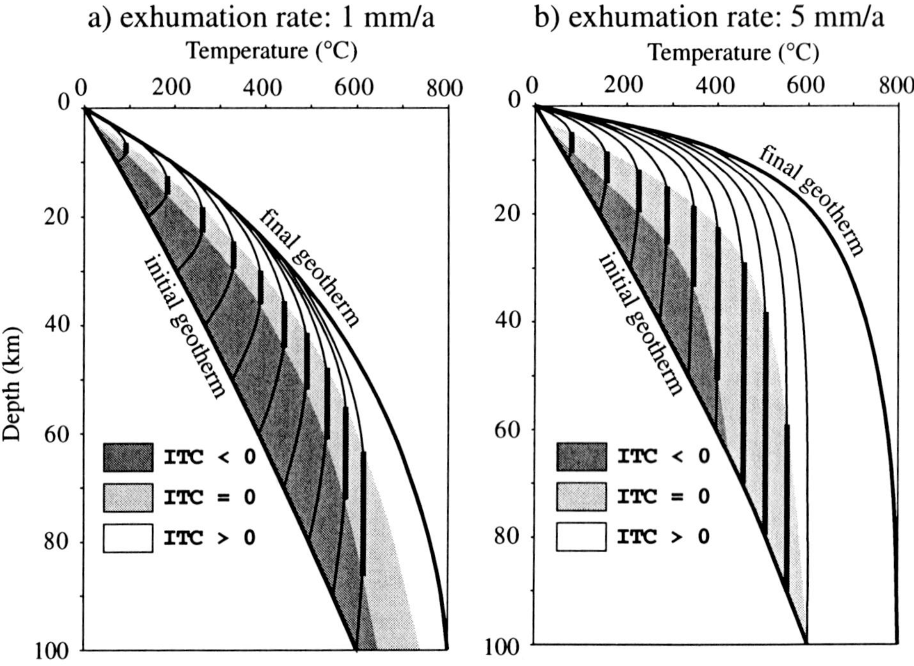

Fig. 5. Results of Model II, where the lower boundary condition at 100 km depth is an increasing temperature from 600 to 800°C. Other parameters are the same as model I. The increasing temperature at the base of the slab precludes that steady state temperature distribution is reached at the end of the exhumation. Comparison of Figs. 4 and 5 shows that an increasing temperature at the base of the slab influences the P-T-t paths of rocks only slightly.

Although the shapes of the P-T paths for rocks exhuming from crustal levels in our models are similar to P-T paths modeling exhumation of a thickened crust with a constant basal heat flux (e.g., England and Thompson, 1984), the boundary conditions are quite different. In our models the constant temperature at the lower boundary condition results in a variable heat flux at the base of the crust, or more precisely, at *any* depth within the exhuming rock column. In fact, the LGS above the lower boundary condition decreases during exhumation until the geotherm reaches a steady state. This has the important consequence that rocks that start to exhume from directly above this constant temperature boundary condition cannot heat during decompression as long as the LGS remains positive. Mancktelow and Grasemann (1997) demonstrated, by comparing analytical solutions for constant heat flux and for constant temperature lower boundary conditions, that the thermal states at crustal levels are similar for both boundary conditions. However, they did not consider exhumation from mantle depths. In the latter case, we have shown that the model results are sensitive to the chosen lower boundary condition.

Draper and Bone (1981) used a constant temperature lower boundary condition to investigate the thermal evolution and preservation of blueschist terranes. Although they came to the similar conclusion that rocks exhuming from lower crustal level experience the maximum temperature rise, they did not consider the exhumation of rocks directly from above the lower boundary condition. Therefore they came to the different conclusion that all rocks exhuming with constant exhumation rates from greater depth heat during decompression.

4. EXHUMATION RATES AND COOLING RATES IN UHP ROCKS

As discussed above, the exhumation of rocks from UHP depths of 100 km is most likely accompanied by cooling or isothermal decompression. The exhumation rate as denominator in equation (2) means that rapid exhumation causes UHP rocks to approach isothermal decompression, while slow exhumation favors a steady cooling. The different cooling histories of rocks exhuming at different rates are illustrated in Fig. 6. In this plot cooling curves for rocks exhuming from 90 km depth by 1, 2, 3 and 5 mm/a are calculated. The time axis in this plot is normalized to facilitate comparison of the different exhumation rates. All other thermal parameters and boundary conditions are the same as in Model I.

As outlined above, the advection of heat widens the spacing of the isotherms at depth, i.e., it lowers the LGS, and the isotherms are more densely spaced against the surface boundary condition, i.e., the LGS is elevated. This effect increases dramatically with higher exhumation rates. Rocks that move through a temperature field where the distance between isotherms is increasing cool slowly or maintain constant temperature if the exhumation rate is equal to the velocity of the isotherms. Rocks exhuming through closely spaced isotherms record fast cooling (Fig. 6). In Fig. 6, only 10% of the total exhumation history of rock *a* is isothermal; the rest of the cooling history shows steadily increasing cooling rate. Thermochronological methods using closure temperatures below 600°C would record only about the last 40%, or 35 m.y., of the cooling history. This is a dramatic difference from rock *d* exhuming at higher rates; for this rock, 75% of the exhumation history is isothermal and thermochronological methods recording cooling below 600°C would only cover the last 10%, or ~2 m.y.

These considerations have important applications for constraining the exhumation rates and periods of UHP rocks. The cooling of UHP rocks detected by thermochronological methods records only a late stage of the whole exhumation history that began much earlier. Although it may be difficult to determine the exact cooling history of UHP rocks, Fig. 6 can be used as a qualitative guide to the duration of exhumation prior to that constrained by geochronometry. It is thus conceivable that most UHP rocks began to exhume at an early stage of an orogenic cycle. This is in good agreement with geological records and exhumation models suggesting syn-convergent exhumation (e.g., Platt, 1987; Avigad and Garfunkel, 1991; Chopin *et al.*, 1991; Cloos, 1993; Chemenda *et al.*, 1996).

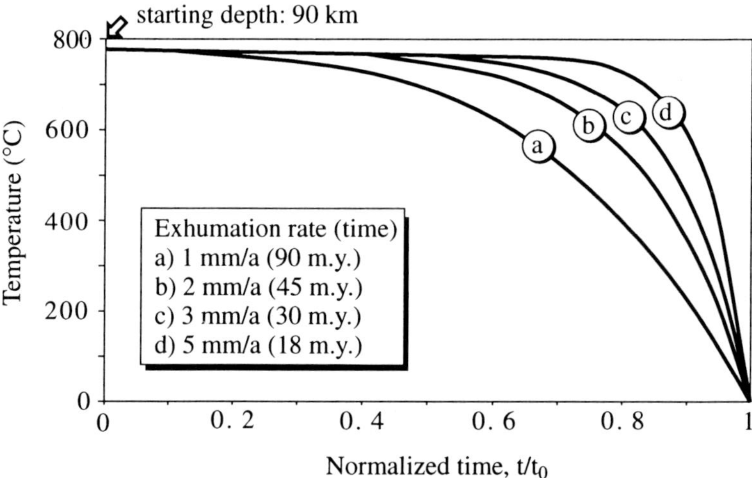

Fig. 6. Cooling curves for rock volumes exhuming from 90 km depth at 1, 2, 3 and 5 mm/a respectively. The time axis is normalized to compare the different time spans needed for the total exhumation. Boundary conditions and thermal parameters the same as Fig. 3. All curves show initially nearly no cooling and cooling rates increase steadily with time. Regardless of the exhumation rate, the exhumation from great depth starts significantly before the onset of cooling.

In summary, for the boundary conditions investigated, UHP rocks can be preserved under a wide range of exhumation rates. By careful examination of the geological data, it should be possible to distinguish between "slow" and "rapid" exhumation of UHP rocks. Slow (~1 mm/a) exhumation (Fig. 4a, 5a and 6) should be characterized by i) a moderate, e.g., ~20-30°C/km near-surface, steady-state geothermal gradient during exhumation through the *crust*; ii) deeply exhumed rocks should show nearly isothermal cooling; iii) shallowly exhumed rocks should exhibit modest temperature increases of 50-100°C, followed by more substantial cooling; and iv) thermochronometers with closure temperatures of 300-600°C should record an age range of 10-20 m.y. In contrast, rapid (~5 mm/a) exhumation (Fig. 4b, 5b and 6) should be characterized by i) an extremely high near-surface geothermal gradient, e.g., ~60-70°C/km at the late stage of exhumation; ii) deeply exhumed rocks should show isothermal cooling to even shallower levels followed by very rapid cooling; iii) shallowly exhumed rocks should exhibit smaller temperature increases of <50°C, again followed by rapid cooling; and iv) thermochronometers with closure temperatures of 300-600°C should record an age range of ~1 m.y.

5. APPLICATION OF THE MODELING TO THE QINLING-DABIE OROGENIC BELT, CENTRAL CHINA

The Su-Lu-Dabie-Hong'an areas contain regional UHP and HP metamorphic rocks developed during a Triassic orogenic event (e.g., Hacker *et al.*, 1995). Current constraints on the P-T-t paths of these rocks are summarized in Fig. 1; the true P-T-t paths were undoubtedly more complicated. Rocks exhumed from the greatest depths, >120 km, exhibit apparent isothermal decompression or cooling, with a marked inflection at around 0.6–0.8 GPa (Liou *et al.*, 1996). High-pressure amphibolite and blueschist-facies rocks reached peak P conditions of ~1.2 GPa, ~550°C and 0.4–0.8 GPa, 300-457°C, respectively, and underwent 25–50°C heating during decompression (Eide, 1993). Among the northernmost UHP rocks in the Dabie Shan, U/Pb zircon dating indicates peak temperatures (700°C) prevailed at ~219 Ma (Ames *et al.*, 1996; Rowley *et al.*, 1997), and $^{40}Ar/^{39}Ar$ phengite and K-feldspar spectra show cooling to temperatures <300°C by ~190 Ma (Hacker and Wang, 1995; Hacker *et al.*, in press).

The Qinling-Dabie-Hong'an P-T paths are consonant with the modeling presented here in that the rocks exhumed from the greatest depth show nearly isothermal decompression, whereas more modestly buried rocks underwent heating during exhumation. If we assume that the exhumation rate was constant, by reference to Fig. 4 and 5, the shapes of the P-T paths of these Chinese rocks suggest an exhumation rate closer to 5 mm/a rather than 1 mm/a. By comparison, assuming that the 219 Ma zircons formed at 3 GPa (Ames *et al.*, 1996) and the 190 Ma ages record cooling to <1 GPa (Hacker and Wang, 1995), yields an exhumation rate of ~2 mm/a.

Strictly, U/Pb zircon ages record the time of zircon crystallization and not necessarily the time of peak pressure. Thus the zircon ages might indicate cooling to temperatures of ~700°C at pressures considerably below maximum (Hacker *et al.*, in press). If so, Fig. 4–6 imply that the main 30 m.y. cooling period followed an equally long period of exhumation from deeper levels.

The P-T paths of Fig. 1 have often been used to argue for a two-stage exhumation history reflecting differential rates of exhumation through mantle and crustal depths (e.g., Liou *et al.*, 1996). Figs. 4–6 reveal, however, that an apparent two-stage cooling history could have been produced by a constant rate of exhumation.

6. CONCLUSIONS

The following points are based on the presented one-dimensional models and thus investigate first-order processes. For quantitative analysis, a two-dimensional approach including additional thermal effects such as induced mantle convection, slab age, fluid production, hydrothermal circulation, and shear heating is required (e.g., Peacock, 1990; Peacock, 1996).

1) The local geothermal slope (LGS) and the rate of change in temperature with respect to change in depth at a given time (ITC) are tools for describing when rocks heat, cool or behave isothermally during exhumation or burial.

2) In the absence of lateral heating, rocks exhuming from great depth always cool or decompress isothermally.

3) The main cooling period follows the main period of exhumation.

4) Shallowly buried rocks follow substantially different P-T paths than deeply exhumed rocks even if all rocks are exhumed by the same orogen-scale process. This observation has important consequences for the interpretation of metamorphic coherent HP terranes with rocks recording different P-T paths.

5) Rocks undergo two-stage cooling histories even for a constant exhumation rate. If the exhumation rate is constant, the P-T paths of all rocks approach a steady state or "final" geotherm at crustal levels, the shape of which is mainly a function of exhumation rate. If it is possible to determine this steady state geotherm by geological methods, exhumation rates can be estimated by reference to the models presented here.

6) After long time scales and corresponding deep exhumation, the geotherms approach a steady state condition at crustal levels. Therefore, reconstruction of P-T-t paths should permit distinction between "fast" and "slow" UHP exhumation processes. For example, fast exhumation is characterized by extremely rapid crustal cooling following small temperature increases or isothermal decompression, whereas slow exhumation is characterized by steady cooling following more modest heating.

ACKNOWLEDGMENTS

N. Mancktelow, M. Thöni, M. Linner and E. Dragnets are thanked for reading an earlier version of the manuscript and for helpful suggestions. A special "thank you" to Simon Peacock for reminding us that a lot of 2–D modeling has been done and that a quantitative understanding of exhumation must go beyond our 1–D approach. L.R. research was funded by grants Ra442/6–2, 9-2 of the German National Science Foundation. B.G. developed the modeling software with funding from project P-11765–GEO of the Austrian Fond zur Förderung der wissenschaftlichen Forschung.

REFERENCES

Ames, L., Zhou, G., and Xiong, B. (1996) Geochronology and geochemistry of ultrahigh-pressure metamorphism with implications for collision of the Sino-Korean and Yangtze cratons, central China, *Tectonics* **15**, 472–489.

Anderson, R.N., Delong, S.E., and Schwarz, W.M. (1978) Thermal model for subduction with dehydration in the downgoing slab, *Journal of Geology* **86**, 731–739.

Andrews, D.J. and Sleep, N.H. (1974) Numerical modelling of tectonic flow behind island arcs, *Geophysical Journal of the Royal Astronomical Society* **38**, 237–251.

Avigad, D. and Garfunkel, Z. (1991) Uplift and exhumation of high-pressure metamorphic terrains: the example of the Cycladic blueshist belt (Aegean Sea), *Tectonophysics* **187**, 1–15.

Benfield, A.E. (1949) A problem of the temperature distribution in a moving medium, *Quarterly of Applied Mathematics* **6**, 439–443.

Blythe, A.E. (this volume) Active tectonics and ultrahigh-pressure rocks, in B.R. Hacker and J.G. Liou (eds.), *When Continents Collide: Geodynamics and Geochemistry of Ultrahigh-Pressure Rocks*, Kluwer Academic Publishers, Dordrecht.

Burg, J.P., Davy, P., Nievergelt, P., Oberli, F., Seward, D., Diao, Z., and Meier, M. (1997) Exhumation during crustal folding in the Namche-Barwa syntaxis, *Terra Nova* **9**, 53–56.

Carslaw, S. and Jaeger, J.C. (1959) *Conduction of Heat in Solids*, Oxford University Press, New York.

Chemenda, A.I., Mattauer, M., and Bokun, A.N. (1996) Continental subduction and a mechanism for exhumation of high-pressure metamorphic rocks: new modeling and field data from Oman, *Earth and Planetary Science Letters* **143**, 173–182.

Chopin, C., Henry, C., and Michard, A. (1991) Geology and petrology of the coesite-bearing terrain, Dora Maira massif, western Alps, *European Journal of Mineralogy* **3**, 263–291.

Cloos, M. (1982) Flow melanges: numerical modeling and geologic constraints on their origin in the Franciscan subduction complex, *Geological Society of America Bulletin* **93**, 330–345.

Cloos, M. (1993) Lithospheric buoyancy and collisional orogenesis: subduction of oceanic plateaus, continental margins, island arcs, spreading ridges, and seamounts, *Geological Society of America Bulletin* **105**, 715–737.

Craw, D., Koons, P.O., Winslow, D., Chamberlain, C.P., and Zeitler, P. (1994) Boiling fluids in a region of rapid uplift, Nanga Parbat massif, Pakistan, *Earth and Planetary Science Letters* **128**, 169–182.

Davies, J.H. and von Blanckenburg, F. (1995) Slab breakoff: a model of lithosphere detachment and its test in the magmatism and deformation of collisional orogens, *Earth and Planetary Science Letters* **129**, 85–102.

Draper, G. and Bone, R. (1981) Denudation rates, thermal evolution, and preservation of blueschist terrains, *Journal of Geology* **89**, 601–613.

Eide, E.A. (1993) Petrology, geochronology, and structure of high-pressure metamorphic rocks in Hubei province, east-central China, and their relationship to continental collision. Stanford, p. 235. Stanford University, California.

Eide, L., McWilliams, M.O., and Liou, J.G. (1994) $^{40}Ar/^{39}Ar$ geochronologic constraints on the exhumation of HP-UHP metamorphic rocks in east-central China, *Geology* **22**, 601–604.

England, P. and Molnar, P. (1990) Surface uplift, uplift of rocks, and exhumation of rocks, *Geology* **18**, 1173–1177.

England, P.C. and Richardson, S.W. (1977) The influence of erosion upon the mineral facies of rocks from different metamorphic environments, *Journal of the Geological Society of London* **134**, 201–213.

England, P.C. and Thompson, A.B. (1984) Pressure-temperature-time paths of regional metamorphism; I, Heat transfer during the evolution of regions of thickened continental crust, *Journal of Petrology* **25**, 894–928.

Ernst, W.G. (1988) Tectonic history of subduction zones inferred from retrograde blueshist P-T paths, *Geology* **16**, 1081–1085.

Ernst, W.G. and Peacock, S.M. (1996) A thermotectonic model for preservation of ultrahigh-pressure phases in metamorphosed continental crust, in G.E. Bebout, D.W. Scholl, S.H. Kirby, and J.P. Platt (eds.), *Subduction Top to Bottom*, American Geophysical Union, Washington, D.C., pp. 171–178.

Hacker, B.R. and Peacock, S.M. (1994) Creation, preservation, and exhumation of coesite-bearing, ultrahigh-pressure metamorphic rocks, in R.G. Coleman and X. Wang (eds.), *Ultrahigh Pressure Metamorphism*, Cambridge University Press, Cambridge, United Kingdom.

Hacker, B.R., Ratschbacher, L., Webb, L., and Dong, S. (1995) What brought them up? Exhumation of the Dabie Shan ultrahigh-pressure rocks, *Geology* **23**, 743–746.

Hacker, B.R., Ratschbacher, L., Webb, L., Ireland, T., Walker, D., and Dong, S. (in press) U/Pb zircon ages constrain the architecture of the ultrahigh-pressure Qinling-Dabie Orogen, China, *Earth and Planetary Science Letters*.

Hacker, B.R. and Wang, Q.C. (1995) Ar/Ar geochronology of ultrahigh-pressure metamorphism in central China, *Tectonics* **14**, 994–1006.

Henry, C., Michard, A., and Chopin, C. (1993) Geometry and structural evolution of ultra-high-pressure and high-pressure rocks from the Dora-Maira Massif, Western Alps, Italy, *Journal of Structural Geology* **15**, 965–981.

Hovius, N., Stark, C.P., and Allen, P.A. (1997) Sediment flux from a mountain belt derived by landslide mapping, *Geology* **25**, 231–234.

Hsui, A.T. and Toksöz, M.N. (1979) The evolution of thermal structures beneath a subduction zone, *Tectonophysics* **60**, 337–349.

Liou, J.G., Zhang, R.Y., Eide, E.A., Maruyama, S., Wang, X., and Ernst, W.G. (1996) Metamorphism and tectonics of high-P and ultrahigh-P belts in Dabie-Sulu Regions, eastern central China, in A. Yin and T.M. Harrison (eds.), *The Tectonic Evolution of Asia*, Rubey Volume IX, Cambridge University Press, Cambridge, United Kingdom, pp. 300–343.

Mancktelow, N.S. and Grasemann, B. (1997) Time-dependent effects of heat advection and topography on cooling histories during erosion, *Tectonophysics* **270**, 167–195.

Michard, A., Chopin, C., and Henry, C. (1993) Compression versus extension in the exhumation of the Dora-Maira coesite-bearing unit, Western Alps, Italy, *Tectonophysics* **221**, 173–193.

Minear, J.W. and Toksöz, M.N. (1970) Thermal regime of a downgoing slab and new global tectonics, *Journal of Geophysical Research* **75**, 1397–1419.

Molnar, P. and Gray, D. (1979) Subduction of continental lithosphere: some constraints and uncertainties, *Geology* **7**, 58–62.

Nie, S., Yin, A., Rowley, D.B., and Jin, Y. (1994) Exhumation of the Dabie Shan ultrahigh-pressure rocks and accumulation of the Songpan-Ganzi flysch sequence, central China, *Geology* **22**, 999–1002.

Okay, A.I. (1993) Petrology of a diamond and coesite-bearing metamorphic terrain: Dabie Shan, China, *European Journal of Mineralogy* **5**, 659–675.

Peaceman, D.W. and Rachford, H.H. (1955) The numerical solution of parabolic and elliptic differential equations, *Journal of the Indian Mathematical Society* **3**, 28–41.

Peacock, S.M. (1990) Numerical simulation of metamorphic pressure-temperature-time paths and fluid production in subducting slabs, *Tectonics* **9**, 1197.

Peacock, S.M. (1996) Thermal and petrologic structure of subduction zones, in G.E. Bebout, D.W. Scholl, S.H. Kirby, and J.P. Platt (eds.), *Subduction Top to Bottom*, American Geophysical Union, Washington, D.C., pp. 119–134.

Peacock, S.M., Rushmer, T., and Thompson, A.B. (1994) Partial melting of subducting oceanic crust, *Earth and Planetary Science Letters* **121**, 227–244.

Platt, J.P. (1987) The uplift of high-pressure low-temperature metamorphic rocks, *Philosophical Transactions of the Royal Society of London* **A321**, 87–103.

Rowley, D.B., Xue, F., Tucker, R.D., Peng, Z.X., Baker, J., and Davis, A. (1997) Ages of ultrahigh pressure metamorphism and protolith orthogneisses from the eastern Dabie Shan: U/Pb zircon geochronology, *Earth and Planetary Science Letters* **151**, 191–203.

Rubie, D.C. (1984) A thermal-tectonic model for high-pressure metamorphism and deformation in the Sesia Zone, western Alps, *Journal of Geology* **92**, 21–36.

Ruppel, C. and Hodges, K.V. (1994) Pressure-temperature-time paths from two-dimensional thermal models: prograde, retrograde, and inverted metamorphism, *Tectonics* **13**, 17–44.

Sandiford, M. and Dymoke, P. (1991) Some remarks on the stability of blueschists and related high P - low T assemblages in continental orogens, *Earth and Planetary Science Letters* **102**, 14–23.

Schmaedicke, E., Okrusch, M., and Schmidt, W. (1992) Eclogite-facies rocks in the Saxonian Erzgebirge, Germany; high pressure metamorphism under contrasting P-T conditions, *Contributions to Mineralogy and Petrology* **110**, 226–241.

Spear, F.S., Selverstone, J., Hickmott, D., Crowley, P., and Hodges, K.V. (1984) P-T paths from garnet zoning: A new technique for deciphering tectonic processes in crystalline terranes, *Geology* **12**, 87–90.

Stüwe, K. and Sandiford, M. (1995) Mantle-lithospheric deformation and crustal metamorphism with some speculations on the thermal and mechanical significance of the Tauern Event, Eastern Alps, *Tectonophysics* **242**, 115–132.

Thompson, A.B. and England, P.C. (1984) Pressure-temperature-time paths of regional metamorphism II. Their inference and interpretation using mineral assemblages in metamorphic rocks, *Journal of Petrology* **25**, 929–955.

Thompson, A.B., Schulmann, K., and Jezek, J. (1997) Extrusion tectonics and elevation of lower crustal metamorphic rocks in convergent orogens, *Geology* **25**, 491–494.

van den Beukel, J. and Wortel, R. (1988) Thermo-mechanical modelling of arc-trench regions, *Tectonophysics* **154**, 177–193.

Zhang, R.Y., Liou, J.G., and Cong, B. (1994) Petrogenesis of garnet-bearing ultramafic rocks and associated eclogites in the Su-Lu ultrahigh-P metamorphic terrane, eastern China, *Journal of Metamorphic Geology* **12**, 169–186.

Zhang, R.Y., Liou, J.G., Ernst, W.G., Coleman, R.G., Sobolev, N.V., and Shatsky, V.S. (1997) Metamorphic evolution of diamond-bearing and associated rocks from the Kokchetav massif, northern Kazakhstan, *Journal of Metamorphic Geology* **15**, 479–496.

APPENDIX

The temperature distribution in a homogeneous isotropic solid whose thermal material parameters are independent of temperature is a particular solution (for given boundary conditions) of the heat transfer differential equation. In two dimensions with only a vertical velocity component this equation reduces to:

$$\frac{\partial T}{\partial t} = \kappa\left(\frac{\partial^2 t}{\partial x^2} + \frac{\partial^2 T}{\partial z^2}\right) + \left(\dot{z}\frac{\partial T}{\partial z}\right) + \frac{A(z,t)}{\rho C} \quad (3)$$

where

T = temperature (°C)

t = time (s)
κ = thermal diffusivity (m^2/s)
x, z = horizontal and vertical (i.e. depth) direction (m)
$\dot{z}$ = exhumation rate (m/s)
$A(z,t)$ = volumetric heat production (W/m^3)
C = specific heat (J/kgK)
ρ = density (kg/m^3)

The equation states that the change of temperature with time depends on heat conduction, advection, and production. This equation was solved numerically over a finite difference grid with a vertical distance of 100 km representing the thickness L of the lithosphere. The equally spaced grid resolution is 100 m. The volumetric heat production was modeled with an exponential function where the heat generation decays with depth:

$$A(z,t) = A_0 \exp\left(\frac{-z}{l(t)}\right) \tag{4}$$

where A_0 is the heat generation in the surface layer, l is the variable depth that changes with time at which the heat production drops to 1/e of this surface value. This system is solved numerically in two dimensions using an ADI (alternating-direction-implicit) method with a two-step scheme (Peaceman and Rachford, 1955). Although the mathematical description and solution used in the calculations is two-dimensional, only the vertical velocity component (i.e., the exhumation rate) was considered and thus reduces strictly speaking to a one-dimensional problem. The advection term was solved separately using an upwind differencing scheme, which applies backward or forward differencing depending on the sign of the velocity vector. This procedure is only stable as long as

$$\Delta t \leq \left\{ \frac{|\dot{x}|}{\Delta z^2} + \frac{2x}{(\Delta z)^2} \right\}^{-1} \tag{5}$$

The other boundary conditions are:

$$T = T_s = 0 \tag{6}$$

at the upper boundary where T_s is the constant surface temperature, and

$$\frac{dT}{dX} = 0 \tag{7}$$

at the left and right sides. The lower boundary condition is either a constant temperature T_l for Model I or a linear function $T_l(t)$ for Model II that increases the temperature during the exhumation process. The algorithm was tested and compared with a one-dimensional analytical solution for the time variation of temperature from an initial steady-state geotherm without erosion to the re-establishment of a steady state following onset of erosion at a constant rate (Mancktelow and Grasemann, 1997):

$$T(z,t) = \zeta(z,t)\exp\left(-\frac{\dot{z}}{2\kappa}z - \frac{\dot{z}^2}{4\kappa}t\right) + T(z) \tag{8}$$

where T(z) is the steady state solution:

$$T(z) = \beta\left[1 - \exp\left(-\frac{z}{l}\right)\right] + \gamma\left[1 - \exp\left(-\frac{\dot{z}}{\kappa}z\right)\right] \tag{9}$$

and

$$\beta = \frac{Al^2}{\rho C(\kappa - \dot{z}l)} \tag{10}$$

A solution for the time and position dependence is given by Carslaw and Jaeger (1959):

$$\zeta(Z,T) = \frac{2}{L}\sum_{n=1}^{\infty}\exp\left(-\kappa\omega_n^2 t\right)\sin\omega_n z\int_0^L \zeta(z',t=0)\sin\omega_n z' dz' \tag{11}$$

where

$$\omega_n = \frac{n\pi}{L} \tag{12}$$

Chapter 6

Active Tectonics and Ultrahigh-Pressure Rocks

Ann E. Blythe
Department of Earth Sciences, University of Southern California, Los Angeles, CA 90089-0740, USA, blythe@earth.usc.edu

Abstract: This chapter compares modern exhumation and surface uplift rates with the rates needed for the preservation of ultrahigh pressure (UHP) metamorphic rocks. The highest recorded exhumation rates of ~5–10 mm/a are inferred from isotopic and fission-track analyses in the Himalaya, Southern Alps of New Zealand, and D'Entrecasteaux Islands. Similar rates (~7 mm/a) of surface uplift are measured from leveling surveys in Nepal and correlations of marine terraces in the Southern Alps. In Nepal, however, this surface uplift rate is occurring despite erosion, and the true rate of surface uplift is probably considerably higher. In restraining bends along the San Andreas and Denali strike-slip faults in North America, contraction has produced localized regions with relatively high exhumation rates of 1–5 mm/a. A similar surface uplift rate (3–5 mm/a) was obtained from marine terrace correlations in the King Range of northern California.

Few modern exhumation rates fall within the range thought to be necessary for the preservation of high-pressure minerals during exhumation (10-100 mm/a), even though, because of a long-term shift in global climate, erosion rates appear to have accelerated during the late Cenozoic. Both the Himalaya and the Southern Alps of New Zealand have high erosion rates that are a direct result of modern precipitation and glaciation patterns; these high rates of erosion may act to enhance exhumation rates. Therefore, because the exhumation rates recorded in these two mountain ranges are the highest modern rates yet recorded, researchers studying UHP rocks are urged to consider the possibility that UHP minerals can be preserved at exhumation rates of <10 mm/a.

1. INTRODUCTION

Ultrahigh pressure (UHP) metamorphism of supracrustal rocks is recognized based on mineral assemblages formed at depths of >90 km (e.g., Chopin, 1984; Smith, 1984). These rocks are unique in many aspects: the depths at which they

B. R. Hacker and J. G. Liou (eds.),
When Continents Collide: Geodynamics and Geochemistry of Ultrahigh-Pressure Rocks, 141–160.
© 1998 *Kluwer Academic Publishers. Printed in the Netherlands.*

experienced maximum P/T conditions are significant not only because of the conditions recorded, but also because rocks subducted or buried to those depths are now exposed at the surface. Perhaps the most perplexing problem regarding UHP rocks is the path and mechanisms by which they are transported to the surface (Hacker and Peacock, 1994). A key constraint for any exhumation model is that the mineral assemblages formed during the UHP metamorphic event must be preserved. Draper and Bone (1981) estimated that minimum exhumation rates of 10-100 mm/a are necessary for blueschist-facies rocks not to be overprinted following the cessation of subduction.

This chapter describes modern mechanisms and rates of tectonic and surficial processes to constrain exhumation models of UHP rocks. The focus is on the Himalaya and Southern Alps of New Zealand, where the highest recorded modern exhumation and surface uplift rates occur. Both the Himalayas and the Southern Alps of New Zealand have high erosion rates that are a direct result of modern precipitation and glaciation patterns; these high rates of erosion may act to enhance exhumation rates. Climate appears to have undergone a long-term change during Cenozoic time, however, and modern erosion rates may not be an appropriate gauge of conditions in the past.

2. BACKGROUND

"Uplift" has often been used in the literature to refer to both surface and rock uplift and therefore, it is necessary to carefully define these terms (England and Molnar, 1990). Surface uplift is the upward motion of Earth's surface with respect to the geoid, or more commonly, sea level. Rock uplift, in contrast, is defined as the upward motion of a rock column relative to the earth's surface, with no reference to surface topography. Denudation refers to the amount of the rock column removed from the surface by erosion. Surface uplift, therefore, is the net result of bedrock uplift and denudation. Exhumation is sometimes used interchangeably with denudation, but it can include deformation, such as extensional faulting, as well as erosion to account for rock motion towards the surface (Summerfield, 1991). Throughout this chapter, all rates will be expressed in terms of mm/a (the equivalent of km/m.y.) for ease in comparison.

3. PLATE RATES

In a discussion of modern exhumation and surface uplift rates, an appropriate starting point is with active plate tectonic rates, which in effect, control exhumation and surface uplift rates. The highest modern exhumation and surface uplift rates are measured along convergent plate boundaries and restraining bends in major transform plate boundaries.

Convergence rates vary according to the type and age of lithosphere being subducted, with higher convergence rates occurring where old oceanic lithosphere is undergoing subduction and lower rates in continent-continent collisions. The highest modern convergence rate occurs at the Tonga trench where a maximum of ~240 mm/a has been recorded geodetically (Bevis *et al.*, 1995). This fast rate is produced by a number of factors: the oceanic crust being subducted is old (100-140 Ma) and therefore, cold; the Lau basin, which is a back-arc basin between the Tonga and older Lau ridge, is opening at a maximum rate of ~160 mm/a; and the Tonga ridge, which appears to be internally undeformed, behaves as a rigid microplate (Bevis *et al.*, 1995).

Convergence rates for continent-continent collisions, such as India-Asia (Fig. 1), are lower than convergence rates involving oceanic lithosphere, where much of the motion can be taken up by subduction. The paleomagnetic reversal record

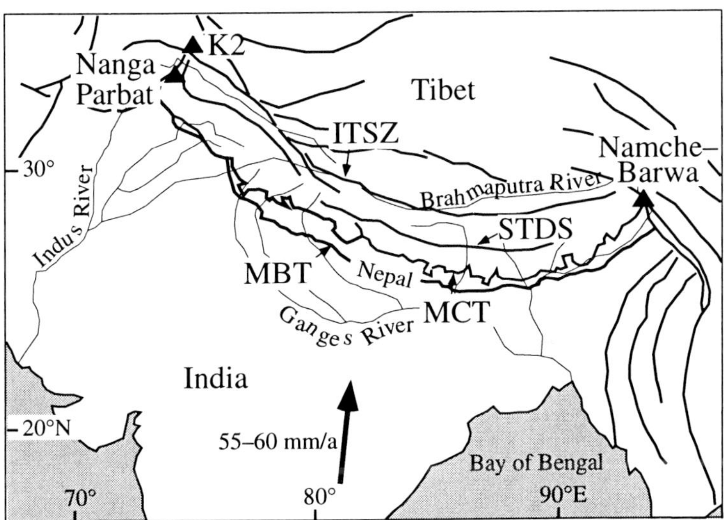

Figure 1. The Himalayan orogen and Tibetan Plateau (gray) and major faults (heavy lines) and drainages (thin lines). MBT, Main Boundary Thrust; MCT, Main Central Thrust; STDS, South Tibetan Detachment System; ITSZ, Indus-Tsangpo Suture Zone. The arrow represents the plate vector for India and Eurasia (Le Pichon *et al.*, 1992).

indicates that the convergence rate for India and Eurasia since collision began at 49 Ma has been 55–60 mm/a (Le Pichon *et al.*, 1992). One-third of this plate convergence rate is taken up on faults within the Himalaya, with the remainder of the India-Asia shortening distributed within the Tibetan plateau and Tian Shan farther north and in China farther east (Molnar and Tapponnier, 1975). Global

Positioning System (GPS) data indicate that the shortening rate for the Nepal Himalaya is 17.5 ± 2 mm/a (Bilham *et al.*, 1997).

In the Himalaya, three major thrust faults (Fig. 1; from north to south, the Indus-Tsangpo suture, Main Central Thrust, and Main Boundary Thrust), have concentrated a significant amount of the convergence between India and Asia (e.g., Gannser, 1981). The Main Central Thrust is thought to have had at least two periods of activity, ~25–15 Ma when sliding occurred along the basal detachment, and from ~9-5 Ma to the present, when significant bedrock uplift occurred along the fault ramp (Macfarlane, 1993; Harrison *et al.*, 1997). Magnetostratigraphic data from foreland basin sediments and a conglomerate derived from the hanging wall suggest that bedrock uplift and denudation of the Main Boundary Thrust began by ~10 Ma and that shortening was concurrent with activity on the Main Central Thrust (Meigs *et al.*, 1995). The South Tibetan Detachment system (STDS), north of the Main Central Thrust, complicates the tectonic setting of the Himalayan orogen with a considerable amount of down-to-the-north motion contemporaneous with shortening on the Main Central Thrust (Burchfiel *et al.*, 1992).

The NE-trending Alpine fault bisecting the South Island of New Zealand is the suture zone in an oblique continental collision occurring at a rate of ~40 mm/a (Walcott, 1978) between the Australian and Pacific plates (Fig. 2). Right-lateral slip of ~480 km on the Alpine fault initiated at ~25 Ma (Kamp, 1986), with convergence beginning at ~9.8 Ma (Stock and Molnar, 1982). Allis (1986) estimated that ~70 km of shortening has occurred across the central part of the South Island in the last 9.8 m.y., giving a crustal shortening rate of ~7 mm/a.

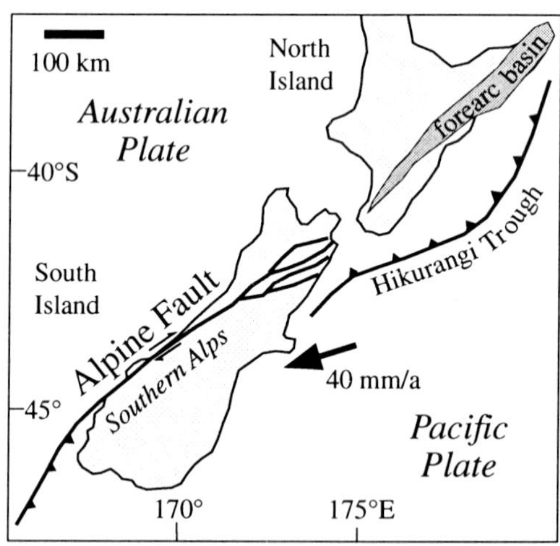

Figure 2. New Zealand. The arrow represents the plate vector for the Pacific and Australian plates across the South Island (from Walcott, 1978).

In California (Fig. 3), fault offsets are reasonably well constrained by a dense array of GPS stations and abundant paleoseismological studies. About 60-70% of the 48 mm/a of total plate motion (DeMets *et al.*, 1990) between the Pacific and North American plates is concentrated along the San Andreas fault, with the majority of the rest of the shear distributed east of the plate boundary, in the Basin-and-Range province (Jordan and Minster, 1988). Offset streams in the Carrizo Plain of central California indicate slip rates of ~32–33 mm/a on the San Andreas fault (Sieh and Jahns, 1984). Significant amounts of extension (the Ridge Basin) and contraction (the Transverse Ranges and Santa Cruz Mountains) have occurred along bends in the San Andreas fault. Although the total exhumation associated with these bends is relatively small (a few km), the rates of exhumation and vertical exhumation for the some of the fault-bounded blocks adjacent to the restraining bends are relatively fast and are discussed below.

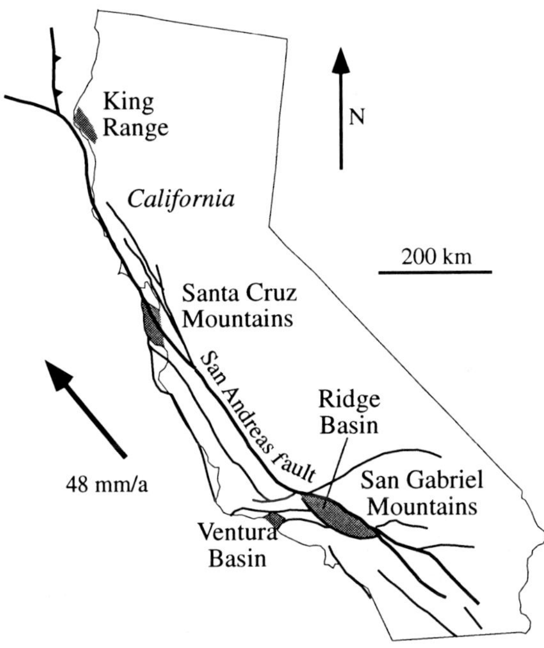

Figure 3. Some of the major faults associated with the San Andreas Fault system. Arrow represents plate vector between Pacific and North American plates (from DeMets *et al.*, 1990).

Although UHP rocks probably form initially in a subduction zone setting, extension, either within the subduction zone or as an overall change in plate motion, has been invoked in many instances to explain the rapid exhumation necessary for preservation of UHP mineral assemblages (e.g., Fossen and Rykkelid, 1992; Michard *et al.*, 1993). Extension as an exhumation mechanism for UHP rocks is attractive because buoyancy of the footwall as the upper plate is removed enhances cooling and erosion rates, leading to more rapid exhumation. The highest modern spreading rate of ~170 mm/a occurs at the Pacific-Nazca ridge (DeMets *et al.*, 1990). A high rate of active extension (30-40 mm/a) within continental crust is occurring near the D'Entrecasteaux islands in the transition from oceanic to continental crust where the Woodlark-D'Entrecasteaux spreading center ends east of Papua New Guinea (Fig. 4; Weissel *et al.*, 1982; Abers, 1997).

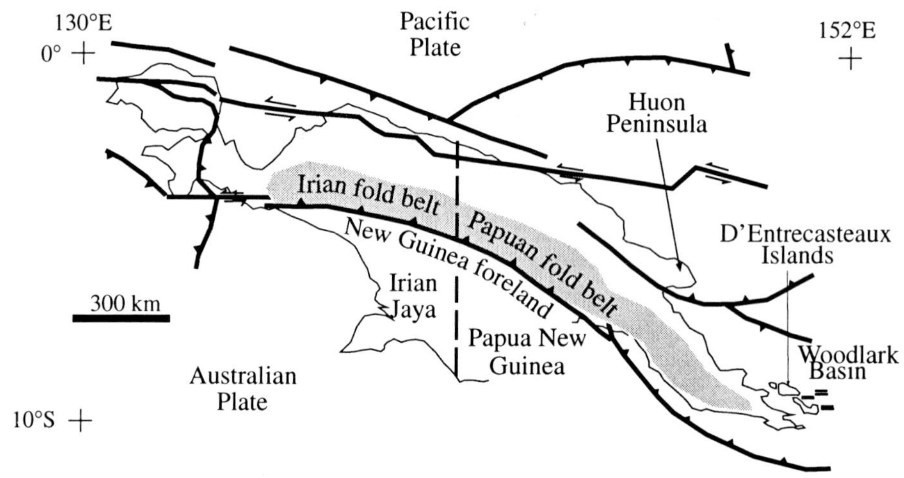

Figure 4. Irian Jaya and Papua New Guinea, including Irian and Papuan Foldbelts, Huon Peninsula, and D'Entrecasteaux Islands.

An additional consideration in a discussion of rates responsible for the exhumation of UHP rocks is whether or not plate rates were faster in the past, and if so, how far back, since most UHP metamorphic rocks, with the exception of 2.0 Ga rocks in Mali (Caby, 1994), were exhumed during the Phanerozoic. Plate rates reconstructed from magnetic stripes in oceanic crust indicate that at least for the last 170 m.y. plate rates were not substantially faster than today. Larsen (1991) proposed that during a mid-Cretaceous (~120-80 Ma) superplume was responsible for a 50-75% increase in the rate of oceanic crust production. He attributed the increase in oceanic crust production to a larger percentage of Pacific ridges undergoing rapid spreading, but estimated that the maximum spreading rate (170 mm/a) at that time was similar to the maximum rates observed today. The geologic record farther back, during the Paleozoic, is not substantially different than the Mesozoic or Cenozoic records, suggesting that plate rates during the Paleozoic were similar to those of today.

4. EXHUMATION RATES

Exhumation rates are commonly inferred from isotopic cooling ages, which measure the time at which a rock cooled through a particular temperature. The methods that measure cooling through the lowest temperatures are $^{40}Ar/^{39}Ar$ analysis of K-feldspar (e.g., McDougall and Harrison, 1988) and zircon and apatite fission-track analysis (e.g., Naeser, 1976). A new method, (U-Th)/He thermochronometry (Farley *et al.*, 1996; Wolf *et al.*, 1997), also shows promise for the study of low-temperature cooling in rocks.

Most of the studies documenting rapid exhumation have used apatite fission-track thermochronometry (see Table 1). Fission tracks are damage zones in a crystal or glass that are formed by fission of ^{238}U (Wagner, 1968). Although fission tracks are stable at low temperature, they shorten and disappear (anneal) with increasing temperature (Naeser and Faul, 1969; Gleadow and Duddy, 1981). Annealing in F-rich apatites (95% of all apatites) occurs over a temperature range of ~60-110°C (Gleadow *et al.*, 1986; Green *et al.*, 1986). This temperature range is known as the partial annealing zone (PAZ) (Gleadow and Fitzgerald, 1987) and forms a pseudostratigraphic horizon that can be used to infer the magnitude of denudation and rock uplift (e.g., Brown, 1991; Fitzgerald *et al.*, 1995). The fission track "age" of a sample is commonly interpreted as the time at which the sample cooled below the closure temperature (or the upper bound of the PAZ) and is determined by measuring the density of fission tracks and the U concentration of the sample (e.g., Naeser, 1976). At a geologic cooling rate of 10°C/m.y., the apatite closure temperature is ~130°C; at a rate of 100°C/m.y., it is closer to 150°C (Gleadow *et al.*, 1983).

Table 1. Summary of Rapid Rock Uplift and Exhumation Rates

Region	Rate, mm/a	Method	Reference
Southern Alps, New Zealand	10	FT	Tippett and Kamp (1993)
Namche Barwa, Tibet	10	U-Th-Pb, FT	Burg *et al.* (1997)
Nanga Parbat, Pakistan	5	FT	Zeitler (1985)
Central Nepal	4.1	$^{208}Pb/^{232}Th$	Harrison *et al.* (1997)
			Foster *et al.* (1994)
D'Entrecasteaux Islands	10	Ar, FT	Baldwin *et al.* (1993)
King Range, CA	4	FT	Dumitru (1991)
Alaska Range	1.5	FT	Fitzgerald *et al.* (1995)
Irian Jaya	1.5	FT	Weiland and Cloos (1996)
San Gabriel Mountains, CA	1.2	FT	Blythe *et al.* (1996)

FT, fission-track analyses; Ar, $^{40}Ar/^{39}Ar$ analyses.

Additional information about the thermal history of a sample can be obtained by measuring the lengths of confined tracks (tracks parallel to the polished surface) in apatite, which constrain the rate and timing of cooling within the PAZ (Gleadow *et al.*, 1986; Green *et al.*, 1989). Samples that cooled rapidly through the PAZ (in less than 5 m.y.) typically have long mean track lengths of ≥14 μm and narrow standard deviations (~1 μm). Samples that cooled more slowly have shorter mean lengths and broader standard deviations. Track-length measurements are commonly used to reconstruct cooling histories between ~60-130 °C (e.g., Dumitru, 1991; Tippett and Kamp, 1993; Fitzgerald *et al.*, 1995; Blythe *et al.*, 1996).

The (U-Th)/He system is based on the production of 8 He atoms for each U-series decay. Apatite has a He closure temperature of ~75°C at a 10°C/m.y. cooling rate, with partial diffusive loss of He occurring over temperatures of ~45–75 °C (Farley *et al.*, 1996; Wolf *et al.*, 1997). This temperature range, termed the helium partial retention zone (PRZ), is analogous to the fission track PAZ (Fig. 5). Thus, (U-Th)/He thermochronology extends cooling histories to even shallower crustal levels and can provide additional constraints on thermal histories indicated by apatite fission-track length models. This technique has not been applied to any area that underwent rapid cooling.

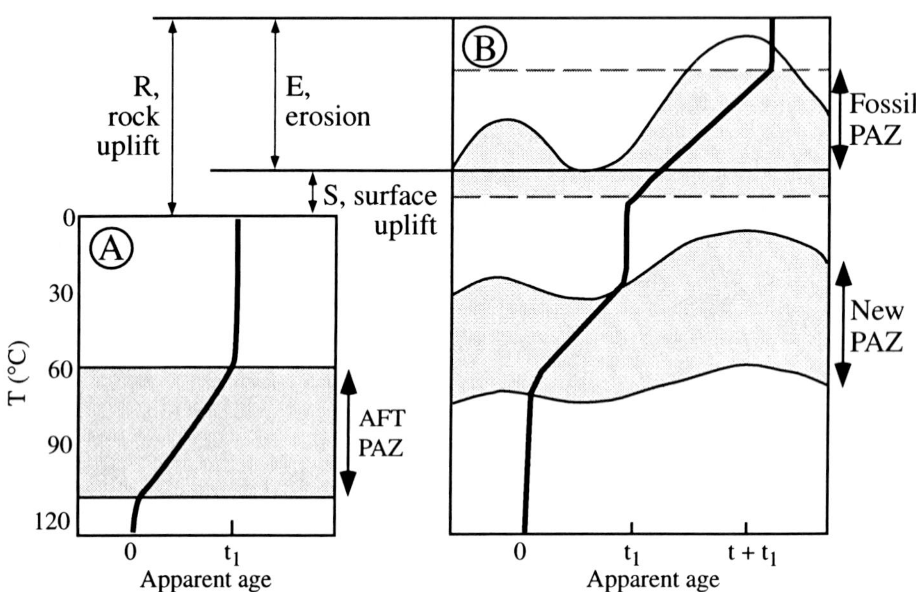

Figure 5. Apatite fission-track partial annealing zone (PAZ). The heavy line traces the fission-track age within the PAZ. a) The PAZ spans a temperature range of 60-110°C. b) Rock uplift and erosion result in surface uplift. The fossil PAZ is also elevated, and a new PAZ forms at depth. The ages obtained along an elevation profile preserve the uplifted fossil PAZ, with ages ranging between the primary cooling age ($t + t_1$) and the age of uplift (t_1). After Fitzgerald and Gleadow (1988).

There are several sources of potential error in any exhumation rate calculation, and it is important to understand how they were treated when evaluating exhumation rates derived in any individual study. The biggest source of potential error in most studies is in the choice of geothermal gradient, which can vary substantially with rock type, the circulation of fluids, and the rate of exhumation. For exhumation rates >1 mm/a (all studies listed in Table 1), a stable geothermal gradient (e.g., 25°C/km) should not be used; instead, thermochronologic data should be evaluated using time-dependent heat transfer models as well as all available independent geologic data (Peacock, 1995). Additionally, topographic relief can greatly disturb near-surface isotherms, leading to overestimation of exhumation rate (Stüwe *et al.*, 1994; Mancktelow and Grasemann, 1997). Another large source of uncertainty in exhumation rate calculations is in the closure temperature of the isotopic system being used: closure temperatures depend on cooling rate.

The fastest estimated exhumation rate, ~10 mm/a (Table 1), was estimated from U-Th-Pb and fission-track analyses in the Namche-Barwa region (Fig. 1) of the easternmost Himalaya (Burg *et al.*, 1997). The difference between this study and the studies described below is that the exhumation rates were calculated using an assumed cooling rate rather than an assumed geothermal gradient. The Namche-Barwa samples yielded U-Th-Pb xenotime ages of 2.9-3.9 Ma, zircon fission-track ages of ~2.6 Ma, and apatite fission-track ages of ~1.1 Ma. Based on a cooling rate of 100°C/Ma, Burg *et al.* (1997) interpreted these ages as representing cooling through 550 ± 150°C (xenotime crystallization), 260 ± 25°C (zircon) and 140 ± 20°C (apatite), respectively. The 10 mm/a exhumation rate was estimated for 3.5–2.5 Ma, with the rate slowing to 3–5 mm/a from 2.1 Ma until the present. These ages, closure temperatures, and exhumation rate require a geothermal gradient of 29 °C/km (upper limit of 46.5°C/km and lower limit of 11.5°C/km, calculated using the errors on the closure temperatures). The required geothermal gradient for 2.6–1.1 Ma ranges from 16–27°C/km (for 5 to 3 mm/a), and for 1.1 Ma to the present, from 24–40°C/km.

Although similar rates were estimated for Nanga Parbat (Fig. 1) in the westernmost Himalaya in Pakistan, the rates were estimated based on much younger zircon and apatite fission-track ages and much higher geothermal gradients. Fluid inclusion studies suggest a geothermal gradient of ~60°C/km in the Nanga Parbat region, which is attributed to rapid unroofing (Winslow *et al.*, 1995). Based on this high geothermal gradient and zircon and apatite fission-track ages of 1.3 and 0.4 Ma, respectively, Zeitler *et al.* (1993) estimated exhumation rates of 3–6 mm/a in the Nanga Parbat-Haramosh Axis; a lower geothermal gradient would have yielded higher exhumation rates.

Foster *et al.* (1994) obtained older apatite and zircon fission-track ages of 2.1 and 4.3 Ma, respectively, from K2 in the Karakorum (Pakistan), and estimated an exhumation rate of 3–6 mm/a from ~5–3 Ma, from an initial geothermal gradient of 25–30°C/km. This geothermal gradient was assumed because an older zircon fission-track age suggested that the total amount of material removed from the top of K2 was ~6 km, and that rapid exhumation in the K2 gneiss began at ~5

Ma. Their thermal models indicated that advection of isotherms caused the geothermal gradient to peak at ~55°C/km at ~2 Ma, and then decrease to ~34°C/km.

Young fission-track ages and high rock uplift rates have also been obtained from the Southern Alps of New Zealand, in the region adjacent to the Alpine fault (Tippett and Kamp, 1993). Rock uplift in the Southern Alps increases to the south along the Alpine fault, with apatite and zircon ages of 0 and ~5–7 Ma, respectively, in the zone closest to the fault. The young zircon and apatite fission-track ages from samples collected adjacent to the NE-trending Alpine fault were used to estimate rock uplift rates of 2–10 mm/a. Note that these rates are rock uplift rates, not exhumation rates, calculated from the equation of Brown (1991):

$$U_r = (T_p - T_s)/G + E_s - E_{im} + \Delta E_{msl}$$

where:

U_r = amount of rock uplift

T_p = pre-uplift paleotemperature

T_s = pre-uplift surface temperature (assumed to be 10°C)

G = pre-uplift geothermal gradient (assumed to be 27.5 ± 2.5 °C/km)

E_s = present sample elevation above mean sea level

E_{im} = initial mean elevation of land surface above mean sea level (assumed to be 0)

ΔE_{msl} = difference between pre-uplift mean sea level and present mean sea level (assumed to be 0).

When the terms assumed to be zero are dropped from the equation, it is clear that the difference between rock uplift and exhumation is that rock uplift also accounts for the current elevation of the samples relative to mean sea level, whereas exhumation does not. Therefore, because only a few of the 140 samples collected in this study were from elevations >1 km, these rock uplift rates can be compared directly with exhumation rates.

Young apatite fission-track ages (<5 Ma) have been reported from a few localities in North America. Fitzgerald *et al.* (1995) obtained ages of 16 Ma (at ~6 km elevation) to 4 Ma (at ~2 km elevation) from Mt. McKinley, the highest peak in North America, which is south of the Denali fault in the central Alaska Range. They used a plot of elevation *vs.* age, which included a sharp inflection at ~6 Ma (elevation of ~4.5 km), to interpret the initiation of rock uplift in the central Alaska Range at ~5–6 Ma, at a rate of ~1.5 mm/a. In this case, the high elevations of the samples make a direct comparison of rock uplift and exhumation rate problematic.

Dumitru (1991) reported young apatite fission-track ages averaging ~1.2 Ma from the King Range (Fig. 3), a restraining bend along the San Andreas fault, just to the south of the Mendocino triple junction. Dumitru estimated that 2–5 km of rock uplift occurred in the King Range over the last 1.2 Ma, suggesting a rock uplift rate of 1.6–4 mm/a, a rate nearly equal to the high rates in the Himalaya and the Southern Alps of New Zealand.

The San Gabriel Mountains (Fig. 3) also appear to have undergone significant, although less rapid, exhumation. The San Gabriel Mountains are part of the central Transverse Ranges, the largest set of ranges in California thought to be caused by contraction along a confining bend in the San Andreas fault. The San Gabriel Mountains lie west of the San Andreas fault and northwest of the Los Angeles basin. Apatite fission-track ages of 3–5 Ma were obtained from a canyon in the hanging wall of the Sierra Madre fault, the active reverse fault on the southern margin of the San Gabriel Mountains (Blythe *et al.*, 1996; Blythe, in preparation). These young ages, which occur at low elevations (~200-400 m), are interpreted to represent exhumation rates of ~1–1.2 mm/a, assuming a geothermal gradient of ~25°C/km. In this case, where exhumation began in the last few million years, and the highest rates are ~1 mm/a, a stable geothermal gradient can be safely assumed.

Rapid exhumation rates are also recorded in the Irian (northwest) and Papuan (southeast) fold and thrust belts of New Guinea (Fig. 4). The Irian and Papuan fold and thrust belts formed by the collision of an island arc with the Australian continental margin (Jacques and Robinson, 1977). Weiland and Cloos (1996) obtained apatite fission-track ages of 2–2.6 Ma from Jurassic and older sedimentary rock on the western margin of the Central Range (the Irian fold and thrust belt), and estimated an exhumation rate of 1.5 mm/a. Their study also documented higher erosion rates from lower elevations on the southwest side of the topographic spine of the fold and thrust belt, which they attributed to higher precipitation rates. The rates of erosion and bedrock uplift are slightly lower to the southeast in the Papuan Fold Belt: Hill and Gleadow (1989) obtained apatite fission-track ages of 2.0-4.0 Ma from samples collected there.

All the studies described above dealt with exhumation or rock uplift rates from regions undergoing contraction. Young $^{40}Ar/^{39}Ar$ ages (ranging from 2–3 Ma for amphiboles to 1–2 Ma for K-feldspars) and apatite fission-track ages (0.4 to 1.0 Ma) were obtained by Baldwin *et al.* (1993) from the D'Entrecasteaux Islands, a region of active extension (Fig. 4). These extremely young ages were used along with thermobarometry (Hill and Baldwin, 1993) to constrain P-T-t paths and to interpret exhumation rates of ~10 mm/a for the last ~4 m.y.

5. SURFACE UPLIFT

Surface uplift can be directly measured over short time periods (tens of years) with leveling surveys, Global Positioning System (GPS) studies and Laser Altimetry data. Over longer time periods (~10^2–10^5 years) surfaces can be dated by a variety of isotopic methods, such as ^{10}Be, ^{26}Al, and ^{3}He dating (e.g., Bierman, 1994). These methods record the exposure time for the surface, and can be used to date surface uplift when applied to marine terraces formed at sea level. Marine terraces containing corals and shell material are commonly dated with U series and amino acid racemization techniques; in a landmark study of the Huon Peninsula in Papua New Guinea, Bloom *et al.* (1974) calculated surface uplift rates of 1–3 mm/a from $^{230}Th/^{234}U$ dates on emergent Pleistocene coral reef terraces.

Modern surface uplift rates can be considered a proxy for exhumation without some of the complications, such as upwarping of isotherms, associated with high denudation rates. One of the problems with comparing surface uplift and exhumation rates, however, is that the isostatic feedback caused by erosional processes that contribute to exhumation rates are missing from surface uplift rates (Montgomery, 1994; Small and Anderson, 1995), and therefore surface uplift rates measured in areas of no erosion underestimate exhumation rates.

Surface uplift rates of >1 mm/a are common in tectonically active regions (see Table 2). As expected, the highest surface uplift rates are from the Southern Alps in New Zealand and the Himalaya. Based on correlation with dated terraces from New Guinea, Bull and Cooper (1986) inferred maximum uplift rates of 7.8 mm/a from flights of marine terraces along the Alpine fault on the South Island of New Zealand. Leveling data from northern Nepal indicate that a zone with a 40-km wavelength centered over the Main Central Thrust is going up 7 ± 3 mm/a relative to the Indian border (Jackson and Bilham, 1994); this surface uplift in Nepal, however, is occurring despite erosion.

Table 2. Summary of Rapid Surface Uplift Rates

Region	Rate, mm/a	Method	Reference
Southern Alps, NZ	7.8	terrace correlation	Bull and Cooper (1986)
Nepal Himalayas	7	leveling	Jackson and Bilham (1994)
King Range, CA	3–5	terrace correlation	Merritts and Bull (1989)
Papua New Guinea	1–3	$^{230}Th/^{234}U$ on terraces	Bloom *et al.* (1974)

In North America, high rates of surface uplift are also recorded in the areas with the highest bedrock uplift rates. In northern California (Fig. 3), uplifted beach terraces in the King Range (where Dumitru (1991) obtained ~1.2 Ma apatite fission-track ages) were used to infer a maximum surface uplift rate of 3–5 mm/a (Merritts and Bull, 1989). Although no surface uplift rates have been determined in the Transverse Ranges of southern California, vertical motion has been measured using stratigraphic offset on faults in the Ventura Basin region (Fig. 3). Based on radiocarbon ages in terrace deposits and soil correlations, Rockwell *et al.* (1984), estimated <0.3–1.3 mm/a vertical slip rates on faults in the Ventura anticline. The fastest moving reverse fault in southern California, however, is the San Cayetano fault, with more than 750 m of fault-plane separation recorded in less than 1 million years (Rockwell, 1988).

6. EFFECTS OF CLIMATE

The study of active tectonics has recently undergone a shift with the recognition that climate might play a much larger role in controlling tectonic processes than previously thought. Beaumont *et al.* (1992) suggested that variation in the rates of erosion in an orogenic belt might affect the style of deformation. Regions undergoing high rates of erosion might in turn have more uplift as the result of isostatic compensation for their reduced load (e.g., Anderson, 1994).

Molnar and England (1990) re-examined the link between late Cenozoic cooling and world-wide increase in mountain range elevation. A late Cenozoic rise in mountain ranges has been interpreted on the basis of the deposition of thick sedimentary sequences, increased incision rates in major mountain belts, and inferred paleo-elevations from plant fossils (e.g., de Sitter, 1952; Trimble, 1980; Hay *et al.*, 1988). Molnar and England (1990) argued that these three lines of evidence-sedimentation rate, incision rate and paleotemperatures derived from paleobotanical evidence-could in fact be explained just as well by global cooling during late Cenozoic time as by a "rise" in mountain ranges. As summarized by Raymo and Ruddiman (1992), global cooling during Cenozoic time is recognized from $^{18}O/^{16}O$ records, the terrestrial fossil record, and the glacial detritus found in many mid-latitude countries. $^{18}O/^{16}O$ records from deep sea sediments indicate that ~3 million years ago, glacial-interglacial cycles had begun (Shackleton, 1984). At the peak of the last glacial period (18,000 years ago), 30% of the Earth's land surface was covered by ice. The cause of global climate change over the last 40 m.y. is unclear, although plateau uplift in Tibet and South America has been suggested as a possible cause of world-wide global cooling during the Cenozoic (Raymo *et al.*, 1988; Ruddiman and Kutzbach, 1989).

7. MECHANISMS OF EROSION

Landscapes are sculpted by fluvial, glacial and hillslope processes that vary in importance with climate. In many places undergoing rapid rock uplift and erosion, such as the Alaska Range, the Southern Alps in New Zealand, and the Himalaya, all three agents of erosion are active within a single drainage basin. Hillslope processes are the most poorly quantified erosional agent in tectonically active regions, although in many areas, such as the Transverse Ranges of southern California, they are the dominant means of erosion (Bull, 1991).

Fluvial erosion is generally thought to be the most important agent of denudation world-wide. Erosion by streams and rivers is controlled by climate, rock type, and relief, although relief appears to be the dominant control (e.g., Ahnert, 1970). One of the difficulties in extrapolating modern fluvial erosion rates to the past is that recent climate changes have affected river erosion rates. Presently, southern Asia and the larger Pacific Ocean islands contribute ~70% of the suspended sediments reaching the oceans (Milliman and Meade, 1983).

In a summary of world-wide fluvial denudation rates, Summerfield and Hulton (1994) listed the Brahmaputra basin in central/southeast Asia as having the highest denudation rate of 0.7 mm/a. The Ganges and Indus river basins, with the next highest denudation rates, had substantially lower values of 0.27 and 0.14 mm/a, respectively. These three rivers draining the Himalayan orogen had low percentages of chemical denudation (<11.5% of the total), suggesting that chemical weathering is not important in rapidly eroding regions.

In many parts of the world, glaciers are a significant agent of erosion. Modern glacial erosion rates compiled by Hallet *et al.* (1996) are summarized in Fig. 6. Only 10% of today's land surface is covered by glaciers (most of that is in the Antarctic and Greenland ice caps), although nearly 30% of the land surface was covered during the glacial maximum, 100,000 years ago. The degree of denudation associated with a glacier depends on the type of glacier, with the highest glacial erosion rates from wet, alpine-type glaciers and the lowest from dry, cold continental glaciers (Hallet *et al.*, 1996).

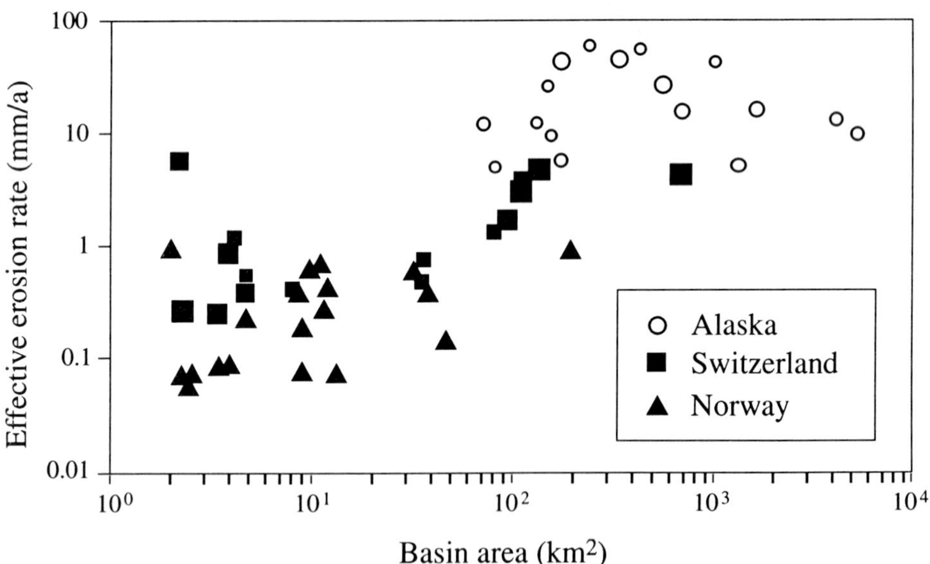

Figure 6. Glacial erosion rates compiled by Hallet *et al.* (1996).

A recent study by Brozovic *et al.* (1997) suggests that glaciers might be the ultimate control on the maximum height of a mountain range. Brozovic and colleagues used digital elevation models to look at the distribution of elevations in the Nanga Parbat region of Pakistan, an area that includes some of the fastest exhumation rates in the world (Zeitler *et al.*, 1993). In plots of slope angle versus elevation for selected regions around Nanga Parbat, a consistent "trough" appeared in slope angles at the elevations glaciers were active, implying that the ultimate control on maximum elevation in this region is glaciation, which acts as a "topographic buzzsaw". Brozovic *et al.* (1997) hypothesized that as the elevation of a specific area increased above the ELA (Equilibrium Line Altitude-the line at which accumulation equals ablation of ice), more glaciers became active, eroding more efficiently that other mechanisms, and decreasing the mean elevation.

8. CONCLUSIONS

Measured bedrock and surface uplift rates world-wide suggest that rates higher than ~10 mm/a over time periods of $>10^3$ years are unlikely. The Himalaya and Southern Alps in New Zealand, areas undergoing active contraction, record exhumation and surface uplift rates of 5–10 mm/a, measured over timescales of 10^6–10^3 years. A similarly high exhumation rate was obtained for the D'Entrecasteaux Islands, Papua New Guinea, an area undergoing active extension. Lower rates of 1–5 mm/a were obtained from other regions, including the King and Transverse Ranges of California, Mt. McKinley in the Alaska Range, and New Guinea. Climatic data suggest that modern erosion rates may be as high as they have ever been, and that these high erosion rates may be enhancing exhumation rates world-wide. Researchers studying UHP metamorphism need to determine whether exhumation rates greater than 10 mm/a are really necessary for the preservation of UHP mineral assemblages.

One of the striking conclusions in the comparison of the shorter term surface uplift and longer term exhumation rates is that these rates are remarkably similar. The "pull of the present" (see discussion in Keller and Pinter, 1996)—the concept that rates are faster in the present than they were in the past—does not appear to be true. New techniques and more careful analyses show that the rates measured over millions of years are similar to rates measured for more short-term processes. Therefore, one of the most important questions researchers must ask is whether the rates and processes that result in the exhumation of UHP metamorphic rocks are unique for the time or conditions at which they occurred. The challenge is to consider climatic conditions as well as plate rates.

ACKNOWLEDGMENTS

The author is partially supported by a NASA grant to D. Burbank. Reviews by Jeff Lee, Lothar Ratschbacher, and Brad Hacker are gratefully acknowledged.

REFERENCES

Abers, G.A. (1997) Shallow dips of normal faults during rapid extension: Earthquakes in the Woodlark - D'Entrecasteaux rift system, Papua New Guinea, *Journal of Geophysical Research* **102**, 15301–15317.

Ahnert, F. (1970) Functional relationships between denudation, relief, and uplift in large mid-latitude basins, *American Journal of Science* **268**, 243–263.

Allis, R.G. (1986) Mode of crustal shortening adjacent to the Alpine fault, New Zealand, *Tectonics* **5**, 15–32.

Anderson, B. (1994) Evolution of the Santa Cruz Mountains, California, through tectonic growth and geomorphic decay, *Journal of Geophysical Research* **99**, 20,161–20,179.

Baldwin, S.L., Lister, G.S., Hill, E.J., Foster, D.A., and McDougall, I. (1993) Thermochronologic constraints on the tectonic evolution of active metamorphic core complexes, D'Entrecasteaux Islands, Papua New Guinea, *Tectonics* **12**, 611–628.

Beaumont, C., Fullsack, P., and Hamilton, J. (1992) Erosional control of active compressional orogens, in K.R. McClay (ed.), *Thrust Tectonics*, Chapman and Hall, London, pp. 1–18.

Bevis, M., Taylor, F.W., Schutz, B.E., Recy, J., Isacks, B.R., Helu, S., Singh, R., Kendrick, E., Stowell, J., Taylor, B., and Calmant, S. (1995) Geodetic observations of very rapid convergence and back-arc extension at the Tonga arc, *Nature* **374**, 249–251.

Bierman, P.R. (1994) Using in situ produced cosmogenic isotopes to estimate rates of landscape evolution: A review from the geomorphic perspective, *Journal of Geophysical Research* **99**, 13885–13896.

Bilham, R., Larson, K., Freymueller, J., and members, P.I. (1997) GPS measurements of present-day convergence across the Nepal Himalaya, *Nature* **386**, 61–63.

Bloom, A.L., Broecker, W.S., Chappell, J.M.A., Matthews, R.K., and Mesolella, K.J. (1974) Quaternary sea level fluctuations on a tectonic coast: New ^{230}Th/^{234}U dates from the Huon Peninsula, New Guinea, *Quaternary Research* **4**, 185–205.

Blythe, A.E., Fielding, E.J., and Burbank, D.W. (1996) Morphology as a function of bedrock uplift and climate: a case study of the Transverse Ranges, Southern California, from apatite fission-track and DEM analyses, *Transactions of the American Geophysical Union, Eos* **77**, F644.

Brown, R. (1991) Backstacking apatite fission-track "stratigraphy": A method for resolving the erosional and isostatic rebound components of tectonic uplift histories, *Geology* **19**, 74–77.

Brozovic, N., Burbank, D.W., and Meigs, A.J. (1997) Climatic limits on landscape development in the northwestern Himalaya, *Science* **276**, 571–574.

Bull, W.B. (1991) *Geomorphic Responses to Climatic Change*, Oxford University Press, New York.

Bull, W.B. and Cooper, A.F. (1986) Uplifted marine terraces along the Alpine fault, New Zealand, *Science* **234**, 1225–1228.

Burchfiel, B.C., Cheng, Z., Hodges, K.V., Liu, Y., Royden, L.H., Deng, C., and Xu, J. (1992) The South Tibetan Detachment System, Himalayan orogen: Extension contemporaneous with and parallel to shortening in a collisional mountain belt, *Geological Society of America Special Paper* **269**, 41.

Burg, J.P., Davy, P., Nievergelt, P., Oberli, F., Seward, D., Diao, Z., and Meier, M. (1997) Exhumation during crustal folding in the Namche-Barwa syntaxis, *Terra Nova* **9**, 53–56.

Caby, R. (1994) Precambrian coesite from northern Mali; first record and implications for plate tectonics in the trans-Saharan segment of the Pan-African belt, *European Journal of Mineralogy* **6**, 235–244.

Chopin, C. (1984) Coesite and pure pyrope in high-grade blueschists of the western Alps: a first record and some consequences, *Contributions to Mineralogy and Petrology* **86**, 107–118.

de Sitter, L.U. (1952) Pliocene uplift of Tertiary mountain chains, *American Journal of Science* **250**, 297–307.

DeMets, C., Gordon, R.G., Argus, D.F., and Stein, S. (1990) Current plate motions, *Geophysics Journal International* **101**, 425–478.

Draper, G. and Bone, R. (1981) Denudation rates, thermal evolution, and preservation of blueschist terrains, *Journal of Geology* **89**, 601–613.

Dumitru, T.A. (1991) Major Quaternary uplift along the northernmost San Andreas fault, King Range, northwestern California, *Geology* **19**, 526–529.

England, P. and Molnar, P. (1990) Surface uplift, uplift of rocks, and exhumation of rocks, *Geology* **18**, 1173–1177.

Farley, K.A., Wolf, R.A., and Silver, L.T. (1996) The effects of long alpha-stopping distances on (U-Th)/He dates, *Geochimica Cosmochimica et Acta* **60**, 4223–4230.

Fitzgerald, P.G. and Gleadow, A.J.W. (1988) Fission-track geochronology, tectonics and structure of the Transantarctic Mountains in northern Victoria Land, Antarctica, *Chemical Geology* **73**, 169–198.

Fitzgerald, P.G., Sorkhabi, R.B., Redfield, T.F., and Stump, E. (1995) Uplift and denudation of the central Alaska Range: A case study in the use of apatite fission track thermochronology to determine absolute uplift parameters, *Journal of Geophysical Research* **100**, 20,175–20,191.

Fossen, H. and Rykkelid, E. (1992) Postcollisional extension of the Caledonide orogen in Scandinavia: structural expressions and tectonic significance, *Geology* **20**, 737–740.

Foster, D., Gleadow, A.J.W., and Mortimer, G. (1994) Rapid Pliocene exhumation in the Karakoram (Pakistan), revealed by fission-track thermochronology of the K2 gneiss, *Geology* **22**, 19–22.

Gannser, A. (1981) The geodynamic history of the Himalaya, in H.K. Gupta and F.M. Delaney (eds.), *Zagros, Hindu Kush, Himalayan geodynamic evolution*, 3, American Geophysical Union, Washington, D.C., pp. 111–121.

Gleadow, A.J.W. and Duddy, I.R. (1981) A natural long-term track annealing experiment for apatite, *Nuclear Tracks* **5**, 169–174.

Gleadow, A.J.W., Duddy, I.R., Green, P.F., and Hegarty, K.A. (1986) Fission track lengths in the apatite partial annealing zone and the interpretation of mixed ages, *Earth and Planetary Science Letters* **78**, 245–254.

Gleadow, A.J.W., Duddy, I.R., and Lovering, J.F. (1983) Fission track analysis: A new tool for the evaluation of thermal histories and hydrocarbon potential, *Petroleum Exploration Association of Australia Journal* **23**, 93–102.

Gleadow, A.J.W. and Fitzgerald, P.G. (1987) Uplift history and structure of the Trans-Antarctic Mountains: New evidence from fission track dating of basement apatites in the Dry Valleys area, Southern Victoria Land, *Earth and Planetary Science Letters* **82**, 1–14.

Green, P.F., Duddy, I.R., Gleadow, A.J.W., Tingate, P.T., and Laslett, G.M. (1986) Thermal annealing of fission tracks in apatite: 1. A qualitative description, *Chemical Geology* **59**, 237–253.

Green, P.F., Duddy, I.R., Laslett, G.M., Hegarty, K.A., Gleadow, A.J.W., and Lovering, J.F. (1989) Thermal annealing of fission tracks in apatite: 4. Quantitative modeling techniques and extensions to geological timescales, *Chemical Geology* **79**, 155–182.

Hacker, B.R. and Peacock, S.M. (1994) Creation, preservation, and exhumation of coesite-bearing, ultrahigh-pressure metamorphic rocks, in R.G. Coleman and X. Wang (eds.), *Ultrahigh Pressure Metamorphism*, Cambridge University Press, Cambridge, United Kingdom.

Hallet, B., Hunter, L., and Bogen, J. (1996) Rates of erosion and sediment evacuation by glaciers: A review of field data and their implications, *Global and Planetary Change* **12**, 213–235.

Harrison, T.M., Ryerson, F.J., Le Fort, P., Yin, A., Lovera, O.M., and Catlos, E.J. (1997) A Late Miocene-Pliocene origin for the central Himalayan inverted metamorphism, *Earth and Planetary Science Letters* **146**, E1–E7.

Hay, W.W., Sloan, J.L., and Wold, C.N. (1988) Mass/age distribution and composition of sediments on the ocean floor and the global rate of sediment subduction, *Journal of Geophysical Research* **93**, 14933–14940.

Hill, E.J. and Baldwin, S.L. (1993) Exhumation of high-pressure metamorphic rocks during crustal extension in the D'Entrecasteaux region, Papua New Guinea, *Journal of Metamorphic Geology* **11**, 261–277.

Hill, K.C. and Gleadow, A.J.W. (1989) Uplift and thermal history of the Papuan Fold Belt, Papua New Guinea: Apatite fission track analysis, *Australian Journal of Earth Sciences* **36**, 515–539.

Jackson, M. and Bilham, R. (1994) Constraints on Himalayan deformation inferred from vertical velocity fields in Nepal and Tibet, *Journal of Geophysical Research* **99**, 13,897–13,912.

Jacques, A.L. and Robinson, G.P. (1977) Continent/island arc collision in northern Papua New Guinea, *Journal of Australian Geology and Geophysics* **2**, 289–303.

Jordan, T.J. and Minster, B. (1988) Crustal deformation in western North America, *Spektrum der Wissenschaft* **1988**, 106–116.

Kamp, P.J.J. (1986) Late Cretaceous-Cenozoic tectonic development of the southwest Pacific region, *Tectonophysics* **121**, 255–281.

Keller, E.A. and Pinter, N. (1996) *Active Tectonics*, Prentice Hall, Upper Saddle River, NJ.

Larson, R.L. (1991) Latest pulse of Earth: Evidence for a mid-Cretaceous superplume, *Geology* **19**, 547–550.

Le Pichon, X., Fournier, M., and Jolivet, L. (1992) Kinematics, topography, shortening, and extrusion in the India-Eurasia collision, *Tectonics* **11**, 1085–1098.

Macfarlane, A.M. (1993) Chronology of tectonic events in the crystalline core of the Himalaya, Langtang National Park, Central Nepal, *Tectonics* **12**, 1004–1025.

Mancktelow, N.S. and Grasemann, B. (1997) Time-dependent effects of heat advection and topography on cooling histories during erosion, *Tectonophysics* **270**, 167–195.

McDougall, I. and Harrison, T.M. (1988) *Geochronology and Thermochronology by the $^{40}Ar/^{39}Ar$ Method*, Oxford University Press, New York.

Meigs, A.J., Burbank, D.W., and Beck, R.A. (1995) Middle-late Miocene (>10 Ma) formation of the Main Boundary thrust in the western Himalaya, *Geology* **23**, 423–426.

Merritts, D. and Bull, W.B. (1989) Interpreting Quaternary uplift rates at the Mendocino triple junction, northern California, *Geology* **17**, 1020–1024.

Michard, A., Chopin, C., and Henry, C. (1993) Compression versus extension in the exhumation of the Dora-Maira coesite-bearing unit, Western Alps, Italy, *Tectonophysics* **221**, 173–193.

Milliman, J.D. and Meade, R.H. (1983) World-wide delivery of river sediment to the oceans, *Journal of Geology* **91**, 1–21.

Molnar, P. and England, P. (1990) Late Cenozoic uplift of mountain ranges and global climate change: Chicken or egg? *Nature* **346**, 29–34.

Molnar, P. and Tapponnier, P. (1975) Cenozoic tectonics of Asia: Effects of a continental collision, *Science* **189**, 419–426.

Montgomery, D.R. (1994) Valley incision and uplift of mountain peaks, *Journal of Geophysical Research* **99**, 13913–13922.

Naeser, C.W. (1976) *Fission Track Dating*, U. S. Geological Survey, Denver.

Naeser, C.W. and Faul, H. (1969) Fission track annealing in apatite and sphene, *Journal of Geophysical Research* **74**, 705–710.

Peacock, S. (1995) Calculating exhumation rates from thermochronologic data, *Transactions of the American Geophysical Union, Eos* **76**, F569.

Raymo, M.E. and Ruddiman, W.F. (1992) Tectonic forcing of late Cenozoic climate change, *Nature* **359**, 117–122.

Raymo, M.E., Ruddiman, W.F., and Froelich, P.N. (1988) Influence of late Cenozoic mountain building on ocean geochemical cycles, *Geology* **16**, 649–653.

Rockwell, T.K. (1988) Neotectonics of the San Cayetano fault, California, *Geological Society of America Bulletin* **100**, 500–513.

Rockwell, T.K., Keller, E.A., Clark, M.N., and Johnson, D.L. (1984) Chronology and rates of faulting of Ventura River terraces, California, *Geological Society of America Bulletin* **95**, 1466–1474.

Ruddiman, W.F. and Kutzbach, J.E. (1989) Forcing of late Cenozoic northern hemisphere climate by plateau uplift in southeast Asia and the American southwest, *Journal of Geophysical Research* **94**, 18409–18427.

Shackleton, N.J. (1984) Oxygen isotope evidence for Cenozoic climate change, in P. Brenchley (ed.), *Fossils and Climate*, Wiley, London, pp. 27–34.

Sieh, K.E. and Jahns, R.H. (1984) Holocene activity on the San Andreas fault at Wallace Creek, California, *Geological Society of America Bulletin* **95**, 883–896.

Small, E.E. and Anderson, R.S. (1995) Geomorphically driven late Cenozoic rock uplift in the Sierra Nevada, California, *Science* **270**, 277–280.

Smith, D.C. (1984) Coesite in clinopyroxene in the Caledonides and its implications for geodynamics, *Nature* **310**, 641–644.

Stock, J. and Molnar, P. (1982) Uncertainties in the relative positions of the Australia, Antarctica, Lord Howe, and Pacific plates since the late Cretaceous, *Journal of Geophysical Research* **87**, 4697–4714.

Stüwe, K., White, L., and Brown, R. (1994) The influence of eroding topography on steady-state isotherms. Application to fission-track analysis, *Earth and Planetary Science Letters* **124**, 63–74.

Summerfield, M. (1991) *Global Geomorphology: an Introduction to the Study of Landforms*, Wiley, New York.

Summerfield, M. and Hulton, N.J. (1994) Natural controls of fluvial denudation rates in major world drainage basins, *Journal of Geophysical Research* **99**, 13,871–13,883.

Tippett, J.M. and Kamp, P.J.J. (1993) Fission track analysis of the Late Cenozoic vertical kinematics of continental pacific crust, South Island, New Zealand, *Journal of Geophysical Research* **98**, 16119–16148.

Trimble, D.E. (1980) Cenozoic tectonic history of the Great Plains contrasted with that of the southern Rocky Mountains; a synthesis, *Mountain Geologist* **17**, 59–69.

Wagner, G.A. (1968) Fission track dating of apatites, *Earth and Planetary Science Letters* **4**, 411–415.

Walcott, R.I. (1978) Geodetic strains and large earthquakes in the axial tectonic belt of North Island, New Zealand, *Journal of Geophysical Research* **100**, 8221–8232.

Weiland, R. and Cloos, M. (1996) Pliocene-Pleistocene asymmetric unroofing of the Irian fold belt, Irian Jaya, Indonesia: apatite fission-track thermochronology, *Geologic Society of America Bulletin* **108**, 1438–1449.

Weissel, J.K., Taylor, B., and Karner, G.D. (1982) The opening of the Woodlark Basin, subduction of the Woodlark spreading center, and the evolution of Melanesia since mid-Pliocene time, *Tectonophysics* **87**, 253–277.

Winslow, D.M., Chamberlain, C.P., and Zeitler, P.K. (1995) Metamorphism and melting of the lithosphere due to rapid denudation, Nanga Parbat Massif Himalaya, *Journal of Geology* **103**, 395–409.

Wolf, R.A., Farley, K.A., and Silver, L.T. (1997) Assessment of (U-Th)/He thermochronometry: the low-temperature history of the San Jacinto mountains, California, *Geology* **25**, 64–68.

Zeitler, P.K. (1985) Cooling history of the NW Himalaya, Pakistan, *Tectonics* **4**, 127–151.

Zeitler, P.K., Chamberlain, C.P., and Smith, H. (1993) Synchronous anatexis, metamorphism, and rapid denudation at Nanga Parbat (Pakistan Himalaya), *Geology* **21**, 347–350.

Chapter 7

K-Ar ($^{40}Ar/^{39}Ar$) Geochronology of Ultrahigh Pressure Rocks

Stéphane Scaillet
Laboratoire de Géochronologie, INSU-UPS-IPGP, Département des Sciences de la Terre, Bât. 504, Université Paris Sud, Orsay, F-91405, France, scaillet@geophy.geol.u-psud.fr

Abstract: Although $^{40}Ar/^{39}Ar$ dating studies have been successful in setting age constraints on many ultrahigh pressure (UHP) occurrences world-wide, argon isotopic data from UHP rocks have on occasion proven difficult to interpret in terms of cooling and exhumation ages. One principal reason is that the closure behavior of common, high-K rock-forming minerals (phengite, phlogopite, and biotite) is poorly understood in UHP contexts, with age variations occurring at the scale of a single outcrop—or even a single mineral—that are often difficult to relate to the actual pressure-temperature-time path experienced by these rocks. The second reason is that these minerals can be contaminated with excess ^{40}Ar, commonly resulting in apparent ages well above those measured by other isotopic systems on the same rocks (e.g., Rb/Sr, U/Pb, Sm/Nd). After reviewing a number of structural-physical aspects bearing on argon transport phenomena at UHP, this chapter outlines the main crystal-related compositional, structural, and pressure effects that are influential in the argon isotopic record of UHP micas. Other important phenomena of argon-fluid-mineral partition behavior are then explored in connection with their relevance for the transport and incorporation mechanisms of excess ^{40}Ar in UHP rocks.

Based on examples selected from the western Alps, it is shown that the $^{40}Ar/^{39}Ar$ isotopic record of UHP rocks can be understood with special reference to their prograde and peak-metamorphic evolution during subduction to mantle depths. Beyond a critical devolatilization depth (between 40-50 km, depending on the thermal structure of the subduction zone), eclogite-facies and UHP rocks enter the fluid-deficient region, making them to evolve largely as a closed system in which the residual fluids and the argon isotopes are internally buffered by continuous UHP fluid-mineral equilibria. According to the starting mineralogy, initial fluid content, and rheology of the parental rock, internal recycling of the fluids and Ar isotopes (whether internally produced and/or inherited from the protolith) can produce complex isotopic signatures with characteristic grain-scale disequilibrium features symptomatic of excess ^{40}Ar contamination and/or protracted internal redistribution

B. R. Hacker and J. G. Liou (eds.),
When Continents Collide: Geodynamics and Geochemistry of Ultrahigh-Pressure Rocks, 161–201.
© 1998 *Kluwer Academic Publishers. Printed in the Netherlands.*

of radiogenic ^{40}Ar during eclogite-facies and UHP deformation. Two end-member cases of isotopic evolution are illustrated—one involving preservation of the pre-Alpine isotopic signature in recycled, anhydrous (high-grade) pre-UHP lithologies (granulite, pegmatite, and amphibolite); the other showing the effects of coupled small-scale isotopic and fluid recycling during deformation of Alpine whiteschists. Finally, the potential of argon as a tracer of fluid-melt-rock interactions and fluid recycling in UHP rocks is emphasized based on existing data from the coesite-bearing unit of the southern Dora Maira nappe.

1. INTRODUCTION

Beyond their identification and tectonic characterization, a major problem in the study of UHP (ultrahigh pressure) metamorphic rocks is to explain how fast and by what mechanism(s) deeply subducted rocks return to Earth's surface from mantle depths. The exposure of UHP oceanic- and continental-derived crustal slices implies incomplete sinking of the subducted slab into the mantle, a mechanism possibly favored by inhibited density changes during partial eclogitization of the subducted crust (e.g., Hacker, 1996; Austrheim *et al.*, 1997). However, the large-scale geodynamic processes by which very large and coherent tracts of oceanic and continental UHP crust are exhumed are still conjectural. Mechanical coupling with the upper (hanging wall) plate and faulting of the HP-UHP crustal slices back to the surface along the active subduction plane has been proposed (e.g., Hsü, 1991; von Blanckenburg and Davies, 1995), possibly aided by buoyant rise and concomitant erosion of the crustal wedge (e.g., Platt, 1986; Chemenda *et al.*, 1995). Large-scale extensional excision of the thickened crust during arrested or continued convergence (e.g., Philippot, 1990; Ballèvre and Merle, 1993; Hacker and Peacock, 1994; Harley and Carswell, 1995), or due to thermal-mechanical weakening of deep crustal levels is possible, as is large-scale subvertical lateral extrusion along oblique convergent plate boundaries (Hacker *et al.*, 1996).

Radiometric dating is fundamental in constraining these different models and the timescales involved in these processes. Indeed, the time-integrated budget of HP-UHP processes must be considered on a global scale so we can address the compatibility in rates of convergence (the horizontal component of motion deduced from plate tectonic reconstructions) with the exhumation rates associated with the huge vertical displacements mandated by these spectacular UHP occurrences. Moreover, time constraints are needed to quantify the mass-transfer budget of subduction processes and their consequences in terms of crustal/elemental recycling in convergent zones. Next, the time calibration of the evolving thermal structure and the heat budget of HP-UHP orogens is an equally important aspect to consider in terms of kinetics of metamorphic reactions, crustal rheology, and magma generation at depth. Finally, HP and UHP metamorphic events in a single orogen may be long-lasting and/or polycyclic, with time gaps as great as 40 m.y. between peak events (e.g., Boundy *et al.*, 1996), an aspect not incorporated in most current thermo-mechanical models of subduction-collision zones (e.g., von Blanckenburg and Davies, 1995).

In the western Alps, attempts at dating HP and UHP rocks by the $^{40}Ar/^{39}Ar$ technique were hampered by the presence of excess ^{40}Ar giving apparent ages well above those obtained on the same samples using other techniques such as Rb/Sr on phengite, U/Pb on zircon, Sm/Nd on garnet, and Lu/Hf on garnet (Tilton *et al.*, 1989; 1991; other references in Scaillet, 1996; Duchêne *et al.*, 1997). In other instances, the presence of excess ^{40}Ar was inferred from sample-scale $^{40}Ar/^{39}Ar$ inhomogeneities (Arnaud and Kelley, 1995), or from outcrop-wide $^{40}Ar/^{39}Ar$ age variations beyond conventionally expected differences in argon closure temperatures (e.g., Ruffet *et al.*, 1995; Inger *et al.*, 1996). The same situation has been recently reported for eclogitic phengites from the Dabie Shan and Sulu UHP provinces in China (Li *et al.*, 1994; Hacker *et al.*, 1996), the HP nappes of the Seward Peninsula in Alaska (Hannula and McWilliams, 1995), and the Bergen Arcs in western Norway (Boundy *et al.*, 1997), suggesting that excess ^{40}Ar is not just a minor nuisance but a widespread feature of HP to UHP rocks (Hammerschmidt *et al.*, 1995). However, $^{40}Ar/^{39}Ar$ ages that are consistent with independently derived radiometric and/or lithostratigraphic constraints can also be found in these and other HP-UHP provinces.

Why excess ^{40}Ar is recorded in some rocks, but not all, is not fully understood. This situation is clearly detrimental to the application of the $^{40}Ar/^{39}Ar$ (or K/Ar) technique in cases where there is no unequivocal evidence for excess ^{40}Ar, or when conflicting data from other radiometric techniques are fraught with problems of inheritance, retention, and/or isotopic disequilibrium that are equally difficult to interpret (e.g., Tilton *et al.*, 1989, 1991, 1997; Thöni and Jagoutz, 1992; Ames *et al.*, 1996). Inverse $^{36}Ar/^{40}Ar$ vs. $^{39}Ar/^{40}Ar$ isochron plots (Heizler and Harrison, 1988) are of little help in resolving the effects of excess ^{40}Ar in samples where the radiogenic yield is so high that it precludes the use of the non-radiogenic ^{36}Ar isotope as a tracer of the excess argon contaminant via the $^{36}Ar/^{40}Ar$ ratio (e.g., Hacker and Wang, 1995; Scaillet, 1996; Boundy *et al.*, 1997). A characteristic feature of discordant $^{40}Ar/^{39}Ar$ phengite age spectra that has been repeatedly documented in HP-UHP settings is their concave-downward (hump) shape. The fact that this feature has been interpreted to reflect inherited argon and/or earlier, rejuvenated, retention ages (this chapter, Wijbrans and McDougall, 1986; Hammerschmidt and Frank, 1991; Scaillet *et al.*, 1992; Hannula and McWilliams, 1995), while $^{40}Ar/^{39}Ar$ spectra with excess ^{40}Ar can appear equally discordant (Scaillet *et al.*, 1992; Hacker and Wang, 1995; Scaillet, 1996; Boundy *et al.*, 1997) or flat (Li *et al.*, 1994) indicates that the $^{40}Ar/^{39}Ar$ spectrum shape cannot be unambiguously used to infer the presence of excess argon in UHP minerals. Thus, if the $^{40}Ar/^{39}Ar$ technique is to be applied with greater confidence to the geochronology of UHP metamorphic processes, it is essential that we develop more objective criteria to recognize the nature and origin of excess ^{40}Ar, based on a firm understanding of the argon isotopic behavior of HP-UHP rocks. Although $^{40}Ar/^{39}Ar$ data from UHP rocks are largely in a reconnaissance stage today, the recent application of micro-analytical $^{40}Ar/^{39}Ar$ laserprobe dating techniques has greatly expanded our understanding of the isotopic behavior of Ar-bearing minerals in HP settings (Scaillet *et al.*, 1992; Scaillet, 1996; Boundy *et al.*, 1997; Pickles *et al.*, 1997).

This chapter delineates and rationalizes the driving mechanisms and main compositional, structural, and field effects influential in the argon isotopic record of HP-UHP rocks. It will be shown that the argon isotopic behavior of UHP rocks can be understood with special reference to the prograde and peak-metamorphic evolution of these rocks. The closed behavior imposed by the reduced fluid availability during HP-UHP recrystallization can explain diverse origins and mechanisms of excess ^{40}Ar incorporation (inherited *vs.* internally recycled), with the resulting argon isotopic signature reflecting the fluid processes attending UHP metamorphism.

2. AR EXCHANGE AND TRANSPORT PHENOMENA IN HP-UHP METAMORPHISM

The closure temperature concept of Dodson (1973) provided the basis to a wealth of thermochronological studies in a wide variety of metamorphic contexts (see overviews by McDougall and Harrison, 1988; Zeitler, 1989; Hodges, 1991). In conjunction with the pressure-temperature (*P-T*) calibration of metamorphic HP-UHP trajectories, temperature-time data derived from different decay schemes and mineral/daughter pairs having distinct closure temperatures (*e.g.*, Sm/Nd on garnet, Rb/Sr on white mica, and $^{40}Ar/^{39}Ar$ on amphibole and micas) can be used to infer vertical translations and exhumation rates of deeply subducted crustal sections (e.g., Boundy *et al.*, 1996; Chavagnac and Jahn, 1996).

The concept of closure temperature has been applied, however, at the cost of some systematization of the closure behavior into a well-behaved sequence of mineral closure temperatures (*e.g.*, Sm/Nd garnet > Rb/Sr white mica > K/Ar white mica ≥ Rb/Sr biotite > K/Ar biotite). Significant deviations from any preconceived closure trend are not unexpected owing to grainsize and compositional-structural effects on diffusion behavior, even among a given mineral family (Lee, 1995; Dahl, 1996a; 1996b). Moreover, in the absence of relevant kinetic data for most isotopes at the metamorphic conditions of interest, it is uncertain whether and to what extent currently field-calibrated, medium *P/T* grade, closure sequences can be directly extrapolated to quite different HP-UHP field situations. Although the closure temperatures of Sm/Nd in garnet and crystallization U/Pb ages in refractory minerals like zircon are reasonably well constrained to be in excess of 600°C, thereby potentially giving peak UHP metamorphic ages (e.g., Chavagnac and Jahn, 1996, but see Erambert and Austrheim, 1993), the situation is more problematic for argon in UHP phases (notably phengite) for which a wide range in closure temperature is expected (see below and Scaillet *et al.*, 1992). Also commonly overlooked when applying the diffusion-based theory of cooling ages is that exchange phenomena of radiogenic isotopes in geologic systems depend on parameters other than diffusion including, *e.g.*, intracrystalline defects, stress, recrystallization, local fluid-rock exchange, inheritance and isotopic disequilibrium.

This section first reviews the basic mechanisms and structural-physical aspects bearing on argon transport phenomena in HP-UHP systems, and addresses the importance of crystal-chemical effects on the closure behavior in the minerals most commonly analyzed by the $^{40}Ar/^{39}Ar$ technique in HP-UHP contexts, the di- and trioctahedral sheet silicates, biotite, phlogopite, and phengite. Other important phenomena of argon partitioning and mixing behavior are then explored in connection with their possible relevance for the transport and incorporation mechanisms of excess ^{40}Ar at UHP. The specific implications of these different processes for the argon isotopic record of HP-UHP rocks are further discussed in a subsequent section in terms of *P-T-t* evolution, with special emphasis placed on the kinetics and fluid evolution during subduction of the crust to mantle depths.

2.1 Closure Behavior, Transport Rates and Lengthscales of Ar Isotope Exchange in HP-UHP Rocks

Argon transport in metamorphic rocks can proceed via four principal mechanisms (Fig. 1): 1) solid-state volume (lattice) diffusion within crystals (*VD*); 2) diffusion along wet or dry grain boundaries ($GBD_{wet/dry}$); 3) solute diffusion (or aqueous diffusion) in an intergranular fluid phase (*SD*); and 4) solute advection (grain-edge fluid flow) if the rock locally contains fluids or volatiles released *in situ* by devolatilization reactions. In rocks undergoing crystal-plastic flow and dynamic recrystallization, fluids and solutes (± argon) can also be involved in local, cm-scale fluid-phase diffusional creep or solution-transfer creep (Stocker and Ashby, 1973; Philippot and van Roermund, 1992; Nakashima, 1995; Shimizu, 1995), a form of pressure-solution operating via dissolution at mineral interfaces under high stress and solute transfer at low-stress interfaces driven by local gradients in chemical potential. These different transport mechanisms vary in efficiency over several orders of magnitude: fluid flow $\sim 10^{-3}$–10^{1} m/yr >> $SD_{aqueous} \sim 10^{-4}$–$10^{-5}$ cm^2/s >> $GBD_{wet} \sim 10^{-8}$-10^{-13} cm^2/s > $GBD_{dry} \sim 10^{-14}$–10^{-21} cm^2/s ≥ $VD \sim 10^{-15}$–10^{-22} cm^2/s (Walther and Wood, 1984; Rubie, 1986). By way of illustration, these values span a range of mean travel distance $x \sim (Dt)^{1/2}$ of 10^7-10^6 cm ($SD_{aqueous}$) >> 10^4–10^{-1} cm (GBD_{wet}) > 10^1–10^{-3}

cm (GBD_{dry}) ~10^{1}-10^{-4} cm (*VD*) for a characteristic 10^7 yr duration. Accordingly, the lengthscale of exchange phenomena and the closed *vs*. open behavior of the Ar-mineral-rock system will be dramatically influenced by the presence or absence of fluids.

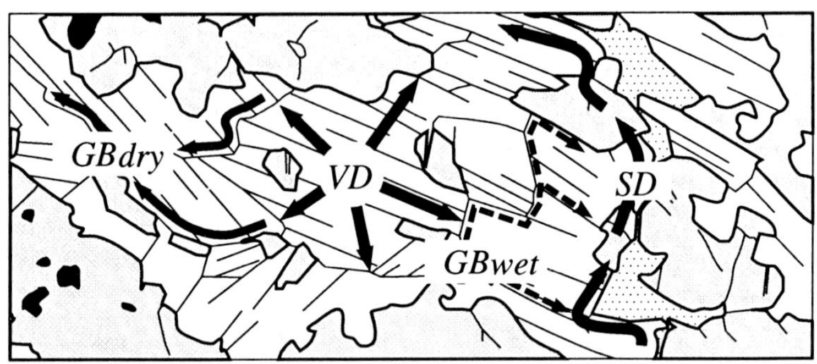

Figure 1. Idealized sketch summarizing main transport phenomena in metamorphic rocks. Abbreviations: *VD*, volume diffusion; *GBwet/dry*, grain-boundary diffusion along wet/dry mineral interfaces; *SD*, solute diffusion in a fluid phase, here shown as filling a microfracture (dotted pattern).

Petrologic and field-based studies indicate that HP-UHP metamorphism is nominally anhydrous (fluid-absent in the thermodynamic sense, with $a_{H_2O} < 1$) in most cases for a wide range of mafic and felsic lithologies, a low water activity being one of the prerequisites for the incomplete reversion of coesite to quartz (Hacker and Peacock, 1994; Liou and Zhang, 1996; Mosenfelder and Bohlen, 1997). Although aqueous and carbonic fluids are common participants in fluid-mineral UHP equilibria (e.g., Massonne, 1995), they often occur on a local scale and in restricted amounts equivalent to the stoichiometric content of bound water and volatiles in high-grade hydrous silicates (e.g., Schertl *et al.*, 1991; Sharp *et al.*, 1993; Philippot *et al.*, 1995). There is now extensive petrographic, fluid-inclusion, Nd-Sr, and C-O-H stable-isotopic evidence to suggest that the lengthscale of fluid transport attending HP-UHP metamorphic transformations does not extend greatly beyond the outcrop scale or the scale of lithologic layering (e.g., Rubie, 1986; Philippot and Selverstone, 1991; Klemd *et al.*, 1992; Nadeau *et al.*, 1993; Philippot *et al.*, 1995), even in cases of deformation-assisted fluid transport (Philippot and van Roermund, 1992; Philippot, 1993; Früh-Green, 1994). The transport efficiency of argon is commensurate with this restricted lengthscale of solute and/or diffusion-driven transfer, and depends greatly on the local fluid availability in the grain boundary network (Mattey *et al.*, 1994; Scaillet, 1996; Boundy *et al.*, 1997). In the absence of a free fluid phase, or any adsorbed monomolecular H_2O layer in the disordered structure of grain boundaries, dry grain-boundary diffusion is too slow (Rubie, 1986) to provide a diffusion pathway faster than lattice diffusion at temperatures common in HP-

UHP settings (500-800°C), making the rocks virtually closed on a sample scale even over timescales of 10^6–10^7 yr (Rubie and Thompson, 1985; Joesten, 1991). Given these constraints, the argon isotopic record of a rock or mineral might actually not reflect the temperature and the time at which it closed to argon loss or gain by diffusion, but instead, the extent to which (i) it has isotopically exchanged with the intergranular fluid phase (if any), and (ii) the mobility and isotopic composition of this fluid phase during the exchange. Clearly speaking, a bulk-closed behavior will result if there are little or no fluids to assist in driving off the argon evolved (or inherited) at prograde or peak-metamorphic conditions.

One first important consequence of the bulk-closed behavior imposed by the absence of a free fluid phase is that it may result in violation of the zero argon grain-boundary concentration (or infinite dilution boundary condition) implicitly assumed when applying Dodson (1973) and Turner (1968) closure/loss models. Figure 2 illustrates the argon radial loss profiles, and the resulting model $^{40}Ar/^{39}Ar$ release spectra, calculated for grains equilibrating in a closed system with a static grain-boundary fluid with an initial atmospheric ratio and fluid/solid Ar partition factor ζ of 0.2 (see Albarède (1976) and caption of Fig. 2 for more details on this parameter). By considering the transient evolution of Ar within the crystal and the isotopic exchange equilibrium between mineral and fluid, this simple model predicts that completely equilibrated (*i.e.* 'open') minerals can produce flat age spectra and homogeneous intracrystalline $^{40}Ar/^{39}Ar$ distributions bearing no geologic relevance to either the original closure age of the sample (t_1 in Fig. 2) or that of the re-equilibration event (t_2). This idealized example of 'self-contamination' may have broad application. A concrete illustration is provided by granulites found metastably preserved as tectonic lenses in the eclogitized pre-Alpine basement from the Sesia Zone at Alpe Toso, western Alps (Droop *et al.*, 1990; Lardeaux and Spalla, 1991). These rocks evolved mostly as a 'dry' closed system throughout the Alpine HP metamorphism (1.4–1.6 GPa; 550°C), locally maintaining textural and small-scale (domainal) equilibrium of the original assemblage orthopyroxene + garnet + plagioclase + brown amphibole + biotite ± ilmenite ± quartz, with incipient breakdown of plagioclase into very fine-grained zoisite + kyanite ± quartz ± jadeite pseudomorphs and the development of pale-green reaction rims on amphibole (Ungaretti *et al.*, 1983). Laser $^{40}Ar/^{39}Ar$ age mapping on two of these pre-Alpine biotites revealed homogeneous apparent ages around 190 Ma (Fig. 3), midway between the Late Variscan age of the granulite metamorphism in this portion of the European continental crust (280-320 Ma, Pin and Vielzeuf, 1983; Brodie *et al.*, 1989) and the 70-120 Ma eclogite-facies event in the Sesia Zone (Stöckhert *et al.*, 1986; Inger *et al.*, 1996; Reddy *et al.*, 1996). Such a homogenous intracrystalline isotopic distribution would likely produce flat $^{40}Ar/^{39}Ar$ plateau via bulk degassing techniques, with no apparent age indication of the Alpine HP overprint even though the eclogite facies temperatures were as much as 250°C in excess of the commonly quoted temperature for argon retention in biotite (300-350°C).

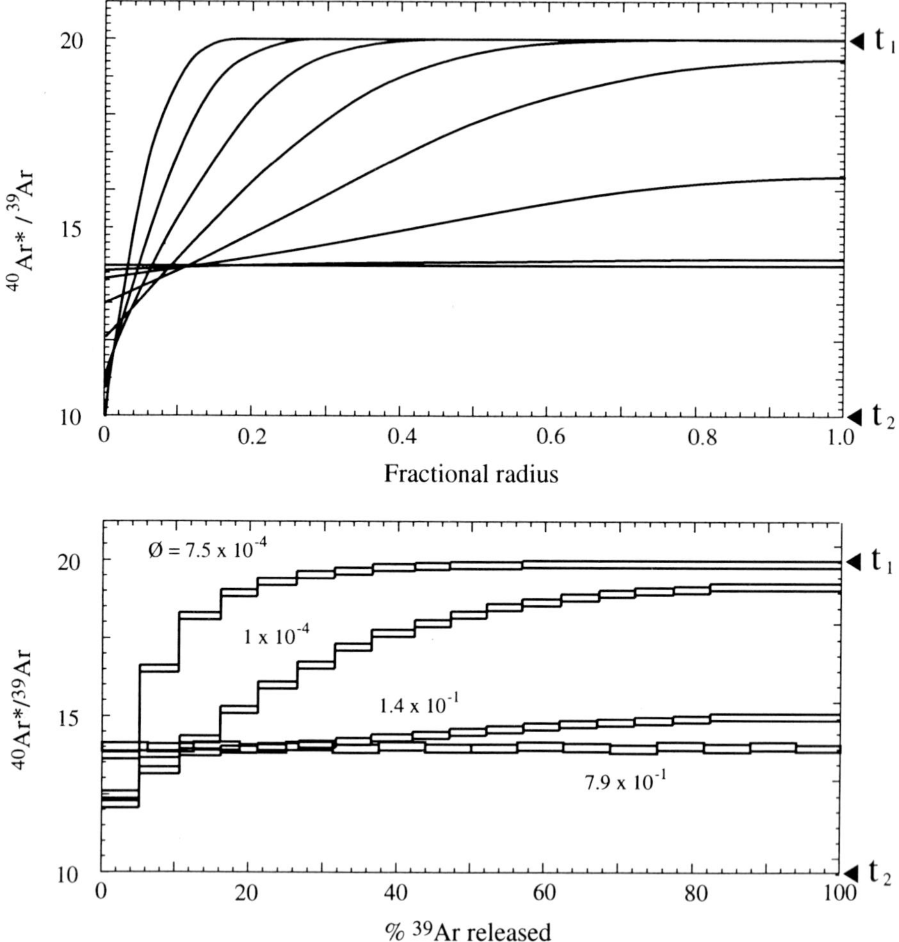

Figure 2. Model of closed system fluid-mineral isotopic exchange by lattice diffusion (from Albarède, 1976). *Top*: $^{40}Ar^{*}/^{39}Ar$ vs. fractional radius loss/equilibration profiles calculated for spherical grains of radius a and age t_1 for different magnitudes $\theta = Dt/a^2$ (D, diffusivity) of isothermal exchange at time t_2. Note the progressive relaxation of the radial $^{40}Ar^{*}$ concentration gradients toward a flat equilibrium profile midway between t_1 ($^{40}Ar^{*}/^{39}Ar = 20$) and t_2 ($^{40}Ar^{*}/^{39}Ar = 10$). The exchange equilibrium coefficient, ζ, between solid and fluid concentration is: $3K_{solid/fluid}(V/V_{porosity}) = 2$, where $K_{solid/fluid}$ ~0.017 is the solubility constant for Ar between solid and fluid, $V_{porosity}$, the porosity of the rock (0.2 vol%), and V, the volume of the grains ($4/3\pi a^3$, $a = 1.6$ μm). *Bottom*: model $^{40}Ar^{*}/^{39}Ar$ stepwise release spectra calculated for some relevant equilibration profiles at increasing θ up to 7.9 x 10^{-1}, where a plateau is obtained that lies well above the age t_2 of the equilibration event. Note that if $\zeta >> 1$, the system remains perfectly closed, yielding a plateau at t_1 for any value of θ.

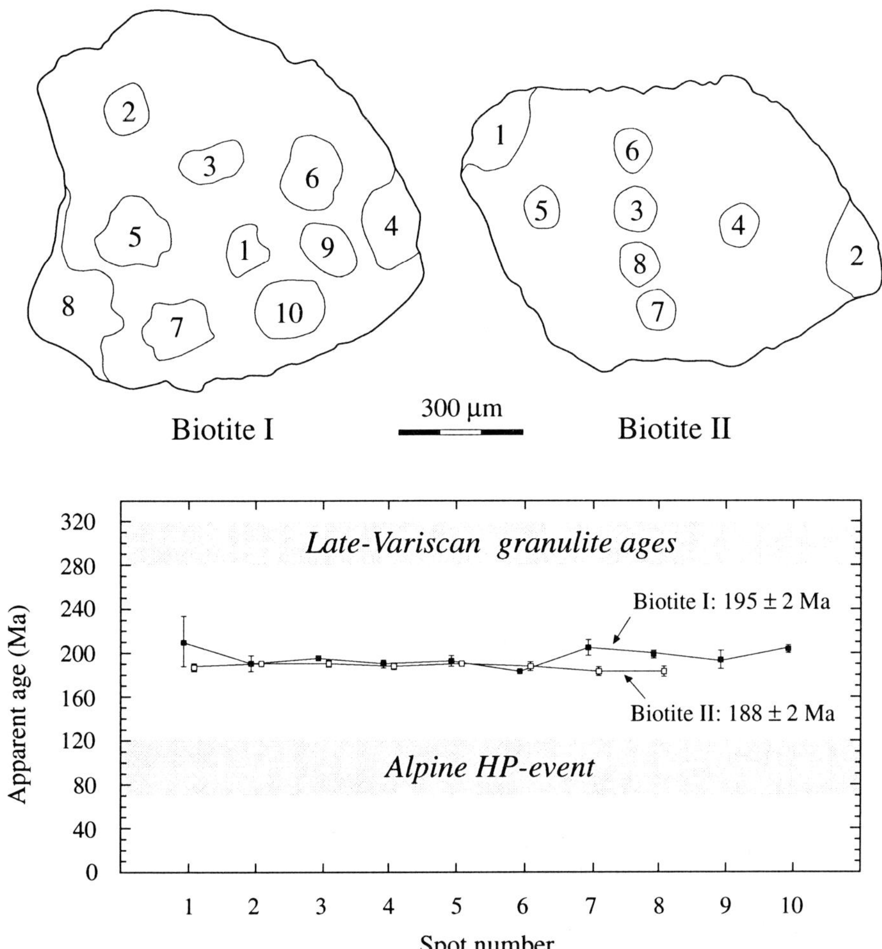

Figure 3. $^{40}Ar/^{39}Ar$ laserprobe spot-ages of pre-Alpine biotites from basic granulites eclogitized during the Eo-Alpine HP event in the Sesia Zone at Alpe Toso, western Alps (Droop *et al.*, 1990). Note the homogeneous age distribution inside both grains at around ~190 Ma, roughly midway between the Late-Variscan granulite ages in this portion of the western European continental crust and the age of the HP event (1.4–1.6 GPa; 550°C). Mean integrated dates indicated for each grains. These data indicate that the granulite biotites survived the HP overprint without complete resetting by maintaining an equilibrium argon edge concentration with the grain boundary fluid. Compare with the equilibrium profile of Fig. 1.

This field-based example of sample-scale closed behavior can be idealized a step further to imagine that a high-K mineral growing in a 'dry' rock with a zero argon partial pressure (*i.e.*, with $\zeta >> 1$) will retain all its K-derived ^{40}Ar upon progressive radiogenic buildup, thereby preserving its original growth age *regardless of its temperature of crystallization and its nominal Dodsonian closure temperature upon cooling*. Accordingly, regardless of grainsize and compositional variations, grain-to-grain differences in crystallization ages could result on a sample scale that would be impossible to explain via the closure temperature concept. Thus, great restraints should be imposed on the use of nominal isotopic closure temperatures unless it can be shown that the mineral-rock system was indeed open to large-scale fluid-rock exchange and fluid flow during exhumation and cooling (see also Villa and Puxeddu, 1994).

A further shortcoming inherent to using nominal closure temperatures is that they neglect possible differences in argon diffusion rates imposed by variations in crystal chemistry and cell parameters. Figure 4 shows the dramatic $^{40}Ar/^{39}Ar$ age difference recorded by high-Si HP phengites from two adjacent metapelitic and gneissic layers in the northern Dora Maira eclogitic nappe, western Alps (1.4–2.5 GPa, T = 550 ± 50°C, Scaillet *et al.*, 1992). These samples share basically the same *P-T* history and grainsize; the age difference can be uniquely equated with their variable Mg/Fe ratio imposed by the bulk rock composition. Scaillet *et al.* (1992) argued that the age difference reflects contrasting argon retention properties arising from structural modifications of the interlayer mica sheet structure imposed by their variable Fe/Mg octahedral ratio. This interpretation has been given compelling crystal-chemical justification by Dahl (1996a), who showed that a number of other isomorphous substitutions, including notably $(OH)F_{-1}$, are potentially capable of lowering argon diffusion (thus increasing closure *T*) in Mg- and F-substituted micas via an increase in ionic packing density and a shortening of the cross-interlayer ⟨K-O⟩ bondlengths.

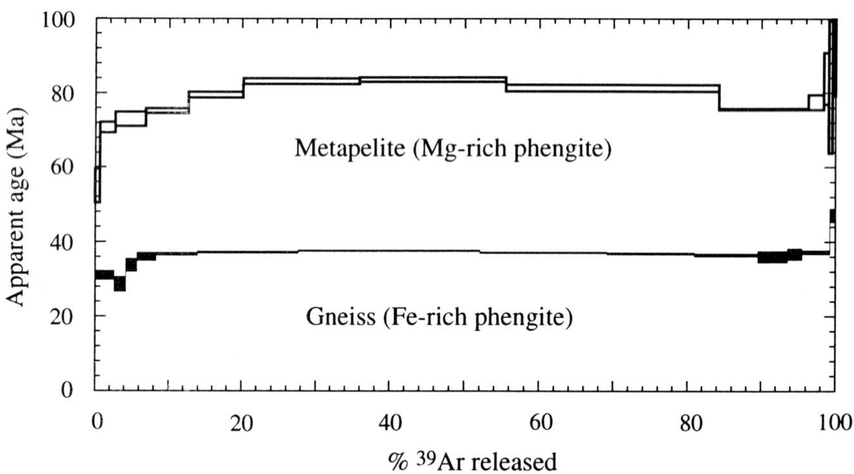

Figure 4. Gneiss-metapelite HP phengite $^{40}Ar/^{39}Ar$ age spectra from the eclogitic northern Dora Maira nappe, western Alps (redrawn from Scaillet *et al.*, 1992). The samples come from adjacent lithologies in the same outcrop. Note the huge age difference illustrating the Fe/Mg compositional control on the isotopic composition of the phengites.

The results of Scaillet *et al.* (1992) allow two opposite interpretations (*i.e.*, variable contamination with excess ^{40}Ar vs. true differences in closure *T*), and have not been fully duplicated so far outside the Dora Maira (Monié and Chopin, 1991) and Monte Rosa (Chopin and Monié, 1984) HP-UHP nappes in the western Alps. Scaillet *et al.* (1992) speculated on different *P-T-t* trajectories to quantitatively explain the great difference (≥ 60 m.y.) in closure age recorded between their Fe- and Mg-phengite endmembers. They concluded that the spread could be accounted for by a two-step evolution involving a separate greenschist-facies thermal overprint after the peak HP event to selectively reset the less-retentive Fe-rich phengites at 30-40 Ma. Another possibility they considered was that the Mg-rich phengites had closure *T* in excess of 500 ± 50°C, and that their age gradients could reflect partial resetting of earlier 120-130 Ma crystallization ages produced via solid-state cationic re-equilibration during decompression. Alternatively, the age discordance could reflect incorporation of excess ^{40}Ar at high-*P* in adjacent lithologies, with the more retentive Mg-phengites retaining preferentially this common initial component upon cooling and subsequent removal of the high ambient argon pressure (Dahl, 1996a). All three explanations are equally probable given the persistent uncertainty surrounding the age of the HP-UHP event in the Dora Maira nappe (Tilton *et al.*, 1989, 1991, 1997; Gebauer *et al.*, 1997). Regardless of which is correct, the interpretations collectively and unambiguously argue that mineral and bulk rock composition strongly influence the argon isotopic record of HP phengites. That similar age patterns have not been documented elsewhere may simply reflect the singular *P-T-t* trajectory of the Dora Maira nappe in the critical *T* range for argon retention in the phengite group that probably forced dramatic enhancement of the closure effect during exhumation and overprinting (Scaillet *et al.*, 1992; Scaillet, 1996). Alternatively, the $Fe(Mg)_{-1}$ effect may have been blurred beyond analytical uncertainties in other HP-UHP occurrences by other competing factors (see below), or by significantly different retrograde *P-T* histories.

Argon retentivity may also be affected by a pressure effect on argon diffusion rates. In contrast with earlier hydrothermal work dismissing a pressure control on argon diffusion in phlogopite up to 1.5 GPa water pressure (Giletti and Tullis, 1977), Harrison *et al.* (1985) found that argon diffusivity in biotite at 700°C and 0.1 GPa decreased by one order of magnitude at P_{H_2O} = 1.4 GPa. From these data they determined an activation volume $(\Delta V)_{700°C}$ of 1.397 J bar^{-1} mol^{-1} and concluded that pressure should not enhance the Ar-biotite closure temperature significantly above 350-400°C for normal crustal geotherms. This cannot be true in UHP rocks that are subducted along steeply depressed geotherms in the range of 7-8°C/km or even less (e.g., Cyclades: 5°C/km, Massonne, 1995), and along which the thermally induced increase in diffusivity cannot counteract the inhibiting pressure effect as efficiently as in lower *P/T* metamorphic gradients. Unfortunately, there are no hydrothermal data that constrain argon diffusion in the *P-T* range of interest for UHP rocks. Lister and Baldwin (1995) proposed using the activation volume of Harrison *et al.* (1985) to account for an increase in argon retentivity (and closure temperature) with pressure. This is probably ill-advised because Harrison *et al.*'s activation volume pertains to their biotite at 700°C only; the pressure and temperature dependence of the activation volume for argon diffusion in this (and other) mineral(s) is probably sufficiently complex to deserve specific experimental investigation before the pressure effect on closure *T* can be fully appreciated.

According to recent expansion and compressibility data (Guggenheim *et al.*, 1987; Comodi and Zanazzi, 1995), the mica interlayer cell volume (V_i) is expected to remain fairly constant along *P/T* gradients of ~15–20°C/km (Dahl, 1996a), which is again far greater than HP-UHP gradients of 5–8°C/km. Thus, UHP should produce significant contraction of the mica interlayer (the path of 'faster' diffusion in micas) that could significantly inhibit argon diffusion (enhance closure) at these conditions. This effect was pointed out by Dahl (1996a) who noted that synthetic high pressure phengite products (Comodi and Zanazzi, 1995) show contraction of the interlayer spacing (c^*) and reduction in interlayer ionic-porosity, $Z = 39.8\%$ (*vs.* 45.5% for common muscovite), a measure of the atomic packing density log-arithmetically correlated with diffusivity and 'hole' space in crystal and melt structures (Fortier and Giletti, 1989; Carroll and Stolper, 1993; 1996b; Dahl, 1996a). These structural changes are confirmed by cell refinements on natural muscovite compressed to 2 GPa (Catti *et al.*, 1994) and HP-UHP phengites synthesized in the KMASH system by Massonne and Schreyer (1986), who showed that substitution of Si for ^{IV}Al with *P* up to 2.5 GPa results in contraction of the basal spacing and unit cell volume, accompanied by a decrease in the monoclinic angle (β) with rising Mg content. The structural changes induced by the simultaneous tetrahedral (α) and octahedral (ω) rotations result in corrugation of the basal plane associated with a strong modification of the interlayer spacing, reflected in a marked reduction of the difference between the ⟨K-O⟩ bondlengths between the inner and outer coordination spheres of the basal oxygens (cf. also Catti *et al.*, 1994). Given the universal Si-rich character of HP-UHP white micas, these structural effects should be common. Dahl (1996a) predicted that the relatively shallow intersection of HP-UHP metamorphic paths with the *dP/dT* isochores of constant V_i could give rise to a wide array of effective closure ages for a given mica composition. The inferred pressure effect might well account for some of the $^{40}Ar/^{39}Ar$ age scatter commonly seen in HP-UHP suites, according to their specific *P-T-t* trajectories.

In concluding this section, it must be emphasized that there exists ample field and crystal-data evidence that argon diffusion kinetics (hence closure temperature, *if* and *where* it applies) are influenced by mineral composition, especially in mica and amphibole (Dahl, 1996a; 1996b). Because the ionic attractive-repulsive effects on cell parameters and lattice spacing are complex and dominantly controlled by isomorphous substitutions involving elements that are seldom analyzed in routine electron-microprobe analyses (*e.g.*, F, Cl), these compositional effects have probably passed largely undetected so far and cannot be rejected solely on the basis of apparent lack of $Fe(Mg)_{-1}$ control. There is also a need for more reliable diffusion and crystal-data to address the pressure effect on closure behavior in phengite via the reduction in interlayer spacing monitored through both the ionic-porosity model (Dahl, 1996a) and cell refinements of HP-UHP synthetic and natural analogues (Massonne and Schreyer, 1986; Catti *et al.*, 1994; Comodi and Zanazzi, 1995). Lastly, the coupled ^{40}Ar and (Al, Si, Na, Fe, Mg) cation zonation commonly documented in Si-rich HP phengites (Wijbrans and McDougall, 1986; Hammerschmidt and Frank, 1991; Scaillet *et al.*, 1992; Boundy *et al.*, 1997) suggests that argon transfer in HP-UHP micas is at least as sluggish as the kinetics of solid-state cationic exchange phenomena, which have blocking temperatures arguably in excess of 500°C (see "Isotopic resetting and phengite re-equilibration", below, and Dempster, 1992). The possibility that UHP mica closure temperatures may reach values well above what is commonly considered (300-400°C) cannot be discounted and merits full consideration.

2.2 Excess Argon Solution Behavior in HP-UHP Minerals

Another important aspect to consider is the partitioning of argon between fluid and minerals, and among different silicate structures. Owing to its inertness and large radius (1.58 Å), argon partitions preferentially into the less dense fluid phase. However, the rather modest enthalpy of solution of argon in common silicates (e.g., Harrison and McDougall, 1981) also implies that sizeable amounts of argon can enter and equilibrate with silicate solid structures (in addition to being occluded and stably trapped in lattice voids), as indicated by the wide range in excess concentrations found in common metamorphic minerals (see reviews in Dalrymple and Lanphere, 1969; McDougall and Harrison, 1988). Possible sites for the incorporation of excess argon in solids include (in addition to fluid inclusions): i) Shottky and Frenkel point defects (cation and anion vacancies); ii) extended defects (dislocations, stacking faults, antiphase boundaries) and cracks; iii) structural holes and channels into which alkali metal ions and noble gases can readily fit; and iv) reaction-driven microporosity along lamellar exsolution structures and tubular nanopores (mostly in feldspar and amphibole: e.g., Hacker and Christie, 1991; Smelik, 1995; Walker *et al.*, 1995). The tightly bound nature of the excess ^{40}Ar component in most high-grade and HP-UHP minerals points to trapping dominantly on lattice defects activated at high temperature (*i.e.*, anion and cation vacancies).

Argon solubility data in common micas are almost completely lacking (e.g., Onstott *et al.*, 1991). However, the experimental results of Karpinskaya *et al.*

(1961) and Karpinskiy *et al.* (1966) provide some clues to the solution behavior of argon in one mineral of relevance to the $^{40}Ar/^{39}Ar$ dating of HP-UHP rocks—muscovite. Their experiments conducted in the stability range of muscovite (checked against no observable loss of structural water upon completion of the runs), and pure argon atmosphere at confining P and T up to 0.45 GPa and 760°C, suggest that the solution of argon in muscovite obeys Henry's law with an enthalpy of solution (ΔH_{sol}) of 2.26 x 10^{-2} J/mol, and a pre-exponential term of 1.6 x 10^{-8} mole g^{-1} bar^{-1} (Albarède, 1976). Compared to ΔH_{sol} values of ~12 kJ/mol and ~2 kJ/mol determined from interatomic potential energy calculations for argon dissolution in the hydroxyl and alkali sites of muscovite (Albarède, 1976), these data suggest that extraneous argon is preferentially sited on hydroxyl vacancies in this muscovite. These experimental and theoretical determinations are of uncertain applicability, however, with regard to the large structural-compositional changes and the *T-O-T* structural ordering and 3*T* polytypism induced at HP-UHP conditions in phengitic white micas relative to pure muscovite (e.g., Massonne and Schreyer, 1986; 1987; Guidotti *et al.*, 1989; 1989; Amisano-Canesi *et al.*, 1994; Catti *et al.*, 1994; Faust and Knittle, 1994; Sassi *et al.*, 1994; Comodi and Zanazzi, 1995).

The argon solution behavior in HP and UHP melts and glasses is, in contrast, better known. The argon solubility in basaltic, rhyolitic, orthoclase, and albite HP (and UHP) melts and silica glasses has been shown to be extremely composition dependent, with Ar and Xe solubility favored in felsic melts relative to more mafic compositions because of polymerization of the melt structure induced by increasing Si/Al ratio and a coordination change of ^{IV}Al (White *et al.*, 1989; Montana *et al.*, 1993). Moreover, the argon solubility in these melts and glass structures correlates linearly with ionic porosity (Carroll and Stolper, 1991; 1993). Importantly, although the high pressure argon solubility data in the investigated melts display Henry's law (*i.e.*, linear) dependence on pressure up to 2.5 GPa (White *et al.*, 1989; Montana *et al.*, 1993), recent data on UHP silica and anorthite melts reveal a saturation behavior for argon in $CaAl_2Si_2O_8$ melts, and even a pronounced drop in argon solubility in SiO_2 melts above 5 GPa (Chamorro-Perez *et al.*, 1996). This marked change, documented in both melt compositions at the same pressure, is thought to correlate with an increase in silicon coordination with pressure accompanied by mechanical closure of the six-membered SiO_4 rings above 5 GPa. This structural collapse, possibly associated with a shift toward a more compact, 'ordered' 4– and 3–membered ring structure, is attested to by the 3–fold reduction in compressibility of the melt above 5 GPa (e.g., Chamorro-Perez *et al.*, 1996), and is paralleled by a decrease in ionic diffusivity in silica melts at densities above 3–3.5 g/cm^3 (Kubicki and Lasaga, 1991).

The compositional-structural and pressure effects documented for the argon solution in felsic and mafic HP to UHP melts could be even more pronounced in more tightly close-packed mineral structures given the contraction in lattice cell parameters of dense HP and UHP silicates at mantle pressures (e.g., Massonne and Schreyer, 1986; Faust and Knittle, 1994; Comodi and Zanazzi, 1995 and previous section). Experimental data are badly needed for quantitative comparison of the solution behavior of argon at UHP conditions among different silicate structures, or even among a given silicate family (*e.g.*, the mica group). This is unfortunate because regional and/or local $^{40}Ar/^{39}Ar$ age discordances occasionally seen within certain sample suites are sometimes taken to imply the presence of excess ^{40}Ar without consideration of this effect. The ionic-porosity model for the solubility of rare gas in melts (Carroll and Stolper, 1993), if extended to solids, predicts the following trend in Ar solubility within the mica group (based on Dahl's (1996a) ionic porosities, *Z*): muscovite/phengite ($Z =$ 45.5–46%) $\leq$ F-phlogopite (45.8%) < eastonite (47.6%) < phlogopite (47.6%) < annite (49.1%). This trend is qualitatively consistent with the empirical observation that 'loose', open, silicate structures (*e.g.*, cordierite, $Z = 48\%$) can accommodate larger quantities of extraneous argon than more tightly packed ionic crystals (Smith and Schreyer, 1962). If the solubility and kinetic behavior of argon in solids can be collectively equated via their common dependence on chemical-structural effects as empirically monitored through the ionic-porosity model, then muscovite should be more 'impervious' to extraneous ^{40}Ar contamination than common biotite (45.8% vs. 47-49%). Interestingly, $^{40}Ar/^{39}Ar$ data on ^{IV}Al-deficient Alpine biotites and HP-phengites from the Sesia Zone, western Alps, reveal the following age sequence (Ruffet *et al.*, 1995): 140 $Ma_{biotite}$ > 70 $Ma_{phengite}$, at odds with the commonly assumed progression: T_{bio} < T_{phe}; from this it was concluded that excess ^{40}Ar was responsible for the observed reverse discordance. Thus, it would appear that even in cases of thorough excess contamination of the coexisting (high-K) biotite, HP phengite is rather immune to ^{40}Ar incorporation and fully capable of retaining meaningful $^{40}Ar/^{39}Ar$ ages (e.g., compare with the 70 Ma U/Pb, Rb/Sr, and $^{40}Ar/^{39}Ar$ data of Inger *et al.*, 1996).

Preliminary $^{40}Ar/^{39}Ar$ data of Hammerschmidt *et al.* (1995) from the coesite-bearing pyrope-quartzite of the southern Dora Maira nappe (3.7 GPa, 800°C) also bear on argon partitioning among high pressure micas. Although phlogopite at this locality is retrogressive after pyrope, and postdates primary phengite (cf. Chopin, 1984; Schertl *et al.*, 1991), the phlogopite is far older (~ 1 Ga, hence heavily contaminated) than the UHP Mg-phengite (104 Ma). Again, despite having been exposed to a post-crystallization high external argon partial pressure that raised the apparent age of the (K-rich) phlogopite to impossibly high values, the phengite was not—or only minimally influenced by excess argon (Scaillet, 1996). These data illustrate that a UHP phase like phengite need not necessarily act as preferential sink for argon as sometimes suggested (e.g., Li *et al.*, 1994; Arnaud and Kelley, 1995; Ruffet *et al.*, 1995; Inger *et al.*, 1996; Reddy *et al.*, 1996). Likewise, age differences taken to reflect the variability in ^{40}Ar

contamination in a given mineral suite can be the product of temporal variations in initial argon coupled with variable retention characteristics among the coexisting phases (*e.g.*, the sequence: phengite (610-560 Ma) > biotite I (420-150 Ma) > biotite II (90-15 Ma) ≥ K-feldspar (85–0!! Ma) documented by Arnaud and Kelley (1995) in the statically overprinted Brossasco UHP metagranite, southern Dora Maira nappe, 700°C, 3 GPa). Such sequences obviously should not be confused with possible Ar/mineral affinity relationships to infer trends in Ar solubility behavior (in this instance with phengite as the main excess-catcher phase).

These few, but significant, examples illustrate the danger in assuming well-behaved trends in mineral closure and/or Ar uptake behavior to infer the pathology of excess argon without considering crystal-chemical and other more context-specific effects (*e.g.,* the argon isotopic inheritance just mentioned in the pre-Alpine Brossasco UHP metagranite). The solubility and partitioning of argon in UHP is not quantified and deserve detailed experimental investigation. As with diffusion, laboratory compressibility and structural crystal-data (e.g., Catti *et al.*, 1994; Faust and Knittle, 1994; Comodi and Zanazzi, 1995) could help in assessing (and correcting for) the effect of pressure on the kinetic/solubility argon behavior of dense UHP silicates via the ionic-porosity model. This is especially critical given the strong reduction in interlayer spacing c^* and the associated reduction in ionic-porosity noted above in synthetic UHP phengite run products.

A final matter of concern is whether the pressure release of stored elastic strain energy could affect the argon solution behavior in metastable UHP phases during unloading. $^{40}Ar/^{39}Ar$ laserprobe analyses of shielded and 'quenched' UHP solid-phase inclusions, such as those preserved in UHP pyrope crystals (including both cœsite and phengite, Schertl *et al.*, 1991; Philippot *et al.*, 1995), are desired to assess whether and how much excess ^{40}Ar, if any, can be trapped and retained in such inclusions. Lastly, the kinetics of argon contamination (uptake and solubility) should receive more attention to address the mechanisms of argon-fluid-solid exchange equilibria at pressures in excess of 2.5–3.0 GPa. Sorption (physisorption) of argon driven by excess surface-free energy of crystals growing in a high ambient argon pressure certainly has lower barrier potentials (and different rate constants) than post-crystallization trapping of argon by solid-state diffusion from the external fluid, a possible explanation for the contrasting argon behavior exhibited by the UHP Mg-phengite and phlogopite mentioned above (see "Excess argon as a UHP-fluid-melt tracer isotope", below, and Hammerschmidt *et al.*, 1995).

The main conclusion at this point is that the inherently lower ionic-porosity and reduced interlayer spacing of HP-UHP phengites are likely to favor this mineral over biotite and/or phlogopite in terms of radiogenic argon retention and 'shielding' against excess contamination. These apparent solubility relationships imply that the loosely packed biotite structure could provide, at sufficiently high modal abundances, an efficient buffer against an increase in argon pore-pressure in two-mica HP-UHP assemblages, thereby enhancing the scope for $^{40}Ar/^{39}Ar$ dating of phengite in such rocks; metamorphic biotite has long been known to readily accommodate excess argon, *unlike coexisting white mica* (e.g., Brewer, 1969; Roddick *et al.*, 1980; Dallmeyer and Gee, 1986). Proved occurrences of excess argon in phengites almost invariably concern rocks lacking biotite (e.g., Li *et al.*, 1994; Scaillet, 1996; Boundy *et al.*, 1997), or represent typically a case of protolith inheritance in which the high local argon pore-pressure might have outpaced the biotite buffering capacity (e.g., Arnaud and Kelley, 1995; Hannula and McWilliams, 1995; Tilton *et al.*, 1997).

2.3 Ar Mixing Behavior and Mobility in HP-UHP Metamorphic Fluids

Metamorphic fluids are frequently invoked as efficient agents to drive-off, advect, and carry the free grain-boundary ^{40}Ar in metamorphic rocks, and the $^{40}Ar/^{39}Ar$ literature is filled with claims of fluid-assisted argon transport phenomena—most of which exclude consideration of whether such a fluid phase is present in the petrologic and fluid inclusion record (e.g., Cumbest *et al.*, 1994). Although they represent only a minor component of high-grade HP-UHP rocks (generally less than 2.0-0.5% of bound volatiles), observational evidence of a free fluid phase in mafic and felsic HP-UHP rocks is common in the form of fluid inclusions and hydrous HP-silicate-bearing veins (*e.g.*, phengite-omphacite veins). Eclogite and UHP-facies veins displaying equilibrium mineralogies similar to the enclosing rocks commonly are segregations produced by ponding of non-wetting fluids and are different from late-stage metasomatic infiltration veins that cut across metamorphic layering. These veins commonly retain outcrop-scale heterogeneities in fluid-inclusion and C-O-H isotopic compositions, implying restricted fluid transport (Burg and Philippot, 1991; Philippot and Selverstone, 1991; Nadeau *et al.*, 1993; Philippot, 1993). The fluids generally range from moderately to strongly saline aqueous brines (e.g., Philippot, 1993; Philippot *et al.*, 1995), and less frequently include CO_2, CO_2 + N_2, and H_2O immiscible high-density mixtures (Klemd *et al.*, 1992; Andersen *et al.*, 1993). The common high salinity of HP-UHP aqueous fluids reported from eclogites and UHP rocks (with up to 84 wt% equivalent NaCl) raises the question of Ar solubility in such fluids. The presence of dissolved electrolytes, such as NaCl, is known to decrease the solubility of argon in water at near-surface conditions in the presence of vapor (the 'salting-out' effect, Smith and Kennedy, 1983), and preliminary data of Harrison *et al.* (1994) suggest that there may be a saturation value of about $^{40}Ar/Cl \sim 10^{-3}$ in mid-crustal aqueous brines.

Similar behavior at HP-UHP conditions could substantially affect the argon mixing and transport properties in common supercritical aqueous brines. Non-ideal mixing increases with increasing *P* and decreasing *T* for a particular mixture. Neutral molecular species with similar properties (*e.g.*, CO_2, N_2, and Ar) mix nearly ideally over a wide range of *P-T-X* conditions, while those differing in their electrokinetic properties, *e.g.*, H_2O-CO_2, H_2O-CH_4 (and H_2O-Ar), form mixtures that can depart significantly from ideality (Ferry and Burt, 1982), especially at low H_2O molar fractions in H_2O-Ar mixtures (Walther, 1992). How the dissociation of fluids at very high *P* can affect excess mixing properties of H_2O-Ar (± NaCl) or more complex brines (by forming charged ions and changing electrostatic interactions among charged and/or polarized molecules) remains uncertain, however, in terms of solvus and unmixing relationships (Belonoshko and Saxena, 1992). Argon, with a zero dielectric constant and low polarizability, behaves much like neutral molecular CO_2, as opposed to H_2O, whose properties depend strongly on electrostatic intermolecular potentials (Saxena and Fei, 1987). Fluid-mineral equilibria in 2 GPa mafic eclogites of the Tauern Window, Austria, led Selverstone *et al.* (1991) to conclude that addition of dissolved salts in complex brines at increasing *P* expands the region of H_2O-CO_2 immiscibility, making it possible that eclogite-facies metamorphism (and UHP metamorphism, Philippot *et al.*, 1995) occurs predominantly in the two-phase field. They further demonstrated that the different wetting characteristics between CO_2 and salty/briny aqueous fluids could act against pervasive fluid flow due to the absence of long-range interconnectivity of these fluids and their segregation as H_2O-NaCl and CO_2 (± Ar?) immiscible pairs.

The well-documented property of CO_2 to form non-wetting fluids raises the possibility that the argon could, at high concentrations, behave in a similar fashion by forming isolated pockets at grain-edge triple-junctions, or at least by modifying the fluid-solid interfacial free energies of H_2O-Ar mixtures and Ar-bearing aqueous brines to significantly increase their wetting angle, an effect apparently supported by ancillary data on wetting angle characteristics for pure Ar and H_2O-Ar mixtures in hot-pressed monomineralic aggregates (Holness, 1993; Holness and Graham, 1995). The data are too sparse to speculate further, but coupled with their reduced mobility, these characteristics would make aqueous brines and carbonic fluids poor conveyers of dissolved argon over a significant lengthscale at HP-UHP conditions (Scaillet, 1996).

A consequence of this complex fluid-mixing behavior is that the small-scale heterogeneous distribution and segregation of HP-UHP metasomatic fluids is likely to be mirrored by complex excess ^{40}Ar spatial patterns at the mineral scale, making it difficult—or requiring particular conditions of inheritance, in situ fluid generation, or local fluid flow—to achieve isotopic equilibrium (e.g., Früh-Green, 1994). In this particular regard, it is significant that reported occurrences of contamination by (proved) excess ^{40}Ar invariably reflect grain-scale isotopic disequilibrium in phengites (Scaillet *et al.*, 1992; Arnaud and Kelley, 1995; Scaillet, 1996; Boundy *et al.*, 1997). The only exception is the homogeneous 104–180 Ma HP phengites of Ruffet *et al.* (1995) from the eclogitized Monte Mucrone metagranitoid in the Sesia Zone. Unfortunately, the petrographic setting of these particular samples was not reported by the authors, precluding an evaluation of their data in connection with the HP deformation and fluid processes described in these rocks (Früh-Green, 1994). Fluid-induced reaction hardening due to localized fluid release and HP deformation in these rocks is known to have enhanced diffusive mass transfer, resulting in a relative homogenization of phase compositions, oxygen isotope exchange and the formation of coarse aggregates of jadeite + garnet + quartz + phengite. This particular case aside, the potentially complex fluid-mixing and reduced transport properties of Ar-bearing HP-UHP fluids are expected to enhance and complicate the bulk closed system behavior of HP and UHP rocks already imposed by sluggish grain-boundary and solid-state argon kinetics, *even in cases where fluids are locally present along grain boundaries* (Boundy *et al.*, 1997).

3. FLUID RECYCLING, FLUID-ROCK INTERACTIONS, AND THE AR ISOTOPIC RECORD OF HP-UHP ROCKS

Based on the different aspects of argon transport and exchange phenomena described above, the argon isotopic record of UHP rocks can be understood with special reference to the prograde *P-T-t* path of these rocks and the fluid fate during subduction. Figure 5 shows two prograde, steady-state, underthrusting paths calculated for the top of the downgoing slab at convergence rates of 3 and 10 cm/yr (see other parameters in the caption, and Peacock, 1990; 1991). Also shown are peak-metamorphic *P-T* estimates from various blueschist-eclogite facies and coesite-bearing units in the western and central Alps. Although these prograde *P-T-t* paths strictly apply only to the top of the subducted oceanic crust, and thus may not accurately reflect the hotter trajectory of rocks belonging to the former continental margin, they provide a valid depiction of the subduction-related origin of HP-UHP oceanic and continental rocks without requiring inordinate crustal thickening (Peacock, 1990; 1991). The *P-T* array displayed by the HP-UHP assemblages recorded throughout the Alps, although they are not all of the same age, suggests a thermal structure similar to that achieved by subduction of ≥ 50 Ma old, cold lithosphere (cf. Philippot, 1993).

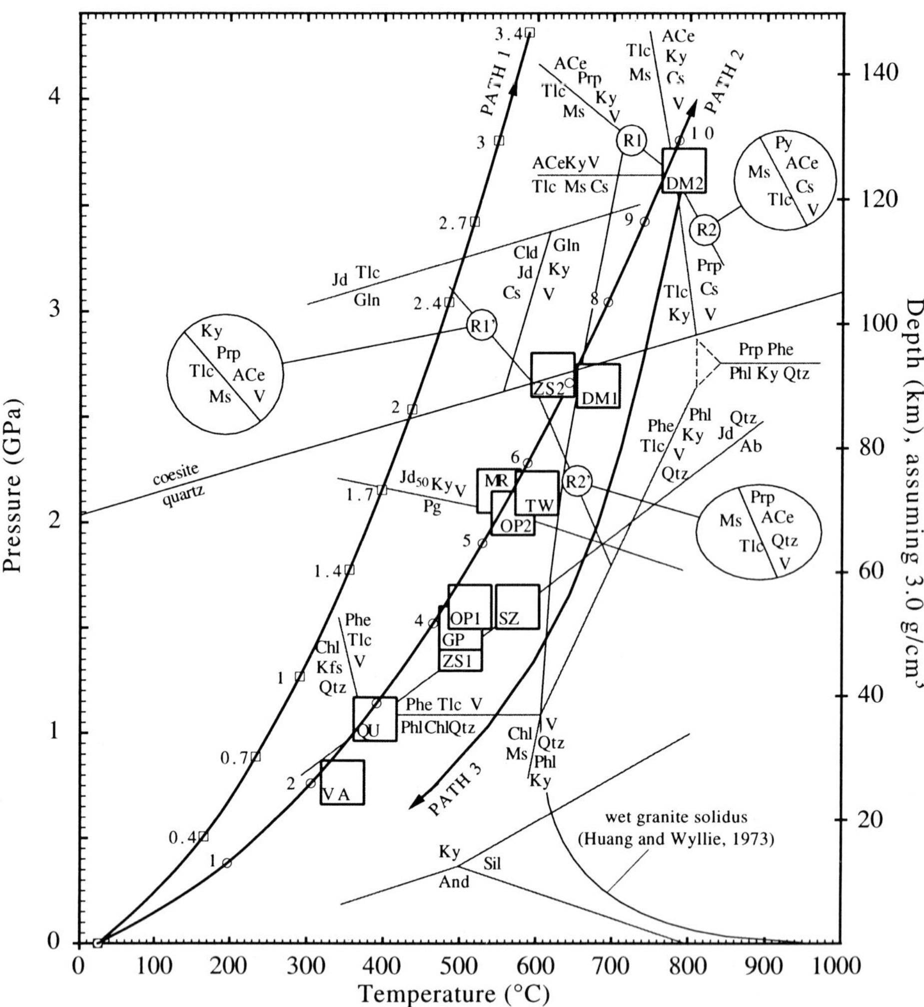

Figure 5. Selected KMASH mineral equilibria (Massonne and Schreyer, 1989; Droop *et al.*, 1990; Reinecke, 1991; Schertl *et al.*, 1991; Sharp *et al.*, 1993) and peak conditions of high-P units in the western and central Alps. Paths 1 and 2 are prograde underthrusting paths calculated with average thermal parameters for convergence rates of 10 (Path 1) and 3 cm/yr (Path 2), shear stresses of 40 (Path 1) and 80 MPa (Path 2), and subduction dip of 25°. Numerals next to squares (path 1) and circles (path 2) give time (m.y.) since start of subduction. Path 3 is exhumation path of the coesite-bearing unit of the southern Dora Maira nappe (Schertl *et al.*, 1991). Notice wide divariant field between reactions involving phengite + talc as calculated from mineral compositions (R1', R2', Reinecke, 1991) and thermodynamic data on endmember phases (R1, R2, Sharp *et al.*, 1993). VA, Vanoise; QU, Queyras blueschist metaophiolite; ZS1, Zermatt-Sass eclogite metaophiolite; GP, Gran Paradiso continental nappe; OP1, Cottian, Graian, and Voltri eclogitic metaophiolites; SZ, Sesia Zone; OP2, Pennine metaophiolite, Zermatt-Sass (Droop *et al.*, 1990); TW, Tauern Window; MR, Monte Rosa continental nappe (Massonne and Schreyer, 1989); ZS2, Zermatt-Sass coesite-bearing metaophiolite; DM1, southern Dora Maira eclogites; DM2, Dora Maira pyrope-quartzite coesite-bearing unit. Abbreviations after Kretz (1983) except for ACe, Al-celadonite, and V, vapor.

3.1 Internal Volatile Buffering and Isotopic Recycling During Subduction

Rocks underthrust along these paths will sequentially cross a number of blueschist- and eclogite-forming reactions before eventually reaching UHP conditions in the *P-T* stability field of coesite and/or diamond (Fig. 5). In virtually all petrogenetic systems (KFMASH, CASH, etc.), these prograde transformations are dominated by devolatilization fluid-mineral equilibria. Devolatilization reactions have greater enthalpy of reaction than solid-solid reactions, and thus are kinetically favored in heat-flow controlled regional metamorphism (Fisher, 1978; Ridley, 1985), allowing them to reach large degrees of equilibrium overstepping even at moderate rates of convergence and subduction (Fig. 5). Buoyant rise of the early fluids released by dewatering and devolatilization will first occur via fracture-driven permeability, shifting with depth to grain-boundary and capillary flow, causing extensive hydration of the hangingwall plate in the first ~30-40 km of burial (Peacock, 1993). Further fluid loss from volatile-rich lithologies (*e.g.*, pelagic and terrigenous sediments, hydrothermally altered seafloor rocks) undergoing incipient blueschist- and eclogite-facies metamorphism will subsequently occur via episodic hydrofracturing and cataclastic flow driven by the 5–10% volume loss attending eclogite-forming reactions (Hacker, 1996), and progressive buildup of the pore-fluid pressure locally reaching lithostatic pressure if fractures already exist. Down to this depth, any argon initially present or released with the devolatilizing fluids (by the breakdown of diagenetic sheet silicates and/or low-grade high-K minerals) will be lost and driven off.

At a critical, irreversible, step along the prograde path, subducted rocks will make a transition from large-scale devolatilization and fluid transfer to reduced fluid flow; the depth of this transition depends on rheology, deformation mechanism, and thermal structure of the subduction zone (Peacock, 1991; Philippot, 1993). Large-scale fluid flow and withdrawal can be expected to operate down to the low-*T* part of the blueschist/eclogite transition in serpentinite-melanges accreted to the base of the mantle wedge, while restricted fluid transfer will dominate eclogites formed at approximately the same depth (~40-50 km) within the subducted slab (Giaramita and Sorensen, 1994). In cold paleo-subduction thermal structures such as recorded in the Alps, most of the downgoing slab will be dehydrated beyond those depths, well before crossing melting reactions at UHP conditions (Fig. 5). In the intervening *P-T* space between the early devolatilization stage and melting reactions, the subducted rocks will evolve largely in the water-deficient region (Koons and Thompson, 1985) because extensive stripping of the primary fluids will cause subsequent dehydration reactions to shift toward the low-*T* side of their respective H_2O-saturated fluid-mineral equilibrium boundary. This may occur at widely different depths and temperatures depending on the starting mineralogy, reaction progress, specific *P-T* location of the relevant fluid-mineral equilibrium, and the rheology of the parental lithology (e.g., Hacker, 1996). This will potentially result in contrasting fluid behavior and sharp gradients in water activity (a_{H_2O}) in intermixed felsic-mafic-carbonate and metasedimentary-metaigneous rock suites or interlayered sequences. One important consequence of entering the fluid-absent region is that extensive fluid transfer (and loss) will be no longer favored because the activity of the residual fluids in equilibrium with blueschist- and eclogite-facies assemblages will be internally buffered by the new HP mineralogy.

This means that rocks in the eclogite-facies field will evolve largely as a closed argon-isotopic system with rising *P* and *T*. The shift from large- to local-scale volatile recycling along the prograde path will impede large-scale homogenization of the argon isotopes formed (or retained) in HP and pre-metamorphic (igneous) minerals in anhydrous mafic eclogites comprising most of the downgoing slab (Hacker, 1996). Limited fluid flow will probably be delayed to greater depths in capping metasediments and seafloor-altered rocks still undergoing progressive dehydration reactions near the overriding mantle interface (Philippot, 1993). Another important effect of entering the water-deficient region is significant expansion of the *P-T* stability of HP-UHP hydrous phases in these rocks (Koons and Thompson, 1985), leading to a maximized preservation of H_2O during prograde subduction until fluid-absent melting occurs (Poli and Schmidt, 1997). The main carrier phases of bound water in HP-UHP silicates are, with increasing *P/T*: lawsonite, zoisite, amphibole, micas (chlorite, biotite, phengite, phlogopite) and talc in mafic, felsic, and ultrabasic rocks. These common rock-forming hydrous phases may survive to pressures in excess of 3–7 GPa for temperatures above 700-950°C in fluid-absent conditions characteristic of subduction zones (Poli and Schmidt, 1995; Domanik and Holloway, 1996; Poli and Schmidt, 1997), contributing significantly to volatile recycling at convergent margins (Poli and Schmidt, 1995; Sorensen *et al.*, 1997). Talc and phengite are especially important in this respect: they form UHP-index phase associations of great interest for $^{40}Ar/^{39}Ar$ dating in both felsic and metaluminous lithologies, and they provide efficient internal buffers of a_{H_2O} via continuous $Al_2Si_{-1}Fe_{-1}$ and $FeMg_{-1}$ exchange equilibria spanning a wide *P-T* field up to at least 4 GPa (Fig. 5, Massonne and Schreyer, 1987; 1989; Massonne and Szpurka, 1997).

Because argon mobility is locally controlled by the availability and short-range transport of the residual, post-devolatilization, HP-UHP fluids, the noble gas evolution will parallel the fluid fate during subduction, and most hydrated (juvenile) lithologies will be progressively purged of their initial argon. However, at some critical depth a dual behavior of argon and aqueous fluids can be expected to locally arise as a by-product of the internal fluid buffering by HP-

UHP fluid-mineral equilibria. This is because the water stored in hydrous HP-UHP silicates will leave the residual and/or internally produced ^{40}Ar to be locally recycled according to the fluid budget of the metamorphic transformations in progress in a given rock volume during subduction. This parallel between the fluid and argon behavior during prograde subduction bears directly on the possible origins of excess ^{40}Ar in UHP rocks (inherited versus internally produced or recycled), and is illustrated below by two endmember cases of isotopic evolution (juvenile versus recycled anhydrous protolith).

3.2 Kinetic Hindrance and Inherited Excess Argon

Along a *P-T-t* path crossing an equilibrium boundary, a reaction will only start when the critical degree of ΔP or ΔT overstepping required for nucleation is reached (Rubie and Thompson, 1985). Solid-solid reactions require larger amounts of overstepping to proceed than devolatilization reactions because they generally involve large interfacial and strain free energies, and have smaller changes in excess Gibbs free energy even at high degrees of overstepping. Moreover, metamorphic transformations involving heterogeneous solid-solid reactions among non-contiguous grains (*e.g.*, dispersed reactant phases in coarse-grained igneous rocks) are diffusion controlled rather than interface controlled. Hence, their rate-limiting step is the transport of cations along grain boundaries, and the driving forces (equilibria overstep) for reaction are greatly affected by whether a free fluid phase can assist in the transformation.

High-grade rocks such as felsic granulites, sillimanite-grade gneisses, and granitoids are the metastable anhydrous equivalent of lower grade, pelitic and quartzofeldspathic mid-crustal and/or supracrustal series. Because their dehydrated mineralogy is already purged of its equilibrium (*i.e.* high-grade) volatile-bearing fluid phase, these rocks cannot react to lower-T hydrous assemblages during subsequent metamorphism unless exotic fluids are introduced. Moreover, they cannot release fluid at HP/LT conditions. In the absence of deformation and fluid infiltration, these rocks are unable to re-equilibrate to a new stable HP/LT assemblage via dehydration and fluid-absent (water-deficient) prograde reactions to any significant extent, with the lack of fluid release further preventing efficient cationic transfer for subsequent, higher-grade solid-solid reactions. This will invariably result in their almost complete preservation, or in corona-textured rocks characterized by markedly irregular transformations locally controlled by grain-scale gradients in a_{H_2O} (e.g., see chapter by Austrheim, this volume)

These endmember cases of kinetic hindrance are common in HP-UHP suites (Austrheim, this volume) and include, *e.g.*: oceanic- and continental-derived gabbros (e.g., Biino and Pognante, 1989; Wayte *et al.*, 1989; Pognante, 1991; Zhang and Liou, 1997), metagranitoids (Koons *et al.*, 1987; Biino and

Compagnoni, 1992), and the Sesia Zone granulites (Lardeaux and Spalla, 1991). The static, metastable preservation of these dry lithologies throughout the HP and UHP metamorphism is most commonly mirrored by a nearly complete retention of the parental isotopic signature in their primary high-grade minerals, much like the Alpe Toso biotites described earlier (Fig. 3). The preservation of the protolith isotopic signature is much less frequently observed in polymetamorphic metapelites, which is again paralleled by the common eradication of primary textures and minerals in these rocks, as opposed to their higher-grade coarse-grained igneous counterparts which are inherently much more resistant to deformation because their dominant network-supporting phases (K-feldspar, plagioclase) have much greater mechanical strength than quartz and mica. Figure 6 shows an uncommon example of inheritance from a garnet + chloritoid + phengite + rutile-bearing HP metapelite from the polymetamorphic eclogite unit of the northern Dora Maira nappe in the western Alps (P = 1.4–2.5 GPa, T = 550 ± 50°C). The phengites of this metapelite retain discordant ages up to 200 Ma, most probably reflecting partial inheritance from the pre-Alpine (Variscan?) age of the protolith. Significantly, this is the only sample containing large (5–10 mm) garnets distinctly displaying pre-Alpine inclusions of lobate (partially resorbed) quartz and ilmenite conspicuously absent in common metapelites dated in this area between 60-130 Ma, and exhibiting the typical garnet-hosted Alpine inclusions, chloritoid + phengite + rutile (Scaillet *et al.*, 1992).

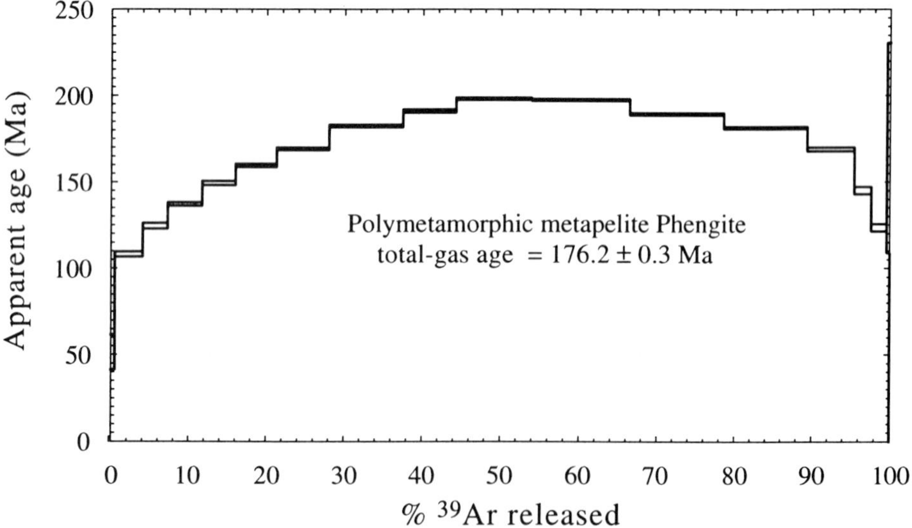

Figure 6. $^{40}Ar/^{39}Ar$ phengite age spectrum from a chloritoid + garnet-bearing metapelite from the polymetamorphic eclogitic basement of the northern Dora Maira nappe, western Alps. This sample displays inherited excess ^{40}Ar and contains garnet-hosted pre-Alpine inclusions of quartz and ilmenite conspicuously absent in common metapelites dated in this area between 60-130 Ma.

Interestingly, even in cases of pervasive deformation and recrystallization, a mineral like muscovite is exceedingly resistant to resetting as illustrated in Fig. 7 for a pegmatite-muscovite specimen from the polymetamorphic eclogite unit of the northern Dora Maira nappe. This sample originally consisted of a single, 1–cm long, porphyroclastic muscovite crystal set in a finely recrystallized matrix of phengite + nematoblastic albite ± microcline + tourmaline and quartz exhibiting a characteristic mortar texture resulting from the complete transposition of pre-Alpine pegmatite into schistose country rocks during the Alpine event. Although the muscovite displays conspicuous signs of internal deformation in the form of kinks and undulatory extinction, its pre-metamorphic isotopic composition was not eradicated during the Alpine deformation and eclogite metamorphism. The $^{40}Ar/^{39}Ar$ age spectrum of the matrix phengites displays markedly younger, mixed, ages due in part to incomplete separation of minute muscovite clasts from the mica concentrate, with values of 60-70 Ma that grossly approach the 35–40 Ma phengite $^{40}Ar/^{39}Ar$ ages typically found in Alpine felsic (gneissic) rocks of this area (Scaillet *et al.*, 1992).

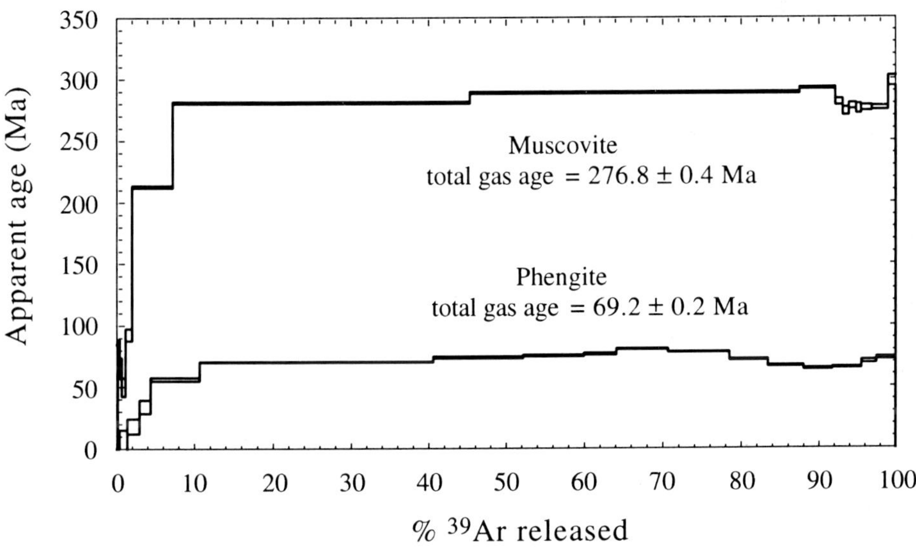

Figure 7. $^{40}Ar/^{39}Ar$ age spectrum illustrating the lack of isotopic resetting in a pre-eclogitic muscovite porphyroclast from a sheared pegmatite in the polymetamorphic eclogitic basement of the northern Dora Maira nappe, western Alps. Matrix phengites display younger discordant $^{40}Ar/^{39}Ar$ apparent ages.

Overall, in the absence of pervasive deformation one should not be surprised to find that newly formed HP-UHP micas in texturally well-preserved and statically overprinted igneous bodies are extensively contaminated with the radiogenic ^{40}Ar internally derived from the protolith, making these rocks probably the least likely to yield $^{40}Ar/^{39}Ar$ age constraints on the UHP event (Arnaud and Kelley, 1995; Tilton *et al.*, 1997).

3.3 Deformation and In Situ Local Ar Recycling

Deformation is unfortunately not a sufficient condition to attain isotopic equilibrium (and resetting) during HP-UHP metamorphism. Strain-induced recrystallization may actually result in highly heterogeneous isotopic distributions at the grain scale because the fluids internally produced and buffered isotopically by the (pre-existing) isotopic composition of the host rock can be recycled in situ.

Figure 8 shows the $^{40}Ar/^{39}Ar$ age spectrum of HP phengites from an eclogitized metabasite from the polymetamorphic northern Dora Maira nappe. These phengites were collected from a concordant segregation vein formed during Alpine eclogitization and mylonitization of the (amphibolite) protolith. Although the sample displays no hint of the protolith mineralogy, these Alpine phengites retain strong grain-scale concentration gradients in excess ^{40}Ar derived from the fluid that was internally recycled during the progressive deformation and formation of the synkinematic phengite vein (Scaillet, 1996). Similar C-O-H isotopic and fluid behavior appears to have operated in the nearby Monviso mylonitic eclogite, where *in situ* recycling of the residual, post-devolatilization HP fluids produced segregation veins with different fluid and stable-isotopic compositions during subduction and deformation of juvenile oceanic gabbroic crust (Philippot and Selverstone, 1991; Nadeau *et al.*, 1993; Philippot, 1993). A major difference between these two equivalent field situations of internal eclogite-facies fluid recycling is that, unlike the pre-metamorphic Dora Maira metabasite, the juvenile Monviso ophiolite did not retain nor recycle any initial argon, which was presumably expelled during the ~90% water loss incurred during prograde devolatilization, making it possible to obtain homogeneous $^{40}Ar/^{39}Ar$ phengite plateau ages of 50 Ma (Philippot, 1993). The divergent evolutionary paths between the juvenile hydrated seafloor Monviso gabbro/basalt and the recycled continent-derived mafic Dora Maira protolith clearly show the importance of the initial volatile content (and imposed fluid availability) of the pre-metamorphic assemblage on the final isotopic record of their constitutive HP micas.

The effect of coupled closed-behavior (internal isotopic buffering) and small-scale recycling during ductile deformation is also illustrated by a suite of Mg-rich phengites from three HP-UHP Pennine units of the western Alpine arc (Fig. 9). These particular samples include two from the northern eclogitic Dora Maira nappe (Scaillet *et al.*, 1992), one from the coesite-bearing pyrope-quartzite of the southern Dora Maira nappe (Monié and Chopin, 1991), and one from the Monte

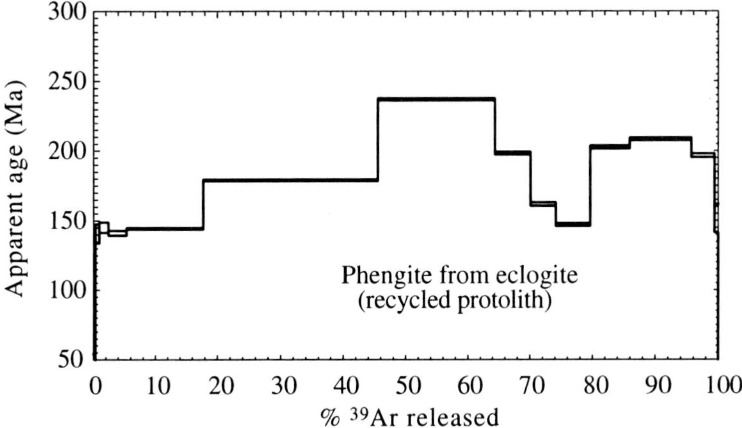

Figure 8. Discordant phengite $^{40}Ar/^{39}Ar$ age spectrum from an Alpine mafic eclogite of the northern Dora Maira nappe, western Alps (redrawn from Scaillet, 1996). Old apparent ages in excess of 130 Ma reflect the incorporation of internally derived excess ^{40}Ar heterogeneously distributed within and between the crystals.

Rosa eclogitic nappe (Chopin and Monié, 1984). The last two samples are among the best preserved and studied HP-UHP phengite + talc occurrences in the western Alps (see op. cit. for full details), and all four samples are characterized by an unusual, highly magnesian host rock composition that is reflected in the near magnesioceladonite endmember composition of the phengites. This unusual composition is thought to have resulted from Mg metasomatism of the enclosing gneisses (Dal Piaz and Lombardo, 1986), along shear zones infiltrated by low-$\delta^{18}O$, high-δD aqueous-rich fluids of probable seawater origin (Sharp *et al.*, 1993). The age of their protolith and deformation is uncertain (Alpine or older), although equivalent, lower-grade Austroalpine units in Hungary suggest that they could be the product of Alpine shearing during upward infiltration of fluids derived from devolatilization of Alpine ophiolites during subduction (Sharp *et al.*, 1993). The early appearance of Mg-chlorite (stable with phengite + quartz in the northern Dora Maira samples) indicates that they must have originated early in the stability field of chlorite + quartz ($P < 1.5$ GPa at $T < 500$°C) along the prograde path en route to eclogite-facies (northern Dora Maira and Monte Rosa) and UHP peak-metamorphic conditions (southern Dora Maira coesite unit). Fluid inclusion data from the Dora Maira pyrope-quartzite bear evidence of Mg-rich fluids in pyrope-hosted healed fractures, possibly indicating that the Mg metasomatism lasted until peak-UHP conditions at this locality (type-2 inclusion of Philippot, 1993). However, Philippot *et al.* (1995) reported fluid-inclusion and mineral composition data pointing to an internal origin of the trapped fluids in the pyrope-quartzite beyond the quartz + chlorite isograd (*i.e.*, 1.5 GPa at ~500°C).

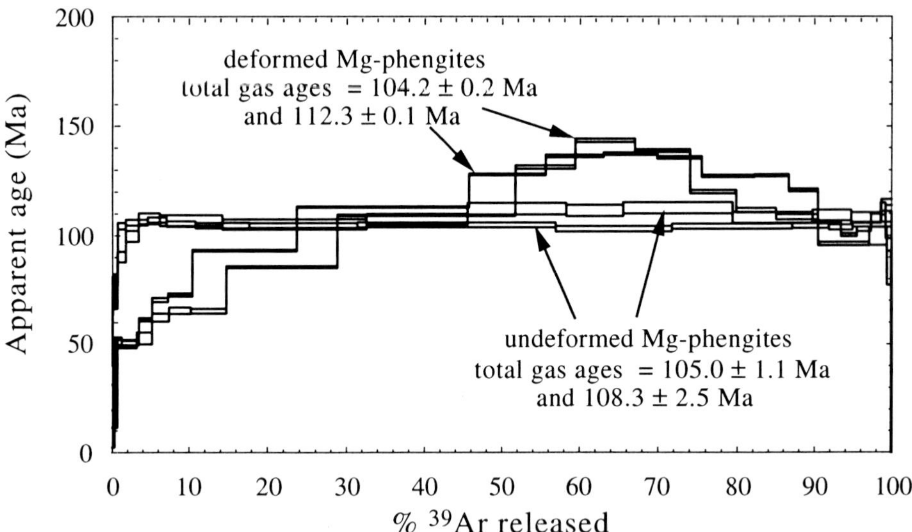

Figure 9. $^{40}Ar/^{39}Ar$ phengite age spectra from Alpine Mg-rich whiteschists from the eclogitic northern Dora Maira nappe (from Scaillet *et al.*, 1992), the Dora Maira coesite-bearing unit (from Monié and Chopin, 1991), and the Monte Rosa eclogitic nappe (from Chopin and Monié, 1984). The two specimens from the northern Dora Maira nappe are ductilely deformed and display virtually the same internally discordant spectrum most probably reflecting inhomogeneous, internal ^{40}Ar redistribution during deformation. The two other samples are undeformed, and all samples display nearly the same integrated (total gas) age, suggesting that they share a common initial HP-UHP isotopic composition corresponding to an apparent age of around 104–112 Ma.

What makes these rocks noteworthy from an isotopic viewpoint is that the two samples from the northern Dora Maira nappe are strongly deformed and display two irregular, yet strikingly reproducible $^{40}Ar/^{39}Ar$ release age spectra, while the southern Dora Maira and Monte Rosa specimens are pristine and yield two homogeneous plateau ages of 105 and 108 Ma. Scaillet (1996) argued that the reproducible discordant spectra and similar total gas age of the two deformed samples (104 and 112 Ma) reflect internal buffering by the starting isotopic composition of the HP Mg-phengites during ductile deformation, resulting in the closed system evolution and fluid recycling at the grain scale in both samples. The integrated (total gas) ages of all four samples are tightly grouped around 104–110 Ma, suggesting a common starting ^{40}Ar composition according to a closed system evolution. Because these rocks should have lost any earlier argon isotopic signature during the early metasomatic event (suggested by the juvenile origin of the early fluids in the prograde path and the O-H stable-isotope homogenization in the pyrope-quartzite, Sharp *et al.*, 1993), and because the primary metasomatizing aqueous fluids could hardly have buffered the phengite isotopic composition at exactly the same value at these different localities, this common starting composition would be best explained by a mechanism of *in situ* argon generation (via radioactive decay with zero initial ^{40}Ar) starting at nearly the same time, 104–110 Ma in these rocks.

Certainly, the significance of this common initial argon will remain controversial and enigmatic until a more ample and systematic laserprobe $^{40}Ar/^{39}Ar$ study is attempted on similar rocks throughout the Pennine nappes in the western Alps (Tilton *et al.*, 1989; 1991; Scaillet, 1996; Tilton *et al.*, 1997). These data lend support, however, to the view that internal fluid buffering and small-scale recycling of argon isotopes in HP-UHP rocks are intimately related phenomena. The role of ductile deformation and fluid behavior on heterogeneous argon distribution at HP conditions is also argued by recent $^{40}Ar/^{39}Ar$ laserprobe data on Norwegian eclogites showing that even where formed in response to pervasive fluid infiltration (*i.e.*, fluid-present conditions), heterogeneous grain-scale ^{40}Ar isotopic disequilibrium can arise and be maintained (or even enhanced) in eclogitic phengites once the HP mineralogy is formed and subsequently deformed (Boundy *et al.*, 1997).

In summary, these examples of coupled fluid behavior and excess argon contamination show that, in most cases, the latter should be expected to display characteristic disequilibrium concentration gradients. In contrast, more homogeneous isotopic compositions with equilibrium ^{40}Ar concentration patterns (and plateau ages) should characterize true closure ages formed via *in situ* radiogenic growth with zero initial ^{40}Ar. Accordingly, those rare cases of proved homogeneous contamination (e.g., the Monte Mucrone phengite samples of Ruffet *et al.*, 1995) require specific and unusual conditions of inheritance, deformation, and internal fluid redistribution (see "Ar mixing behavior and mobility in HP-UHP metamorphic fluids", above). Fluid-inclusion, petrographic, and $^{40}Ar/^{39}Ar$ laserprobe study of these rocks is clearly needed to more fully understand their genesis and the lengthscale of isotopic homogenization attending fluid-assisted mylonitic HP transformations. The recognition that elemental/isotopic recycling during static and/or dynamic HP-UHP recrystallizations is typically associated with strong $^{40}Ar/^{39}Ar$ grain-scale disequilibrium patterns provides us with a valuable guide for identifying the origin (excess *vs.* radiogenic) of argon isotopes in cases where independent information is lacking.

A final question that remains to be answered is why, with the exception of the isotopic inheritance documented above in polycyclic high-grade rocks—and the biotite-gneiss sample 92H23 of Hacker and Wang (1995) from the Dabie Shan UHP orogen, all (proved) occurrences of excess argon in phengite almost invariably occur in mafic eclogites. Note that the reverse is not true—not all eclogite phengites are contaminated. This may reflect the absence of a buffering phase like biotite capable of counteracting any rise in Ar-pore fluid pressure in such rocks (see "Excess argon solution behavior in HP-UHP minerals", above), or the variability of the water-buffering capacity of hydrous (zoisite ± Na-amphibole bearing) versus anhydrous (bimineralic) eclogites, probably coupled with the high kinetic hindrance of these rocks (Rubie, 1990). Altogether, these characteristics would make the phengites from recycled mafic eclogites in continental crust more susceptible to ^{40}Ar excess contamination than others.

3.4 Excess Argon as a UHP Fluid-Melt Tracer Isotope

The common fate of argon and fluid during subduction holds considerable potential for tracing fluid-rock interactions at UHP conditions, and bears directly on the mechanism of melt generation in UHP settings. Progressive burial and heating in the fluid-deficient region and the buffering of volatiles by UHP fluid-mineral equilibria can culminate in fluid-absent, partial melting producing UHP hydrous phases in equilibrium with a H_2O-undersatured melt phase (Fig. 5). In southern Dora Maira, the UHP jadeite-quartzites intimately associated with the pyrope-quartzite suite of the coesite-bearing unit are thought to represent the 'quench' products of in situ UHP melts (Schertl *et al.*, 1991; Philippot, 1993; Sharp *et al.*, 1993; Philippot *et al.*, 1995). These local melts have been suggested to act as a sink for residual volatiles, notably H_2O, to explain the low a_{H_2O} needed to stabilize the pyrope + coesite association at peak UHP conditions (see discussion in Philippot *et al.*, 1995).

Two important consequences derive from the presence of such UHP melts. The high argon solubility in jadeite melts at high pressures (Montana *et al.*, 1993) means that free argon in the grain boundary network would have been preferentially partitioned with H_2O in the melt (now the jadeite-quartzites). This means that if excess argon was ever present in these rocks, we should find it in the jadeite-quartzites rather than the pyrope-bearing ones. Philippot *et al.* (1995) argued that during decompression, a_{H_2O} locally rose in response to fluid oversaturation of the cooling UHP melt, producing hypersaline and water-rich fluids that reacted with pyrope to form secondary phlogopite. This secondary phlogopite contains huge excess argon concentrations (see "Excess argon solution behavior in HP-UHP minerals", above, and Hammerschmidt *et al.*, 1995) that might represent residual argon expelled from the crystallizing melt during decompression.

Because the argon isotopic record is sensitive to fluid behavior during subduction, it can offer the opportunity to track and quantify the extent of fluid-melt-rock interaction at UHP. The Dora Maira example suggests that argon isotopes have great potential for identifying possible sources and signatures of metasomatic UHP fluids, and could provide a critical test for the mechanism of melt generation at UHP. The excess argon geochemistry of UHP metasomatic fluids may be helpful for investigating how their signature can be modified by ultradeep metamorphic processes, hopefully allowing comparison with recycled primary fluids from the incoming slab (e.g., Sharp *et al.*, 1993; Scambelluri *et al.*, 1997).

3.5 Isotopic Resetting and Phengite Re-equilibration

We have focused on mechanisms during prograde subduction and at mantle pressures; however, retrogression during exhumation also influences $^{40}Ar/^{39}Ar$ isotopic patterns. For example, consider the reaction pathway by which HP-UHP phengites re-equilibrate to lower Si contents during exhumation via Tschermak ($Al_2Si_{-1}FM_{-1}$) exchange. The extent of the Tschermak re-equilibration depends on the minerals buffering the Si content of phengite at a given P and T. In the limiting biotite + quartz + K-feldspar assemblage, the relevant equilibrium is (Koons and Thompson, 1985):

$$3 \text{ Muscovite} = 1 \text{ Biotite} + 2 \text{ K-Feldspar} + 3\ Al_2Si_{-1}FM_{-1} + 3 \text{ Quartz} + 2\ H_2O.$$

This is a dehydration reaction (Massonne and Schreyer, 1987) that releases variable amounts of bound water from phengite (in the form OH^-) during decompression, according to the extent of re-equilibration, modal abundance, and composition of the reacting phases. Similarly, in eclogitic metapelites containing the assemblage phengite + paragonite + garnet + quartz, phengite re-equilibration at lower P will proceed via a dehydration reaction whose extent is governed by the NaK_{-1} (paragonite) and Tschermak exchange capacity of phengite, according to its starting composition and modal abundance (Heinrich, 1982). These continuous transformations affect the mica interlayer structure. Stripping of bound OH^- may re-activate argon hosted in the phengite interlayer, with the amount of water produced determining whether the released ^{40}Ar can migrate out of the rock during decompression and overprinting at lower P. The efficiency of these continuous retrograde transformations in altering the argon isotopic composition of the HP-UHP phengite from such rocks will thus depend on its peak-equilibrium Si-content and Tschermak exchange capacity with the relevant buffering assemblage, again implying a strong lithologic control. The driving force for ^{40}Ar loss in this case is cation exchange driven by continuous changes in chemical potential among the different components in the divariant equilibrium assemblage, not volume diffusion of neutral argon in a thermodynamically stable lattice structure. Therefore, the resulting argon resetting pattern may bear little relevance, both in terms of kinetics and grain-scale geometry, to the loss ideally predicted by thermally activated volume-diffusion of argon (Scaillet *et al.*, 1992), imposing further limitations on the applicability of the closure temperature concept (see "Closure behavior, transport rates and lengthscales of Ar isotope exchange in HP-UHP rocks", above).

In contrast, the dominant continuous equilibria adjusting the peak Si-content of UHP phengites toward lower values in the critical KFMASH assemblage, phengite + talc ± pyrope, are (Massonne and Schreyer, 1989):

3 Al-Celadonite + 3 Kyanite + 1 H_2O =
3 Muscovite + 1 Talc + 2 Quartz/Coesite,

and

3 Al-Celadonite + 3 Pyrope + 4 Quartz/Coesite + 4 H_2O =
3 Muscovite + 4 Talc,

which are *hydration* reactions, precluding dehydration-induced Ar loss from operating in these rocks and implying much more unfavorable kinetics and radically different P-T-X_{H_2O} conditions and extent of phengite resetting during decompression down to P_{H_2O} ~1 GPa and T ~500-600°C (Fig. 5). These examples show that potentially different conditions of phengite re-equilibration can prevail in co-facial UHP rocks sharing the same P-T-t exhumation path but differing in bulk composition and mineralogy. In addition to other factors imparting complexity during exhumation (*e.g.*, changes in crustal fluid regimes at shallower depths, deformation, thermal resetting, etc.), attention should be paid to these effects when comparing $^{40}Ar/^{39}Ar$ phengite systematics from different UHP assemblages.

4. CONCLUSIONS

UHP rocks record complex histories apparent in the form of metastable assemblages diversely surviving exhumation, mainly due to slow kinetics. This evolution is reflected in complex chemical and isotopic patterns manifested by sharp age contrasts at the crystal and outcrop scale. The dependence of Ar closure behavior of HP-UHP micas on extrinsic factors such as fluid/solid metasomatic interactions, pressure, and deformation imposes great restrictions upon the application of the closure temperature concept in UHP rocks without evaluation of the processes (reaction pathway, dehydration and fluid budget) at work during subduction and at mantle pressures.

The ^{40}Ar isotopic record faithfully reflects internal fluid-mineral exchange attending prograde subduction and UHP. A strong internal control is predicted to arise from the P-T-t reaction pathway and the volatile budget imposed by the starting mineralogy and the fluid-buffering UHP assemblage specific to a given rock composition. As a consequence, apparent vagaries in regional or outcrop-wide age distributions could arise among compositionally diverse rocks that might actually reflect differences in closure (or crystallization?) ages, and/or isotopic signatures locally imposed by variations in protolith legacy and the extent of fluid-rock interactions during UHP metamorphism. Because of the inherently short lengthscales of isotopic re-distribution usually involved in these processes, the ability of the $^{40}Ar/^{39}Ar$ laserprobe technique to resolve fine-scale isotopic heterogeneities potentially offers more reliable information on the isotopic behavior at UHP than more conventional bulk analytical K/Ar, Sm/Nd, and Rb/Sr techniques. The high level of complexity faced by these different methods in their common problems of resetting, inheritance, and intracrystalline isotope re-distribution still requires caution in assigning more merit to one age or one technique over another.

Adherence to the view that excess ^{40}Ar is an ineluctable feature of UHP metamorphism is a sterile oversimplification preventing advances in understanding processes during HP-UHP metamorphism. There are only roughly a dozen HP and UHP phengite $^{40}Ar/^{39}Ar$ age spectra with *proved* excess ^{40}Ar world-wide (and probably more to come), many others open to more ambiguous interpretation, and a wealth of regional phengite $^{40}Ar/^{39}Ar$ ages consistent with the geologic and U/Pb, Sm/Nd, Rb/Sr isotopic record. The implications of the behavior of excess argon for UHP metamorphism are just being explored. Hopefully, the crystal-chemical aspects and fluid-rock exchange phenomena outlined in this chapter will stimulate a more objective approach to the problem, including:

1) An evaluation of the coupled behavior of Ar + fluids via a petrologic, stable-isotope, $^{40}Ar/^{39}Ar$ laserprobe, and fluid-inclusion study of well-characterized geologic structures (compositional layering, mineral interfaces: e.g., Pickles *et al.*, 1997) and critical mineral assemblages (*e.g.*, the whiteschists scattered throughout the western Alps; the phengite inclusions in pyrope; the eclogite-mylonites of the Monte Mucrone in the Sesia Zone; and the fluid-induced eclogitization spectacularly documented in Norwegian eclogites (Austrheim *et al.*, 1997, this volume; Boundy *et al.*, 1997).

2) The experimental study of crystal-chemical aspects of argon diffusion and solution behavior are in need of urgent quantification via the hydrothermal treatment of different mica compositions at high confining pressures; coupled with the ionic-porosity model (Dahl, 1996a), the experimental calibration of the argon kinetics in UHP minerals will produce more tightly constrained *P-T-t* paths.

Rather than compromising the applicability of the $^{40}Ar/^{39}Ar$ technique, the behavior of excess ^{40}Ar can provide us with the opportunity to track fluid-rock interaction and the fate of fluids and volatiles at UHP.

ACKNOWLEDGMENTS

It is a pleasure for me to thank the editors, Brad Hacker and Juhn Liou, for their invitation to participate in this book, giving me the opportunity to discuss at length many basic aspects and some more personal views on argon systematics in UHP rocks. Brad Hacker and Andy Calvert provided very detailed and helpful reviews that resulted in substantial improvement of the original manuscript. Last but not least, I would like to thank Grazia Vita-Scaillet for her continuous support and for permitting to include her unpublished data on the Alpe Toso biotites in this review.

REFERENCES

Albarède, F. (1976) Géochronologie comparée par la méthode ^{39}Ar - ^{40}Ar de deux régions d'histoire post-hercynienne différente: la Montagne Noire et les Pyrénées orientales. Paris. IPGP Université Paris 6–7, France.

Ames, L., Zhou, G., and Xiong, B. (1996) Geochronology and geochemistry of ultrahigh-pressure metamorphism with implications for collision of the Sino-Korean and Yangtze cratons, central China, *Tectonics* **15**, 472–489.

Amisano-Canesi, A., Chiari, G., Ferraris, G., Ivaldi, G., and Soboleva, S.V. (1994) Muscovite- and phengite-3T: crystal structure and conditions of formation, *European Journal of Mineralogy* **6**, 489–496.

Andersen, T., Austrheim, H., Burke, E.A.J., and Elvevold, S. (1993) N_2 and CO_2 in deep crustal fluids: evidence from the Caledonides of Norway, *Chemical Geology* **108**, 113–132.

Arnaud, N.O. and Kelley, S. (1995) Evidence for excess Ar during high pressure metamorphism in the Dora-Maira (western Alps, Italy), using a Ultra-Violet Laser Ablation Microprobe $^{40}Ar/^{39}Ar$ technique, *Contributions to Mineralogy and Petrology* **121**, 1–11.

Austrheim, H. (this volume) The influence of fluid and deformation on metamorphism of the deep crust and consequences for geodynamics of collision zones, in B.R. Hacker and J.G. Liou (eds.), *When Continents Collide: Geodynamics and Geochemistry of Ultrahigh-Pressure Rocks*, Kluwer Academic Publishers, Dordrecht

Austrheim, H., Erambert, M., and Engvik, A.K. (1997) Processing of crust in the root of the Caledonian continental collision zone: the role of eclogitization, *Tectonophysics* **273**, 129–153.

Ballèvre, M. and Merle, O. (1993) The Combin Fault: compressional reactivation of a Late Cretaceous-Early Tertiary detachment fault in the Western Alps, *Schweizerische Mineralogische und Petrographische Mitteilungen* **73**, 205–227.

Belonoshko, A.B. and Saxena, S.K. (1992) A unified equation of state for fluids of C-H-O-N-S-Ar composition and their mixtures up to very high temperatures and pressures, *Geochimica Cosmochimica et Acta* **56**, 3611–3626.

Biino, G. and Compagnoni, R. (1992) Very-high pressure metamorphism of the Brossasco coronite metagranite, southern Dora-Maira massif, Western Alps, *Schweizerische Mineralogische und Petrographische Mitteilungen* **72**, 347–363.

Biino, G. and Pognante, U. (1989) Palaeozoic continental-type gabbros in the Gran Paradiso nappe (Western Alps, Italy): Early-Alpine eclogitization and geochemistry, *Lithos* **24**, 3–19.

Boundy, T.M., Essene, E.J., Hall, C.M., Austrheim, H., and Halliday, A.N. (1996) Rapid exhumation of the lower crust during continent-continent collision and late extension: Evidence from $^{40}Ar/^{39}Ar$ incremental heating of hornblendes and muscovites, Caledonian orogen, western Norway, *Geological Society of America Bulletin* **108**, 1425–1437.

Boundy, T.M., Hall, C.M., Li, G., Essene, E.J., and Halliday, A.N. (1997) Fine-scale heterogeneities and fluids in the deep crust: a $^{40}Ar/^{39}Ar$ laser ablation and TEM study of muscovites from a granulite-eclogite transition zone, *Earth and Planetary Science Letters* **148**, 223–242.

Brewer, M.S. (1969) Excess radiogenic argon in metamorphic micas from the eastern Alps, Austria, *Earth and Planetary Science Letters* **6**, 321–331.

Brodie, K.H., Rex, D., and Rutter, E.H. (1989) On the age of deep crustal extensional faulting in the Ivrea zone, northern Italy, in M.P. Coward, D. Dietrich, and R.G. Park (eds.), *Alpine Tectonics*, Geological Society of London, London, pp. 203–210.

Burg, J.-P. and Philippot, P. (1991) Asymmetric compositional layering of syntectonic metamorphic veins as way-up criterion, *Geology* **19**, 1112–1115.

Carroll, M.R. and Stolper, E.M. (1991) Argon solubility and diffusion in silica glass: Implications for the solution behavior of molecular gases, *Geochimica Cosmochimica et Acta* **55**, 211–225.

Carroll, M.R. and Stolper, E.M. (1993) Noble gas solubilities in silicate melts and glasses: New experimental results for argon and the relationship between solubility and ionic porosity, *Geochimica Cosmochimica et Acta* **57**, 5039–5051.

Catti, M., Ferraris, G., Hull, S., and Pavese, A. (1994) Powder neutron diffraction study of 2M1 muscovite at room pressure and at 2 GPa, *European Journal of Mineralogy* **6**, 171–178.

Chamorro-Perez, E., Gillet, P., and Jambon, A. (1996) Argon solubility in silicate melts at very high pressures. Experimental set-up and preliminary results for silica and anorthite melts, *Earth and Planetary Science Letters* **145**, 97–107.

Chavagnac, V. and Jahn, B.-m. (1996) Coesite-bearing eclogites from the Bixiling Complex, Dabie Mountains, China; Sm-Nd ages, geochemical characteristics and tectonic implications, *Chemical Geology* **133**, 29–51.

Chemenda, A.I., Mattauer, M., Malavieille, J., and Bokun, A.N. (1995) A mechanism for syn-collisional rock exhumation and associated normal faulting: Results from physical modelling, *Earth and Planetary Science Letters* **132**, 225–232.

Chopin, C. (1984) Coesite and pure pyrope in high-grade blueschists of the western Alps: a first record and some consequences, *Contributions to Mineralogy and Petrology* **86**, 107–118.

Chopin, C. and Monié, P. (1984) A unique magnesiochloritoid-bearing, high-pressure assemblage from the Monte Rosa, Western Alps: a petrologic and ^{40}Ar-^{39}Ar study, *Contributions to Mineralogy and Petrology* **87**, 388–398.

Comodi, P. and Zanazzi, P.F. (1995) High-pressure structural study of muscovite, *Physics and Chemistry of Minerals* **22**, 170–177.

Cumbest, R.J., Johnson, E.L., and Onstott, T.C. (1994) Argon composition of metamorphic fluids: Implications for $^{40}Ar/^{39}Ar$ geochronology, *Geological Society of America Bulletin* **106**, 942–951.

Dahl, P.S. (1996a) The crystal-chemical basis for Ar retention in micas: inferences from interlayer partitioning and implications for geochronology, *Contributions to Mineralogy and Petrology* **123**, 22–39.

Dahl, P.S. (1996b) The effects of composition on retentivity of argon and oxygen in hornblende and related amphiboles: A field-tested empirical model, *Geochimica Cosmochimica et Acta* **60**, 3687–3700.

Dal Piaz, G.V. and Lombardo, B. (1986) Early Alpine metamorphism in the Penninic Monte Rosa-Gran Paradiso basement nappes of the northwestern Alps, *Geological Society of America Memoir* **164**, 249–265.

Dallmeyer, R.D. and Gee, D.G. (1986) $^{40}Ar/^{39}Ar$ mineral dates from retrogressed eclogites within the Baltoscandian miogeosyncline: Implications for a polyphase Caledonian orogenic evolution, *Geological Society of America Bulletin* **97**, 26–34.

Dalrymple, G.B. and Lanphere, M.A. (1969) *Potassium-Argon dating. Principles, Techniques and Applications to Geochronology*, W. H. Freeman, San Francisco.

Dempster, T.J. (1992) Zoning and recrystallization of phengitic micas: implications for metamorphic equilibration, *Contributions to Mineralogy and Petrology* **109**, 526–537.

Dodson, M.H. (1973) Closure temperature in cooling geochronological and petrological systems, *Contributions to Mineralogy and Petrology* **40**, 259–274.

Domanik, K.J. and Holloway, J.R. (1996) The stability and composition of phengitic muscovite and associated phases from 5. 5 to 11 GPa: Implications for deeply subducted sediments, *Geochimica Cosmochimica et Acta* **60**, 4133–4150.

Droop, G.T.R., Lombardo, B., and Pognante, U. (1990) Formation and distribution of eclogite facies rocks in the Alps, in D.A. Carswell (ed.), *Eclogite Facies Rocks*, Blackie, Glasgow, pp. 225–259.

Duchêne, S., Blichert-Toft, J., Luais, B., Télouk, P., Lardeaux, J.-M., and Albarède, F. (1997) The Lu-Hf dating of garnets and the ages of the Alpine high-pressure metamorphism, *Nature* **387**, 586–589.

Erambert, M. and Austrheim, H. (1993) The effects of fluid and deformation on zoning and inclusion pattern in poly-metamorphic garnets, *Contributions to Mineralogy and Petrology* **115**, 204–214.

Faust, J. and Knittle, E. (1994) Equation of state, amorphization, and high-pressure phase diagram of muscovite, *Journal of Geophysical Research* **99**, 19785–19792.

Ferry, J.M. and Burt, D.M. (1982) Characterization of metamorphic fluid composition through mineral equilibria, in J.M. Ferry (ed.), *Characterization of metamorphism through mineral equilibria, Reviews in Mineralogy*, **10**, Mineralogical Society of America, Washington, D.C., pp. 207–262.

Fisher, G.W. (1978) Rate laws in metamorphism, *Geochimica Cosmochimica et Acta* **42**, 1035–1050.

Fortier, S.M. and Giletti, B.J. (1989) An empirical model for predicting diffusion coefficients in silicate minerals, *Science* **245**, 1481–1484.

Früh-Green, G.L. (1994) Interdependence of deformation, fluid infiltration and reaction progress recorded in eclogitic metagranitoids (Sesia Zone, Western Alps), *Journal of Metamorphic Geology* **12**, 327–343.

Gebauer, D., Schertl, H.-P., Brix, M., and Schreyer, W. (1997) 35 Ma old ultrahigh-pressure metamorphism and evidence for very rapid exhumation in the Dora Maira massif, Western Alps, *Lithos* **41**, 5–24.

Giaramita, M.J. and Sorensen, S.S. (1994) Primary fluids in low-temperature eclogites: evidence from two subduction complexes (Dominican Republic, and California, USA, *Contributions to Mineralogy and Petrology* **117**, 279–292.

Giletti, B.J. and Tullis, J. (1977) Studies in diffusion. IV. Pressure dependence of Ar diffusion in phlogopite mica, *Earth and Planetary Science Letters* **35**, 180–183.

Guggenheim, S., Chang, Y.-H., and van Groos, A.F.K. (1987) Muscovite dehydroxylation: High-temperature studies, *American Mineralogist* **72**, 536–550.

Guidotti, C.V., Sassi, F.P., and Blencoe, J.G. (1989) Compositional controls on the a and b cells dimensions of 2M1 muscovite, *European Journal of Mineralogy* **1**, 71–84.

Hacker, B.R. (1996) Eclogite formation and the rheology, buoyancy, seismicity, and H_2O content of oceanic crust, in G.E. Bebout, Scholl, D., Kirby, S.H., Platt, J.P. (ed.), *Dynamics of Subduction*, Monograph, American Geophysical Union, Washington, D.C., pp. 337–246.

Hacker, B.R. and Christie, J.M. (1991) Observational evidence for a possible new diffusion path, *Science* **251**, 67–70.

Hacker, B.R. and Peacock, S.M. (1994) Creation, preservation, and exhumation of coesite-bearing, ultrahigh-pressure metamorphic rocks, in R.G. Coleman and X. Wang (eds.), *Ultrahigh Pressure Metamorphism*, Cambridge University Press, Cambridge, United Kingdom.

Hacker, B.R. and Wang, Q.C. (1995) Ar/Ar geochronology of ultrahigh-pressure metamorphism in central China, *Tectonics* **14**, 994–1006.

Hacker, B.R., Wang, X., Eide, E.A., and Ratschbacher, L. (1996) Qinling-Dabie ultrahigh-pressure collisional orogen, in A. Yin and T.M. Harrison (eds.), *The Tectonic Evolution of Asia*, Cambridge University Press, Cambridge, United Kingdom.

Hammerschmidt, K. and Frank, E. (1991) Relics of high pressure metamorphism in the Lepontine Alps (Switzerland) - $^{40}Ar/^{39}Ar$ and microprobe analyses on white micas, *Schweizerische Mineralogische und Petrographische Mitteilungen* **71**, 261–274.

Hammerschmidt, K., Schertl, H.P., Friedrichsen, H., and Schreyer, W. (1995) Excess argon a common feature in ultrahigh-pressure rocks: a case study on micas from the Dora-Maira massif, western Alps, Italy, *Terra Nova* **7**, 349.

Hannula, K.A. and McWilliams, M.O. (1995) Reconsideration of the age of blueschist facies metamorphism on the Seward Peninsula, Alaska, based on phengite $^{40}Ar/^{39}Ar$ results, *Journal of Metamorphic Geology* **13**, 125–139.

Harley, S.L. and Carswell, D.A. (1995) Ultradeep crustal metamorphism: A prospective view, *Journal of Geophysical Research* **100**, 8367–8380.

Harrison, T.M., Duncan, I., and McDougall, I. (1985) Diffusion of ^{40}Ar in biotite: temperature, pressure, and compositional effects, *Geochimica Cosmochimica et Acta* **49**, 2461–2468.

Harrison, T.M., Heizler, M.T., Lovera, O.M., Wenji, C., and Grove, M. (1994) A chlorine disinfectant for excess argon released from K-feldspar during step heating, *Earth and Planetary Science Letters* **123**, 95–104.

Harrison, T.M. and McDougall, I. (1981) Excess ^{40}Ar in metamorphic rocks from Broken Hill, New South Wales: implications for $^{40}Ar/^{39}Ar$ age spectra and the thermal history of the region, *Earth and Planetary Science Letters* **55**, 123–149.

Heinrich, C.H. (1982) Kyanite-Eclogite to Amphibolite facies evolution of hydrous mafic and pelitic rocks, Adula nappe, central Alps, *Contributions to Mineralogy and Petrology* **81**, 30–38.

Heizler, M.T. and Harrison, T.M. (1988) Multiple trapped argon isotope components revealed by $^{40}Ar/^{39}Ar$ isochron analysis, *Geochimica Cosmochimica et Acta* **52**, 1295–1303.

Hodges, K.V. (1991) Pressure-temperature-time paths, *Annual Reviews of Earth and Planetary Science* **19**, 207–236.

Holness, M.B. (1993) Temperature and pressure dependence of quartz-aqueous fluid dihedral angles: the control of adsorbed H_2O on the permeability of quartzites, *Earth and Planetary Science Letters* **117**, 363–377.

Holness, M.B. and Graham, C.M. (1995) P-T-X effects on equilibrium carbonate-H_2O-CO_2-NaCl dihedral angles: constraints on carbonate permeability and the role of deformation during fluid infiltration, *Contributions to Mineralogy and Petrology* **119**, 301–313.

Hsü, K.J. (1991) Exhumation of high-pressure metamorphic rocks, *Geology* **19**, 107–110.

Inger, S., Ramsbotham, W., Cliff, R.A., and Rex, D.C. (1996) Metamorphic evolution of the Sesia-Lanzo Zone, Western Alps: time constraints from multi-system geochronology, *Contributions to Mineralogy and Petrology* **126**, 152–168.

Joesten, R. (1991) Grain-boundary diffusion kinetics in silicate and oxide minerals, in J. Ganguly (ed.), *Diffusion, Atomic Ordering, and Mass Transport; Selected Topics in Geochemistry, Advances in Geochemistry*, **8**, Springer-Verlag, New York, pp. 345–395.

Karpinskaya, T.B., Ostrovskii, I.A., and Shanin, L.L. (1961) Artificial injection of argon in mica at high pressures and temperatures, *Izvestiia Akademii nauk SSSR. Seriia geologicheskaia* **8**, 99–100.

Karpinskiy, T.B., Shanin, L.L., and Borisevich, I.B. (1966) Artificial injection of argon in mica, olivine and pyroxene, *International Geology Review*, 768–769.

Klemd, R., van den Kerkhof, A.M., and Horn, E.E. (1992) High-density CO_2_N_2 inclusions in eclogite-facies metasediments of the Münchberg gneiss complex, SE Germany, *Contributions to Mineralogy and Petrology* **111**, 409–419.

Koons, P.O., Rubie, D.C., and Frueh-Green, G. (1987) The effects of disequilibrium and deformation on the mineralogical evolution of quartz diorite during metamorphism in the eclogite facies, *Journal of Petrology* **28**, 679–700.

Koons, P.O. and Thompson, A.B. (1985) Non-mafic rocks in the greenschist, blueschist and eclogite facies, *Chemical Geology* **50**, 3–30.

Kretz, R. (1983) Symbols for rock-forming minerals, *American Mineralogist* **68**, 277–279.

Kubicki, J.D. and Lasaga, C.A. (1991) Molecular dynamics and diffusion in silicate melts, in J. Ganguly (ed.), *Diffusion, Atomic Ordering, and Mass Transport, Selected Topics in Geochemistry*, 8, Springer-Verlag, New York, pp. 1–50.

Lardeaux, J.-M. and Spalla, M.I. (1991) From granulites to eclogites in the Sesia zone (Italian Western Alps): a record of the opening and closure of the Piedmont ocean, *Journal of Metamorphic Geology* **9**, 35–59.

Lee, J.K.W. (1995) Multipath diffusion in geochronology, *Contributions to Mineralogy and Petrology* **120**, 60–82.

Li, S., Wang, S., Chen, Y., Liu, D., Qiu, J., Zhou, H., and Zhang, Z. (1994) Excess argon in phengite from eclogite: Evidence from the dating of eclogite minerals by the Sm-Nd, Rb-Sr and $^{40}Ar/^{39}Ar$ methods, *Chemical Geology* **112**, 343.

Liou, J.G. and Zhang, R.Y. (1996) Occurrence of intergranular coesite in ultrahigh-P rocks from the Sulu region, eastern China: Implications for lack of fluid during exhumation, *American Mineralogist* **81**, 1217–1221.

Massonne, H.J. (1995) Experimental and petrogenetic study of UHPM, in R.G. Coleman and X. Wang (eds.), *Ultrahigh Pressure Metamorphism*, Cambridge University Press, Cambridge, pp. 33–95.

Massonne, H.J. and Schreyer, W. (1986) High-pressure syntheses and X-ray properties of white micas in the system K_2O-MgO-Al_2O_3_SiO_2_H_2O, *Neues Jahrbuch Fur Mineralogie Geologie Und Palaontologie Abhandlungen* **153**, 177–215.

Massonne, H.J. and Schreyer, W. (1987) Phengite barometry based on the limiting assemblage with K-feldspar, phlogopite, and quartz, *Contributions to Mineralogy and Petrology* **96**, 212–224.

Massonne, H.J. and Schreyer, W. (1989) Stability field of the high-pressure assemblage talc + phengite and two new phengite barometers, *European Journal of Mineralogy* **1**, 391–410.

Massonne, H.J. and Szpurka, Z. (1997) Thermodynamic properties of white micas on the basis of high-pressure experiments in the systems K_2O-MgO-Al_2O_3_SiO_2_H_2O and K_2O-FeO-Al_2O_3_SiO_2_H_2O, *Lithos* **41**, 229–250.

Mattey, D., Jackson, D.H., Harris, N.B.W., and Kelley, S. (1994) Isotopic constraints on fluid infiltration from an eclogite shear zone, Holsenøy, Norway, *Journal of Metamorphic Geology* **12**, 311–325.

McDougall, I. and Harrison, T.M. (1988) *Geochronology and Thermochronology by the $^{40}Ar/^{39}Ar$ Method*, Oxford University Press, New York.

Monié, P. and Chopin, C. (1991) $^{40}Ar/^{39}Ar$ dating in coesite-bearing and associated units of the Dora Maira massif, western Alps, *European Journal of Mineralogy* **3**, 239–262.

Montana, A., Guo, Q., Boettcher, S., White, B.S., and Brearley, M. (1993) Xe and Ar in high-pressure silicate liquids, *American Mineralogist* **78**, 1135–1142.

Mosenfelder, J.L. and Bohlen, S.R. (1997) Kinetics of the cœsite → quartz transformation, *Earth and Planetary Science Letters* **153**, 133–147.

Nadeau, S., Philippot, P., and Pineau, F. (1993) Fluid inclusion and mineral isotopic compositions (H-C-O) in eclogitic rocks as tracers of local fluid migration during high-pressure metamorphism, *Earth and Planetary Science Letters* **114**, 431–448.

Nakashima, S. (1995) Diffusivity of ions in pore water as a quantitative basis for rock deformation rate estimates, *Tectonophysics* **245**, 185–203.

Onstott, T.C., Phillips, D., and Pringle-Goodell, L. (1991) Laser microprobe measurement of chlorine and argon zonation in biotite, *Chemical Geology* **90**, 145–168.

Peacock, S.M. (1990) Fluid processes in subduction zones, *Science* **248**, 329–337.

Peacock, S.M. (1991) Numerical simulation of subduction zone pressure-temperature-time paths: constraints on fluid production and arc magmatism, *Philosophical Transactions of the Royal Society of London* **A335**, 341–353.

Peacock, S.M. (1993) Large-scale hydration of the lithosphere above subducting slabs, *Chemical Geology* **108**, 49–59.

Philippot, P. (1990) Opposite vergence of nappes and crustal extension in the French-Italian western Alps, *Tectonics* **9**, 1143–1164.

Philippot, P. (1993) Fluid-melt-rock interaction in mafic eclogites and coesite-bearing metasediments; constraints on volatile recycling during subduction, *Chemical Geology* **108**, 93–112.

Philippot, P., Chevallier, P., Chopin, C., and Dubessy, J. (1995) Fluid composition and evolution in cœsite-bearing rocks (Dora-Maira massif, Western Alps): Implications for element recycling during subduction, *Contributions to Mineralogy and Petrology* **121**, 29–44.

Philippot, P. and Selverstone, J. (1991) Trace-element-rich brines in eclogitic veins: implications for fluid composition and transport during subduction, *Contributions to Mineralogy and Petrology* **106**, 417–430.

Philippot, P. and van Roermund, H.L.M. (1992) Deformation processes in eclogitic rocks: evidence for the rheological delamination of the oceanic crust in deeper levels of subduction zones, *Journal of Structural Geology* **14**, 1059–1077.

Pickles, C.S., Kelley, S., Reddy, S.M., and Wheeler, J. (1997) Determination of high spatial resolution argon isotope variations in metamorphic biotites, *Geochimica Cosmochimica et Acta* **61**, 3809–3833.

Pin, C. and Vielzeuf, D. (1983) Granulites and related rocks in Variscan median Europe: A dualistic interpretation, *Tectonophysics* **93**, 47–94.

Platt, J.P. (1986) Dynamics of orogenic wedges and the uplift of high-pressure metamorphic rocks, *Geological Society of America Bulletin* **97**, 1037–1053.

Pognante, U. (1991) Petrological constraints on the eclogite- and blueschist-facies metamorphism and P-T-t paths in the Western Alps, *Journal of Metamorphic Geology* **9**, 5–17.

Poli, S. and Schmidt, M.W. (1995) H2O transport and release in subduction zones: Experimental constraints on basaltic and andesitic systems, *Journal of Geophysical Research* **100**, 22,299–22,314.

Poli, S. and Schmidt, M.W. (1997) The high-pressure stability of hydrous phases in orogenic belts: an experimental approach on eclogite-forming processes, *Tectonophysics* **273**, 169–184.

Reddy, S.M., Kelley, S.P., and Wheeler, J. (1996) A $^{40}Ar/^{39}Ar$ laser probe study of micas from the Sesia Zone, Italian Alps: implications for metamorphic and deformation histories, *Journal of Metamorphic Geology* **14**, 493–508.

Reinecke, T. (1991) Very-high-pressure metamorphism and uplift of cœsite-bearing sediments from the Zermatt-Saas zone, Western Alps, *European Journal of Mineralogy* **3**, 7–17.

Ridley, J. (1985) The effect of reaction enthalpy on the progress of a metamorphic reaction, in A.B. Thompson and D.C. Rubie (eds.), *Metamorphic reactions. Kinetics, textures, and deformation, Advances in Physical Geochemistry*, **4**, Springer-Verlag, Berlin, pp. 80–97.

Roddick, J.C., Cliff, R.A., and Rex, D.C. (1980) The evolution of excess argon in Alpine biotites—A $^{40}Ar/^{39}Ar$ analysis, *Earth and Planetary Science Letters* **48**, 185–208.

Rubie, D.C. (1986) The catalysis of mineral reactions by water and restrictions on the presence of aqueous fluid during metamorphism, *Mineralogical Magazine* **50**, 399.

Rubie, D.C. (1990) Role of kinetics in the formation and preservation of eclogites, in D.A. Carswell (ed.), *Eclogite Facies Rocks*, Blackie, Glasgow, pp. 111–140.

Rubie, D.C. and Thompson, A.B. (1985) Kinetics of metamorphic reactions at elevated temperatures and pressures: an appraisal of available experimental data, in A.B. Thompson and D.C. Rubie (eds.), *Advances in Physical Geochemistry*, **4**, Springer-Verlag, Berlin, pp. 27–79.

Ruffet, G., Feraud, G., Ballevre, M., and Kienast, J.-R. (1995) Plateau ages and excess argon in phengites: an ^{40}Ar-^{39}Ar laser probe study of Alpine micas (Sesia Zone, Western Alps, northern Italy), *Chemical Geology* **121**, 327–343.

Sassi, F.P., Guidotti, C.V., Rieder, M., and De Pieri, R. (1994) On the occurrence of metamorphic 2M1 phengites: some thoughts on polytypism and cristallization conditions of 3T phengites, *European Journal of Mineralogy* **6**, 151–160.

Saxena, S.K. and Fei, Y. (1987) Fluids at crustal pressures and temperatures. I. Pure species, *Contributions to Mineralogy and Petrology* **95**, 370–375.

Scaillet, S. (1996) Excess ^{40}Ar transport scale and mechanism in high-pressure phengites: A case study from an eclogitized metabasite of the Dora-Maira nappe, western Alps, *Geochimica Cosmochimica et Acta* **60**, 1075–1090.

Scaillet, S., Feraud, G., Ballevre, M., and Amouric, M. (1992) Mg/Fe and [(Mg,Fe)Si-Al_2] compositional control on argon behavior in high-pressure white micas: a $^{40}Ar/^{39}Ar$ continuous laser-probe study from the Dora-Maira nappe of the internal western Alps, Italy, *Geochimica Cosmochimica et Acta* **56**, 2851–2872.

Scambelluri, M., Piccardo, G.B., Philippot, P., Robbiano, A., and Negretti, L. (1997) High salinity fluid inclusions formed from recycled seawater in deeply subducted alpine serpentinite, *Earth and Planetary Science Letters* **148**, 485–499.

Schertl, H.P., Schreyer, W., and Chopin, C. (1991) The pyrope-cœsite rocks and their country rocks at Parigi, Dora-Maira Massif, Western Alps: detailed petrography, mineral chemistry and PT-path, *Contributions to Mineralogy and Petrology* **108**, 1–21.

Selverstone, J., Morteani, G., and Staude, J.M. (1991) Fluid channeling during ductile shearing; transformation of granodiorite into aluminous schist in the Tauern Window, Eastern Alps, *Journal of Metamorphic Geology* **9**, 419–431.

Sharp, Z.D., Essene, E.J., and Hunziker, J.C. (1993) Stable isotope geochemistry and phase equilibria of coesite-bearing whiteschists, Dora Maira Massif, western Alps, *Contributions to Mineralogy and Petrology* **114**, 1–12.

Shimizu, I. (1995) Kinetics of pressure solution creep in quartz: theoretical considerations, *Tectonophysics* **245**, 121–134.

Smelik, E.A. (1995) Fluid channelization at the submicroscopic scale in exsolved orthoamphiboles, *European Journal of Mineralogy* **7**, 825–834.

Smith, J.V. and Schreyer, W. (1962) Location of argon and water in cordierite, *Mineralogical Magazine* **33**, 226–236.

Smith, S.P. and Kennedy, B.M. (1983) The solubility of noble gases in water and in NaCl brine, *Geochimica Cosmochimica et Acta* **47**, 503–515.

Sorensen, S.S., Grossman, J.N., and Perfit, M.R. (1997) Phengite-hosted LILE enrichment in eclogite and related rocks: Implications for fluid-mediated mass transfer in subduction zones and arc magma genesis, *Journal of Petrology* **38**, 3–34.

Stocker, R.L. and Ashby, M.F. (1973) On the rheology of the upper mantle, *Reviews of Geophysics* **11**, 391–426.

Stöckhert, B., Jäger, E., and Voll, G. (1986) K-Ar age determinations on phengites from the internal part of the Sesia zone, Western Alps, *Contributions to Mineralogy and Petrology* **92**, 456–470.

Thöni, M. and Jagoutz, E. (1992) Some new aspects of dating eclogites in orogenic belts : Sm-Nd, Rb-Sr, and Pb-Pb isotopic results from the Austroalpine Saualpe and Koralpe type-locality (Carinthia/Styria, southeastern Austria), *Geochimica Cosmochimica et Acta* **56**, 347–368.

Tilton, G.R., Ames, L., Schertl, H.P., and Schreyer, W. (1997) Reconnaissance isotopic investigations on rocks of an undeformed granite contact within the cœsite-bearing unit of the Dora Maira Massif, *Lithos* **41**, 25–36.

Tilton, G.R., Schreyer, W., and Schertl, H.-P. (1991) Pb-Sr-Nd isotopic behavior of deeply subducted crustal rocks from the Dora Maira massif, western Alps, Italy-II: what is the age of the ultrahigh pressure metamorphism, *Contributions to Mineralogy and Petrology* **108**, 22–33.

Tilton, G.R., Schreyer, W., and Schertl, H.P. (1989) Pb-Sr-Nd behavior of deeply subducted crustal rocks from the Dora Maira Massif, western Alps, Italy, *Geochimica Cosmochimica et Acta* **53**, 1391–1400.

Turner, G. (1968) The distribution of potassium and argon in chondrites, in L.H. Ahrens (ed.), *Origin and Distribution of the Elements*, Pergamon Press, New York, pp. 387–398.

Ungaretti, L., Lombardo, B., Domeneghetti, C., and Rossi, G. (1983) Crystal-chemical evolution of amphiboles from eclogitized rocks of the Sesia-Lanzo Zone, Italian Western Alps, *Bulletin de Minéralogie* **106**, 645–672.

Villa, I.M. and Puxeddu, M. (1994) Geochronology of the Larderello geothermal field: new data and the "closure temperature" issue, *Contributions to Mineralogy and Petrology* **115**, 415–426.

von Blanckenburg, F. and Davies, J.H. (1995) Slab breakoff: a model for syncollisional magmatism and tectonics in the Alps, *Tectonics* **14**, 120–131.

Walker, F.D.L., Lee, M.R., and Parson, I. (1995) Micropores and micropermeable texture in alkali feldspar: geochemical and geophysical implications, *Mineralogical Magazine* **59**, 505–534.

Walther, J.V. (1992) Ionic association in H_2O-CO_2 fluids at mid-crustal conditions, *Journal of Metamorphic Geology* **10**, 789–797.

Walther, J.V. and Wood, B.J. (1984) Rate and mechanism in prograde metamorphism, *Contributions to Mineralogy and Petrology* **88**, 246–259.

Wayte, G.J., Worden, R.H., Rubie, D.C., and Droop, G.T.R. (1989) A TEM study of disequilibrium plagioclase breakdown at high pressure: the role of infiltrating fluid, *Contributions to Mineralogy and Petrology* **101**, 426–437.

White, B.S., Brearley, M., and Montana, A. (1989) Argon in silicate liquids at high pressures, *American Mineralogist* **74**, 513–529.

Wijbrans, J.R. and McDougall, I. (1986) $^{40}Ar/^{39}Ar$ dating of white micas from an Alpine high-pressure metamorphic belt on Naxos (Greece): the resetting of the argon isotopic system, *Contributions to Mineralogy and Petrology* **93**, 187–194.

Zeitler, P.K. (1989) The geochronology of metamorphic processes, in J.S. Daly, R.A. Cliff, and B.W.D. Yardley (eds.), *Evolution of Metamorphic Belts, Special Publication* **43**, Geological Society of London, London, pp. 131–147.

Zhang, R.Y. and Liou, J.G. (1997) Partial transformation of gabbro to coesite-bearing eclogite from Yangkou, the Sulu Terrane, eastern China, *Journal of Metamorphic Geology* **15**, 183–202.

Chapter 8

Geochemical and Isotopic Characteristics of UHP Eclogites and Ultramafic Rocks of the Dabie Orogen: Implications for Continental Subduction and Collisional Tectonics

Bor-ming Jahn

Géosciences Rennes, Université de Rennes 1, Campus de Beaulieu, 35042 Rennes Cedex, France, jahn@univ-rennes1.fr

Abstract: Geochemical and Nd-Sr isotope tracer studies of ultrahigh-pressure (UHP) eclogites and associated ultramafic rocks from the Su–Lu and Dabie terranes of central China provide important constraints on tectonic models of continental subduction, collision, and subsequent exhumation in the Dabie orogen. UHP eclogites have three types of occurrence: I) enclaves in ortho- and paragneisses gneisses of granitic composition, II) enclaves in or interlayers with marbles or metaclastic sediments (*e.g.*, garnet-mica schists), and III) enclaves and interlayers with ultramafic rocks. The eclogites are mainly of basaltic composition, but show a wide range of major and trace element abundances, suggesting multiple origins and heterogeneous sources, and in some cases, metamorphic segregation.

Types I and II eclogites show LREE- (light rare earth element) enriched patterns and highly negative ε_{Nd}(220 Ma) values (–6 to –20). The overall geochemical and isotopic data indicate that most of the eclogite protoliths resemble basalts/gabbros and their metamorphic equivalents (amphibolites or basic granulites) of Precambrian gneiss terranes, and they constitute an integral part of ancient continental craton. A minimum age of 1.7 Ga was established for those from the Weihai area (northeastern Su–Lu). Sm-Nd model ages (T_{DM}) are highly variable, ranging from >3 Ga to 800 Ma. This indicates that the protoliths were emplaced in Precambrian times. The eclogites do not represent subducted Tethys Ocean crust. Type III eclogites are genetically related to the associated ultramafic rocks (lherzolite, harzburgite, or pyroxenites). Sr-Nd isotopic data indicate minor contamination by lower crustal materials for the Bixiling, whereas a greater contamination by upper crustal rocks is recorded in the Maowu layered intrusions. Isotopic signatures of ultramafic rocks (*e.g.*, Rizhao, Rongcheng, Donghai) indicate that some rocks evolved from an enriched reservoir (Rongcheng) whereas others are from a slightly depleted mantle source (Rizhao).

B. R. Hacker and J. G. Liou (eds.),
When Continents Collide: Geodynamics and Geochemistry of Ultrahigh-Pressure Rocks, 203–239.
© 1998 *Kluwer Academic Publishers. Printed in the Netherlands.*

The present study provides the following constraints to tectonic models of the Dabie Orogen. 1) Type I and II eclogites have continental protoliths, and are not subducted Tethyan oceanic crust. This is significantly different from the eclogites of oceanic origin from the Alpine and Hercynian orogenic belts. Sm-Nd model age data suggest that the eclogite protoliths were produced at different Precambrian times; some Su–Lu eclogites could have formed in the late Archean, whereas Dabie eclogites are Proterozoic. 2) Type III eclogites are generally of mantle derivation. Different degrees of crustal contamination are observed in the Bixiling and Maowu layered intrusions. The UHP parageneses of these intrusions indicate deep continental subduction, whereas those of some ultramafic rocks (*e.g.*, Rizhao) do not. However, they clearly indicate exhumation of deep-seated mantle slices in the collision zone. 3) The UHP metamorphism took place at ~220 Ma based on a variety of reliable age data (U-Pb, Sm-Nd, Ar-Ar) obtained in different laboratories. Other "aberrant" ages can be interpreted as due to a) lack of isotope equilibrium in UHPM conditions (Sm-Nd, U-Pb systems), or b) excess Ar, particularly in phengite. No ages can be determined from the Rb-Sr systems because most eclogites experienced Rb depletion relative to Sr, leading to unsupported high $^{87}Sr/^{86}Sr$ ratios. However, the lowest $^{87}Sr/^{86}Sr$ (~0.7036) for the Bixiling complex and Rizhao clinopyroxenites may represent the Sr isotopic composition of the subcontinental lithosphere of the Dabie orogen. 4) Identical U-Pb zircon ages for UHP eclogites and apparently non-UHP host gneisses are consistent with the *in situ* origin of the UHP complex of the Dabie orogen.

1. INTRODUCTION

The Qinling–Tongbai–Dabie orogenic belt is the suture zone formed by collision between the Sino-Korean and Yangtze cratons. In its eastern extremity, ultrahigh-pressure (UHP) metamorphic rocks are widely distributed in the Su–Lu and Dabie areas, which have been offset about 500 km by the Tan–Lu Fault (Fig. 1). These UHP metamorphic rocks have been the subject of numerous studies in recent years because they provide an excellent opportunity to study the rare phenomenon of continental subduction and exhumation of deeply subducted rocks. There is a great diversity of opinions on tectonic evolution of the Dabie orogen (e.g., Mattauer *et al.*, 1985; Okay and Sengör, 1992; Yin and Nie, 1993; Li *et al.*, 1994; Maruyama *et al.*, 1994; Xu *et al.*, 1994; Ernst and Liou, 1995; Hacker *et al.*, 1996; Liou *et al.*, 1996; Wang *et al.*, 1996), in large part due to the lack of constraints from geochemical, isotopic tracer and age studies. At present, much controversy still concerns i) the primary age and nature of protoliths for the UHP eclogites, their associated ultramafic rocks and enclosing granitic gneisses, ii) the areal extent of the UHPM rocks, iii) the origin and tectonic affinity of Northern Dabie high-T/low-P orthogneiss and its role in evolution of the Dabie orogen, and iv) the timing of continental collision and mechanism of subsequent exhumation.

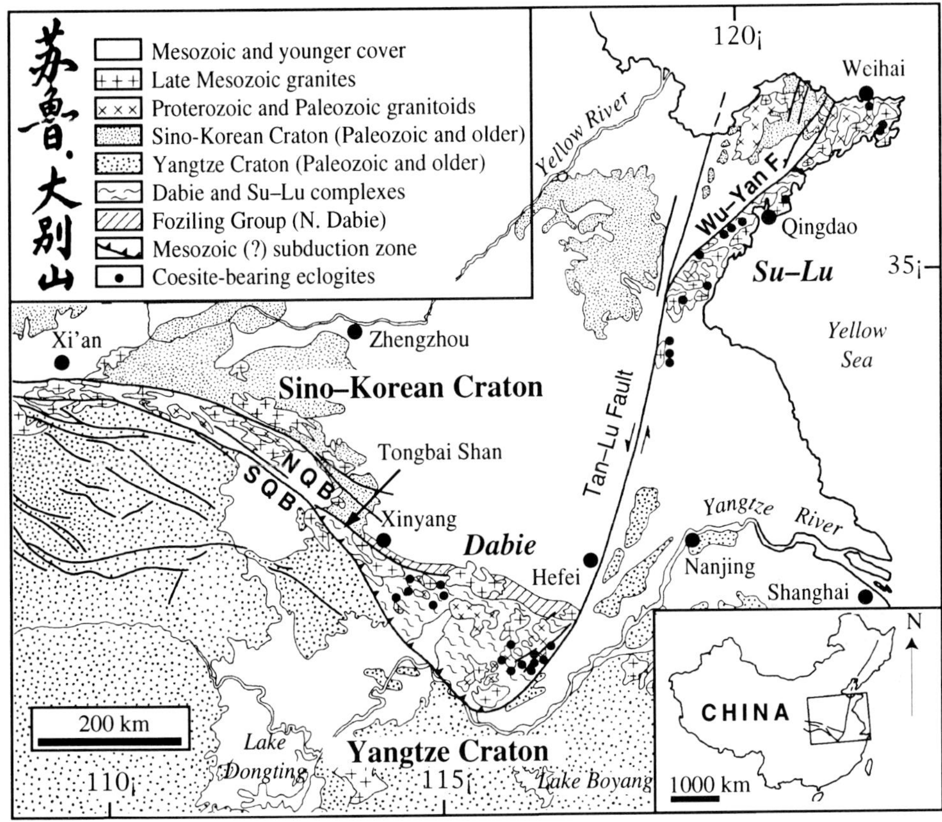

Figure 1. General geologic map of east-central China, showing the Dabie orogen with respect to the Sino-Korean and Yangtze cratons. NQB and SQB are Northern and Southern Qinling Belts, respectively.

Peculiar features that distinguish the Dabie orogen from other orogenic belts are as follows; some of the statements will be substantiated by geochemical and isotopic data presented in this chapter.

1) The Dabie orogen contains the most extensive UHP metamorphic rocks in the world. Evidence of UHP is mainly found in eclogites and ultramafic rocks, but also in metasediments, such as jadeite quartzite (Liou *et al.*, 1997), marbles and calc-silicates (Wang and Liou, 1991; 1993; Schertl and Okay, 1994; Zhang *et al.*, 1996b). Some granitic gneisses also show evidence of UHP (Hirajima *et al.*, 1993; Carswell *et al.*, 1997).

2) The time of UHP metamorphism is ~220 Ma based on different geochronological studies (Ames *et al.*, 1993; Li *et al.*, 1993b; Li *et al.*, ; Hacker and Wang, 1995; Ames *et al.*, 1996; Chavagnac and Jahn, 1996; Rowley *et al.*, 1997).

3) Geochemical and isotope tracer studies indicate that most eclogite protoliths are nor subducted oceanic crust (this paper, Jahn *et al.*, 1994; 1995; 1996), in contrast to eclogites of oceanic origin in Hercynian and Alpine orogenic belts.

4) No ophiolite complexes have been positively identified.

5) Some eclogites show extraordinary geochemical and isotopic characteristics. For example, the eclogites from Weihai (eastern Shandong Peninsula) have the highest $\varepsilon_{Nd}(0)$ values (+170 to +264) ever measured for terrestrial rocks. Geochemical data indicate that severe metasomatism at high pressure and temperature took place at ~1.7 Ga, leading to extreme LREE depletion and subsequent rapid growth of $^{143}Nd/^{144}Nd$ ratios (Jahn *et al.*, 1996). On the other hand, eclogites from Qinglongshan of northern Jiangsu Province show $\delta^{18}O$ values as low as –10 (Yui *et al.*, 1995; Zheng *et al.*, 1996), the lowest recorded in high temperature rocks. Similarly, very low whole-rock $\delta^{18}O$ values (–5.1 to –5.5) have been found in biotite gneisses from the Shima and Taihu areas in the Dabie Shan (Baker *et al.*, 1997). The low values are ascribed to interaction between the basaltic (or granitic gneiss) protoliths and very light meteoric water prior to UHP metamorphism (see also Rumble, this volume).

6) There is an almost total absence of syn-collisional granites in the Dabie orogen. This is an extreme case among the world's "sterile orogens", and contrasts strongly with the "fertile chains" such as the Hercynian with abundant syn-tectonic leucogranites. In the Dabie Shan, the post-orogenic granites were emplaced at ~130–120 Ma (Zhou *et al.*, 1992; Xue *et al.*, 1997), about 100 m.y. after the continental collision.

7) Ultramafic rocks of the Dabie orogen have multiple origins and formed under different conditions. There are syntectonic UHP and post-tectonic non-UHP varieties, with the latter, mainly distributed in the Northern Dabie Complex, showing evidence of crust–mantle interaction (Jahn *et al.*, in preparation-a).

This chapter presents new geochemical and Nd-Sr isotopic analyses on UHP eclogites and their associated ultramafic rocks from the Su–Lu and Dabie areas. These data are used to identify protoliths and provide constraints on the tectonic evolution of the Qinling–Dabie collisional belt.

2. UHP Rock Assemblages and Implications for Continental Subduction

Eclogites are volumetrically minor in the Dabie and Su–Lu UHP areas, but they are the principal rock type that bears evidence of UHP. If the eclogites were produced by metamorphism of a subducted oceanic lithosphere, as in the Hercynian and Alpine orogenic chains (Bernard-Griffiths and Cornichet, 1985; Bernard-Griffiths *et al.*, 1985; Stosch and Lugmair, 1990; Beard *et al.*, 1992;

Thöni and Jagoutz, 1992; Miller and Thöni, 1995), they must have tectonic contacts with their host granitic gneisses. Thus, even if the eclogites show UHP parageneses, it is not necessary that the country-rock gneisses also underwent the same UHP metamorphism. However, if the UHP eclogites can be identified as part of ancient continental crust, then their presence would imply deep subduction of a continental block.

Fig. 2 illustrates the types of mafic–ultramafic rocks formed in UHP conditions. UHP metamorphic rocks of sedimentary origin—such as those of Dora Maira (Chopin, 1984; Tilton *et al.*, 1989; 1991)—are clearly of continental derivation; they are excluded from the present discussion. In the Dabie orogen, eclogites of possible ocean-floor affinity have been reported (Zhai and Cong, 1996), but they appear to be volumetrically insignificant. Eclogites of mantle origin have not been identified in the Su–Lu or Dabie areas, and their presence, if any, does not imply continental subduction. Only eclogites formed from continental mafic rocks (basalts/gabbros) or calcareous silicate sediments (as para-amphibolites) and ultramafic rocks of layered intrusions (*e.g.*, Bixiling and Maowu) have clear crustal signals. Their UHP mineral parageneses provide a definitive signature of continental subduction and UHP metamorphism.

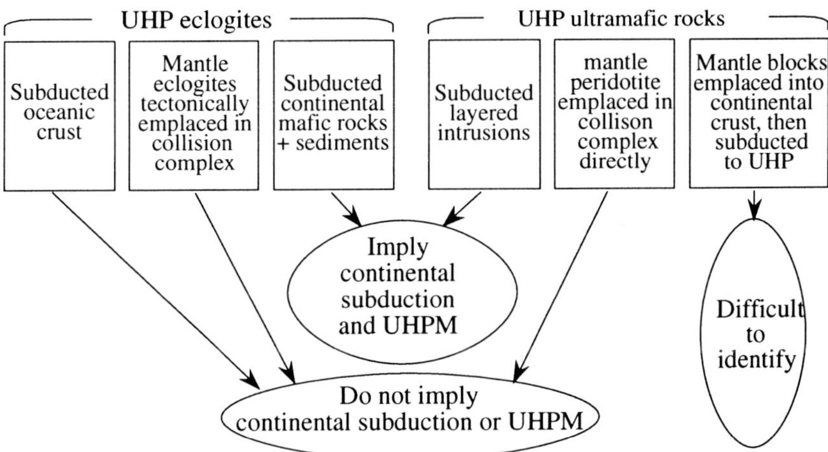

Figure 2. Tectonic implications of a variety of UHP rocks. Most have been identified in the Dabie orogen by means of geochemical and isotope tracer studies. Only subducted continental mafic rocks, including some eclogites possibly transformed from para-amphibolites, and subducted layered mafic-ultramafic intrusions clearly signify subduction of continental crust. Eclogites and peridotites of oceanic affinity and eclogites from kimberlite pipes, though bearing UHP parageneses, do not provide evidence for continental subduction.

The implications of UHP ultramafic rocks may be more controversial. Some may be mantle blocks emplaced tectonically into the continental crust and then subducted to UHP together with the continental crust. This kind of rock may be represented by the Rizhao clinopyroxenites (Hiramatsu and Hirajima, 1995), but such a process is difficult to identify if the rocks have chemical and isotopic compositions comparable to the upper mantle. If, however, the mantle peridotites record crustal contamination in their Nd-Sr-O isotopic compositions and/or petrological features, then their UHP assemblages would imply deep continental subduction. The garnet peridotites of Rongcheng and Donghai (Yang and Jahn, in preparation) may represent this kind of rock.

3. ECLOGITE OCCURRENCES

The geology of the Dabie orogen has been described in numerous recent publications (e.g., Xu *et al.*, 1994; Cong, 1996; Hacker *et al.*, 1996; Liou *et al.*, 1996). Generally, eclogites in the Su–Lu and Dabie UHP areas occur as blocks and lenses within granitic gneisses, ultramafic massifs and marbles. They are homogeneous and massive (*e.g.*, Bixiling), or heterogeneous, foliated and banded with granoblastic or porphyroblastic textures (*e.g.*, Qinglongshan). All the eclogites experienced various stages and extents of retrogression.

The Su–Lu UHP rocks are separated from the Jiaodong Archean terrane to the NW by the Wu–Yan Fault (Fig. 3). UHP parageneses identified in Su–Lu eclogites and ultramafic rocks include coesite, talc, low-Al orthopyroxene, magnesite, and Ti-clinohumite (Yang and Smith, 1989; Enami and Zang, 1990; Zhao *et al.*, 1992; Wang *et al.*, 1993; Yang *et al.*, 1993; Zhang *et al.*, 1994; Liou and Zhang, 1995; Zhang *et al.*, 1996b; 1996c). Recent geochronological studies indicate that most of the granitic gneisses in the Su–Lu UHP area formed in the late Proterozoic at ~700–800 Ma (Li *et al.*, 1993a; Ishizaka *et al.*, 1994; Ames *et al.*, 1996). However, the protolith age of eclogite from the Weihai area has been established at ~1.7 Ga (Jahn *et al.*, 1996), hence its enclosing granitic gneiss could be even older.

The Dabie Shan are composed of three major petrotectonic units: 1) the Northern Dabie Complex, 2) the central Dabie UHP metamorphic complex, and 3) the southern Dabie HP blueschist/greenschist unit (Fig. 4). They are bounded to the south by a foreland fold-thrust belt and to the north by the Foziling Group, several km of greenschist-facies quartzite and biotite-muscovite quartz schist (R.G.S. Anhui, 1987; Xu *et al.*, 1994) commonly interpreted as flysch (Xu *et al.*, 1994) or passive continental margin deposits of the Yangtze Craton (Okay *et al.*, 1993).

The Northern Dabie Complex (NDC) is a high-T/low-P metamorphic unit whose tectonic affinity has long been debated (Okay *et al.*, 1993; Zhai *et al.*, 1994; Zhang *et al.*, 1996a; Xue *et al.*, 1997). The complex consists of a "basement" assemblage of orthogneiss, migmatite, amphibolite, garnet granulite, marble and ultramafic rocks, and an intrusive suite of Cretaceous granites and minor mafic-ultramafic intrusions that range from weakly deformed to gneissic (Wang and Liou, 1991; Okay *et al.*, 1993; Hacker *et al.*, 1996; Xue *et al.*, 1997). The orthogneiss is mainly of TTG composition (trondhjemite-tonalite-granodiorite) and the age of the protolith(s) is potentially Precambrian; our unpublished Sm-Nd isotopic results give T_{DM} model ages of 1.5–1.8 Ga. However, there are also similar-looking Cretaceous orthogneisses, as revealed by recent zircon U-Pb dating (Hacker *et al.*, 1997; Xue *et al.*, 1997; Hacker *et al.*, in press).

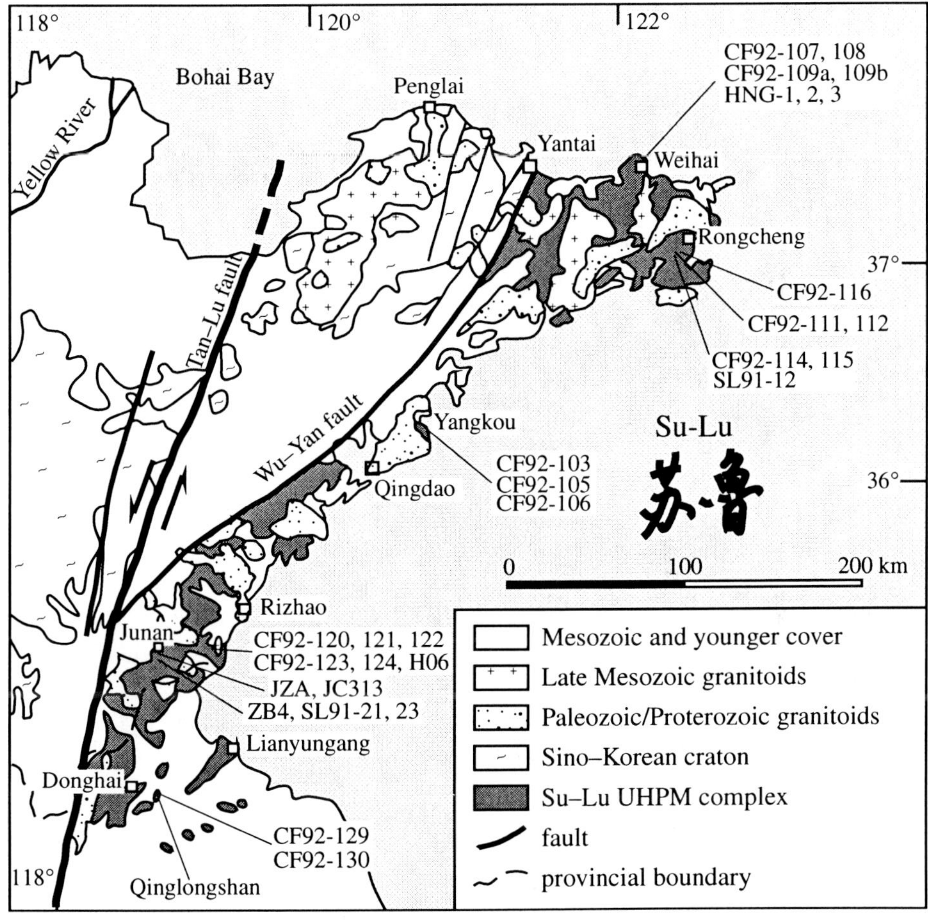

Figure 3. Geologic map of the Su–Lu area showing sample localities.

The presence of Precambrian and Cretaceous orthogneisses make the tectonic interpretation of the NDC even more controversial. Nevertheless, other age data useful for tectonic reconstruction include the following: biotite and hornblende from the NDC yielded $^{40}Ar/^{39}Ar$ ages of 120–130 Ma (Hacker and Wang, 1995) and biotite-plagioclase-whole-rock Rb-Sr isochrons gave ~115 Ma (Potel and Jahn, unpublished). Rb-Sr, Sm-Nd and $^{40}Ar/^{39}Ar$ analyses on mineral fractions of coarse-grained gabbros (Shacun) also show Cretaceous ages (~125 Ma; Hacker and Wang, 1995; Jahn *et al.*, in preparation-b).

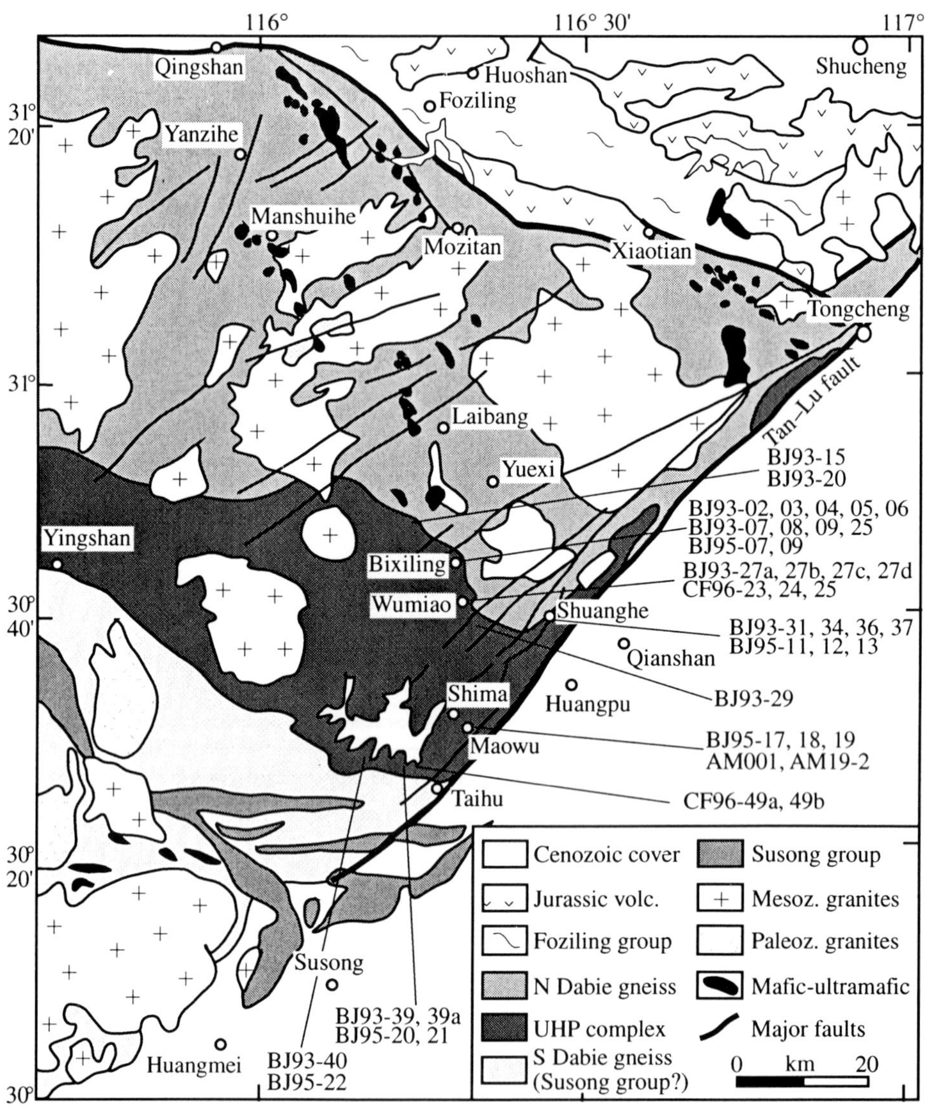

Figure 4. Geologic map of the Dabie Shan showing sample localities.

In the Central Dabie Complex abundant UHP eclogites occur as layers and blocks in granitic gneisses (including paragneisses), as interlayers with metasediments of marbles, and as layered intrusions, such as the Bixiling and

Maowu bodies (Wang *et al.*, 1990; Wang, 1991; Xu *et al.*, 1992; Okay, 1993; Wang and Liou, 1993; Zhang *et al.*, 1995). Okay (1993) subdivided the complex into "hot eclogite" and "cold eclogite", based on the distribution of coesite, sodic amphibole and inferred metamorphic temperatures. Petrologic studies have shown that the eclogites and their host rocks underwent a clockwise P-T evolution with peak conditions of 550–850°C and 2.6–3.2 GPa for coesite-bearing eclogite, and 635 ± 40°C and 2.3 ± 0.3 GPa for the cold eclogite (Liou *et al.*, 1996).

The eclogites can be divided into three types based on their occurrence and the lithology of their host rocks: Type I are enclaves or layers within granitic gneisses; they are the most abundant; Type II are enclaves or interlayers with marbles or calc-silicate rocks; and Type III are members of layered mafic-ultramafic intrusions, such as the Bixiling and Maowu complexes, or simply in association with ultramafic rocks with no clear genetic relationship. All three types of eclogite contain inclusions of coesite and/or quartz pseudomorphs after coesite. Type II eclogites could be of metasedimentary origin (from marl → para-amphibolite → eclogite); they are common in the Dabie Shan, but rarely encountered in Su–Lu outside the Rongcheng area. By contrast, Type III eclogites are more widely distributed in the Su–Lu than in the Dabie areas, but the largest layered complexes, Bixiling and Maowu, are within the Dabie Shan. The protoliths of the Dabie Type I eclogites formed in the Proterozoic based on U-Pb zircon upper intercept ages (Ames *et al.*, 1993) and Sm-Nd model ages (T_{DM}) presented below. By contrast, a Palaeozoic protolith age was determined for Type III eclogite from the Maowu layered intrusion by a zircon U-Pb upper intercept at ~450 Ma (Rowley *et al.*, 1997).

4. GEOCHEMICAL AND ISOTOPIC CHARACTERISTICS OF ECLOGITES

4.1 Elemental Abundances

Eclogites and associated ultramafic rocks were collected from the Su–Lu and Dabie areas at localities shown in Figs. 3 and 4. The results of major and trace element analyses are presented in Tables 1 and 2.

Figure 5 presents the variation diagrams of major oxides *vs.* SiO_2 with N-MORB reference lines for comparison. The SiO_2 content of the eclogites varies from 35–60%, but the majority of the data fall in the basaltic range of 45–50%. The variability of major oxides is impressive; no clear correlation of any oxides with SiO_2 is discernible and little distinction among the three types of eclogite can be made from the major element compositions. However, type II eclogites,

Table 1. Chemical Compositions of Eclogites and Ultramafic Rocks from the Su–Lu Region

Sample	Rock	Locality	SiO_2	Al_2O_3	Fe_2O_3	MnO	MgO	CaO	Na_2O	K_2O	TiO_2	P_2O_5	LOI	Σ	Nb	Zr	Y	Sr	Rb	Co	V	Ni	Cr	Ba	Ga	Cu	Zn	Th	Pb
Type I: enclaves in granitic gneisses																													
CF92–103	EC	Yangkou	53.91	18.59	8.85	0.18	3.25	7.28	4.33	1.22	1.19	0.45	0.62	99.87	6.3	198	25	399	37	17	192	25	42	1540	23	20	119	3	29
CF92–105	EC	Yangkou	53.50	18.42	9.36	0.19	3.46	7.64	4.34	1.22	1.19	0.47	0.30	100.09	6.8	167	26	349	37	20	201	2	14	1405	22	16	121	1	27
CF92–106	EC	Yangkou	47.70	18.12	12.03	0.19	6.49	9.36	3.66	0.83	1.07	0.33	0.11	99.89	2.8	68	14	246	15	40	294	26	44	2660	19	38	115	2	9
CF92–107	EC	Weihai	55.40	14.34	13.86	0.22	7.78	7.52	0.89	0.14	0.27	0.01	0.29	100.72	0.6	5	33	17	8	50	341	73	338	33	12	74	79	3	3
CF92–108	EC	Weihai	60.08	11.43	12.15	0.20	6.88	8.24	1.18	0.10	0.22	0.01	0.25	100.74	1.3	5	30	11	6	46	267	123	277	28	11	76	66	1	3
CF92–109a	GG	Weihai	57.09	16.44	10.23	0.17	3.87	6.37	3.97	0.79	0.80	0.14	0.31	100.18	7.5	112	37	94	20	27	247	29	29	192	27	7	116	1	6
CF92–109b	GG	Weihai	57.25	13.05	14.77	0.28	3.85	6.34	0.54	1.29	1.84	0.46	0.49	100.16	16.1	527	63	191	62	39	182	6	43	459	12	29	124	2	7
CF92–114	EC	Rongcheng	48.14	18.06	14.64	0.24	4.84	10.20	2.41	0.03	1.54	0.49	0.10	100.69	8.7	143	32	163	<1	32	312	2	22	54	21	9	123	2	0
CF92–115	EC	Rongcheng	44.75	15.06	16.88	0.18	6.61	12.91	2.36	0.03	1.58	0.04	0.22	100.62	2.1	38	13	270	1	58	774	20	65	436	19	48	83	2	5
CF92–129	EC	Qinlongshan	53.25	16.47	12.35	0.25	4.47	8.86	2.83	0.06	1.43	0.57	0.12	100.66	3.9	127	32	593	1	44	230	112	501	64	23	22	211	<1	114
CF92–130	EC	Qinlongshan	46.43	20.99	12.27	0.19	4.91	9.33	3.84	0.53	1.02	0.32	0.61	100.44	0.1	61	17	445	12	42	319	40	179	1024	28	42	252	<1	35
Type II: enclaves in or interlayers with marbles																													
CF92–116	GE	Rongcheng	37.59	19.66	15.45	0.12	5.22	18.32	0.58	0.02	1.31	0.03	2.48	100.78	7.6	140	34	48	0	56	222	7	15	205	18	20	57	<1	2
Type III: enclaves in or interlayers with ultramafic rocks																													
CF92–111	GL	Rongcheng	39.95	1.84	7.47	0.11	38.42	1.29	0.00	0.00	0.03	0.01	10.37	99.49	0.4	0	<1	19	0	101	40	2007	2393	14	1	4	41	<1	3
CF92–112	EC	Rongcheng	48.00	17.40	7.47	0.12	10.47	13.64	3.11	0.04	0.19	0.01	0.20	100.65	1.5	5	3	173	2	43	109	290	475	61	14	80	52	1	6
CF92–120	GX	Rizhao	47.53	7.98	6.85	0.10	14.47	21.22	0.28	0.02	1.23	0.01	0.25	99.94	0.6	22	6	112	2	31	163	322	3625	46	9	27	24	<1	6
CF92–121	GX	Rizhao	47.79	7.35	7.70	0.10	15.36	20.63	0.23	0.01	1.37	0.01	0.23	100.78	0.4	18	6	81	0	37	235	308	1142	45	9	62	26	<1	6
CF92–122	GX	Rizhao	46.92	5.87	9.29	0.12	15.76	19.13	0.51	0.17	1.68	0.01	0.96	100.42	1.6	22	6	132	2	41	223	328	771	69	10	83	36	1	3
CF92–123	GX	Rizhao	46.77	6.10	9.56	0.12	15.16	20.61	0.27	0.01	1.72	0.01	0.09	100.42	1.4	23	7	128	1	40	229	294	829	56	11	83	41	2	5
CF92–124	GX	Rizhao	46.77	4.98	9.23	0.12	15.29	20.19	0.41	0.09	2.06	0.01	0.91	100.06	1.7	26	6	152	2	39	213	252	2198	43	9	10	35	1	4
JZA	GX	Junan	47.53	9.32	9.15	0.27	17.32	14.56	0.30	0.01	0.67	0.01		99.14				295	0.76			299	1190						
JC313	EC	Junan	47.96	14.71	11.50	0.24	10.73	10.74	3.14	0.02	0.99	0.02		100.05				96	38.6			189	480						
H06	GX	Rizhao	44.24	12.38	9.35	0.15	14.38	18.10	0.32	0.02	1.35	0.02		100.31				76	0.78			212	1000						
CH5	EC	Rongcheng	46.78	13.74	12.22	0.20	11.50	12.02	2.31	0.02	0.83	0.03	0.94	100.59				195	6	49	208	124	546						
CH17	EC	Rongcheng	46.15	16.89	9.23	0.14	13.15	12.00	1.88	0.02	0.21	0.02	0.85	100.54				133	5	39	65	572	508						
Y1	EC	Rongcheng	49.33	12.39	7.40	0.12	12.85	14.50	2.32	0.06	0.42	0.03	0.53	99.95						28	110	217	652						
D12	EC	Donghai	48.65	19.06	9.90	0.14	8.00	12.57	0.88	0.19	0.85	0.25	0.87	101.36				396		27	182	90	572						
ZC1	EC	Junan	47.47	13.63	11.90	0.23	10.30	11.73	3.37	0.04	0.92	0.34	0.98	100.91				126		21	229	163	508						
Z2	EC	Junan	47.41	16.26	8.13	0.10	13.38	12.48	2.23	0.02	0.23	0.02	0.06	100.32						39	150	366	446						

Oxides in wt%, elements in ppm. AE, amphibolite enclave; AM, amphibolite; CE, cold eclogite; EC, eclogite; GA, garnet amphibolite; Ge, garnet-rich eclogite; GG, garnet granulite; GL, garnet lherzolite; GP, garnet peridotite; GT, garnettite; GX, garnet pyroxenite; OX, orthopyroxenite; W, websterite. Major and trace elements were analyzed by XRF (Philips PW1480 spectrometer) in Rennes. Fe_2O_3 represents total iron oxides. Uncertainties: ±1% to 3% for major elements; ± 5% for trace elements ≥ 20 ppm and ± 10% for those ≤ 20 ppm. JZA, JC313 and H06 are from Yang (1991). CH5, CH17, Y1, D12, ZC1, Z2 are from Zhang et al (1994).

Table 2. Chemical Compositions of Eclogites and Ultramafic Rocks from the Dabie Shan.

Sample	Rock	Locality	SiO_2	Al_2O_3	Fe_2O_3	MnO	MgO	CaO	Na_2O	K_2O	TiO_2	P_2O_5	LOI	Σ	Nb	Zr	Y	Sr	Rb	Co	V	Ni	Cr	Ba	Ga	Cu	Zn	Th	Pb
Type I: enclaves in granitic gneisses																													
BJ93–27a	EC	Wumiao	46.47	16.64	16.67	0.29	3.24	8.77	3.17	0.98	2.51	0.60	0.53	99.87	14.7	104	47	172	29	28	115	13	38	225	29	58	162	2	5
BJ93–27b	EC	Wumiao	46.76	16.29	16.92	0.30	2.84	8.65	2.79	0.78	2.34	0.94	1.14	99.75	14.5	140	56	215	24	20	95	14	28	233	27	41	152	4	7
BJ93–27c	EC	Wumiao	42.89	14.48	22.95	0.41	3.67	8.92	1.27	0.59	2.45	2.31	–0.64	99.30	7.4	47	52	428	24	35	61	7	5	209	21	67	155	9	10
BJ93–27d	EC	Wumiao	40.86	14.93	19.24	0.28	4.77	11.96	2.53	0.06	4.42	1.35	–0.25	100.15	12.3	70	38	226	2	51	195	10	15	8	22	100	133	6	6
CF96–23	EC	Wumiao*	39.98	12.82	20.93	0.32	4.01	10.78	2.28	0.17	4.77	2.79	0.08	98.93	14.6	30	66	496	5	47	128	16	10	56	23	76	165	15	11
CF96–24	AM	Wumiao*	35.77	13.21	17.32	0.27	7.23	9.18	1.74	0.82	5.78	0.15	7.82	99.29	11.5		22	335	23	48	540	91	110	232	24	57	181	1	13
CF96–25	EC	Wumiao*	42.51	16.40	18.93	0.38	3.43	9.05	2.63	1.15	2.49	0.91	1.20	99.08	14.2		81	269	28	30	112	28	28	529	19	142	113	3	16
BJ93–31	EC	Shuanghe	49.34	13.17	17.05	0.23	5.53	9.42	1.98	0.00	2.52	0.29	0.66	100.19	11.9	114	33	118	1	49	537	58	136	36	21	60	130	3	2
BJ93–34	AE	Shuanghe	46.56	16.95	12.48	0.30	7.81	8.36	2.89	1.45	1.39	0.50	0.43	99.12	4.7	141	29	319	27	49	248	129	335	857	23	25	115	2	9
BJ93–36	EC	Shuanghe	46.89	15.89	13.35	0.18	7.38	11.49	2.25	0.12	1.61	0.59	0.14	99.89	1.5	45	19	491	2	45	357	59	175	98	19	38	80	2	5
BJ93–37	EC	Shuanghe	40.89	15.09	17.21	0.22	6.87	11.22	2.37	0.06	3.81	2.13	–0.23	99.64	1.2	37	26	415	1	23	278	<1	16	37	19	24	121	3	3
BJ95–11	EC	Shuanghe	50.45	13.32	17.55	0.20	5.22	8.10	1.68	0.01	2.50	0.31	0.80	100.14	18	169	39	185	2	56	372	51	52	66	20	115	119	2	5
BJ95–12	EC	Shuanghe	47.53	15.77	13.73	0.18	6.46	10.34	2.85	0.50	1.58	0.57	0.43	99.94	2	50	19	449	13	43	356	38	121	332	19	23	97	3	8
BJ95–13	EC	Shuanghe	46.89	16.71	14.61	0.18	7.03	10.61	2.08	0.03	1.82	0.27	0.21	100.44	2	48	17	486	2	47	364	44	86	70	19	41	78	2	4
BJ93–39	CE	Tanshuao	56.53	14.58	11.20	0.23	3.07	6.45	4.57	0.10	1.63	0.64	0.02	99.02	5.7	209	36	372	2	26	205	0	13	108	24	7	143	<1	29
BJ93–39a	CE	Tanshuao	48.96	16.29	10.87	0.17	7.61	10.52	3.49	0.22	1.14	0.31	0.35	99.93	2.9	101	20	428	7	44	278	91	225	169	19	36	95	1	7
BJ93–40	GA	Huangzhen	47.53	16.51	14.94	0.32	4.79	7.14	3.44	0.84	2.21	0.35	1.27	99.34	7.3	184	58	289	17	40	342	11	49	450	22	33	145	4	8
BJ95–20	CE	Zhujiachung	53.08	16.34	10.54	0.22	4.68	7.94	4.30	0.65	1.24	0.30	0.38	99.67	6	189	32	446	16	32	260	39	137	410	21	14	169	1	15
BJ95–21	CE	Zhujiachung	48.82	16.70	11.09	0.20	8.00	8.78	3.38	1.13	1.12	0.31	0.57	100.10	4	111	25	335	28	46	252	109	286	714	17	38	126	1	8
BJ95–22	CE	Huangzhen	50.67	16.75	12.58	0.24	4.65	7.95	4.39	0.12	1.82	0.49	0.32	99.98	4	144	36	302	4	31	220	8	10	84	19	8	149	2	5
CF96–49a	CE	Hualiangting	48.07	16.52	13.60	0.20	7.94	6.43	4.00	0.87	1.57	0.28	0.34	99.82	2.2		26	129	32	55	272	170	338	579	16	40	111	0.7	4.1
CF96–49b	CE	Hualiangting	46.85	16.08	12.12	0.19	9.14	10.31	2.15	0.77	1.15	0.20	0.72	99.68	1.3		20	397	26	50	279	158	372	409	16	78	84	1.2	4.1
Type II: enclaves in or interlayers with marbles																													
BJ93–15	EC	Taolichung	45.51	15.76	13.11	0.20	5.78	13.03	1.44	0.67	1.49	0.14	1.69	98.82	16.8	121	35	169	19	26	248	14	77	340	17	25	97	<1	19
BJ93–29	EC	Xindian	45.24	13.19	10.71	0.14	6.96	12.23	3.75	0.80	0.73	0.04	5.11	98.90	4.4	66	19	260	14	33	242	64	157	272	16	56	91	<1	26
Type III: enclaves in or interlayers with ultramafic rocks																													
BJ93–02	AM	Bixiling	52.17	13.83	11.84	0.30	3.66	7.68	3.37	1.67	2.30	0.71	1.03	98.56	5.7	127	50	177	39	23	229	6	11	487	18	19	99	2	7
BJ93–03	EC	Bixiling	48.53	18.59	7.58	0.14	7.19	13.69	1.91	0.00	0.65	0.09	0.51	98.88	1.1	47	14	458	2	25	158	89	647	31	14	81	49	<1	6
BJ93–04	EC	Bixiling	50.58	16.10	9.09	0.17	8.39	11.26	2.40	0.25	0.73	0.10	0.35	99.42	1.2	33	12	235	7	40	198	45	217	668	15	46	59	<1	2
BJ93–05	EC	Bixiling	48.91	17.57	8.78	0.15	7.70	13.06	2.06	0.26	0.68	0.03	0.39	99.59	0.8	24	12	209	6	33	226	90	429	72	15	48	60	1	2
BJ93–06	EC	Bixiling	49.77	16.65	8.14	0.16	8.96	12.07	2.25	0.20	0.42	0.02	0.44	99.08	0.7	20	10	174	6	35	158	66	109	105	14	41	56	<1	3
BJ93–07	EC	Bixiling	50.77	15.90	8.87	0.17	8.40	11.22	2.41	0.27	0.71	0.11	0.38	99.21	1.2	35	12	231	8	38	201	46	221	820	15	50	57	<1	<1
BJ93–08	GP	Bixiling	43.16	9.75	15.29	0.23	24.63	4.98	0.47	0.00	0.34	0.02	0.29	99.16	1.3	19	4	185	0	127	77	760	1416	15	7	46	110	1	2
BJ93–09	GX	Bixiling	45.57	2.43	16.81	0.20	30.44	1.99	0.14	0.00	0.53	0.01	0.94	99.06	0.9	14	5	91	1	154	116	890	2766	25	5	70	152	2	4
BJ93–25	EC	Bixiling	45.70	13.47	18.56	0.22	7.01	10.98	2.40	0.05	2.34	0.01	–0.49	100.25	3.5	36	26	60	<1	51	794	22	42	68	22	27	124	<1	5
BJ95–07	EC	Bixiling	52.72	14.98	12.39	0.30	3.86	7.00	4.39	0.93	2.37	0.54	0.26	99.74	7	161	45	231	24	22	248	4	14	475	23	12	123	2	6
BJ95–09	EC	Bixiling	49.34	20.61	8.30	0.13	7.63	10.46	2.55	0.16	0.67	0.14	0.41	100.40	1	40	12	488	5	35	126	126	283	164	15	56	56	0	3
BJ93–20	EC	Taolichung	40.32	14.91	22.26	0.32	4.04	7.73	0.80	0.06	11.43	0.06	–0.81	101.12	38	46	51	23	<1	42	1190	35	68	<1	21	51	103	1	3
BJ95–17	OX	Maowu	54.38	3.43	7.04	0.12	33.36	0.36	0.00	0.00	0.19	0.01	0.50	99.39	2	28	6	3	1	78	24	1729	2586	10	4	4	59	<1	<1
BJ95–18	OX	Maowu	52.55	0.75	5.73	0.06	37.17	0.11	0.00	0.00	0.10	0.22	2.34	99.03	1	3	<1	26	1	84	21	2488	2590	13	1	2	55	<1	<1
BJ95–19	WE	Maowu	51.11	9.29	8.33	0.19	29.17	1.51	0.00	0.00	0.22	0.18	0.27	100.27	1	100	18	15	1	76	42	1082	1277	24	6	3	51	22	3
AM001	EC	Maowu*	48.72	15.44	13.52	0.17	6.93	10.08	3.89	0.04	1.81	0.09	0.04	100.73	3	80	7	44	0.6	34	335	98	143	15	144		103	15	
AM19–2	EC	Maowu*	48.93	16.35	11.49	0.17	11.32	8.47	3.74	0.03	0.41	0.03		100.94	4	12	8	103	0.53	30	211	160	621	12	4		68	22	

Major and trace elements analyzed by XRF (Philips PW1480 spectrometer) in Rennes, except for three samples (Wumiao*) whose trace elements were analyzed by ICP-MS at Guangzhou. Uncertainties: ±1% to 3% for major elements; ± 5% for trace elements ≥ 20 ppm and ± 10% for those ≤ 20 ppm. Fe_2O_3 represents total iron oxides. Maowu* are from Fan et al. (1996).

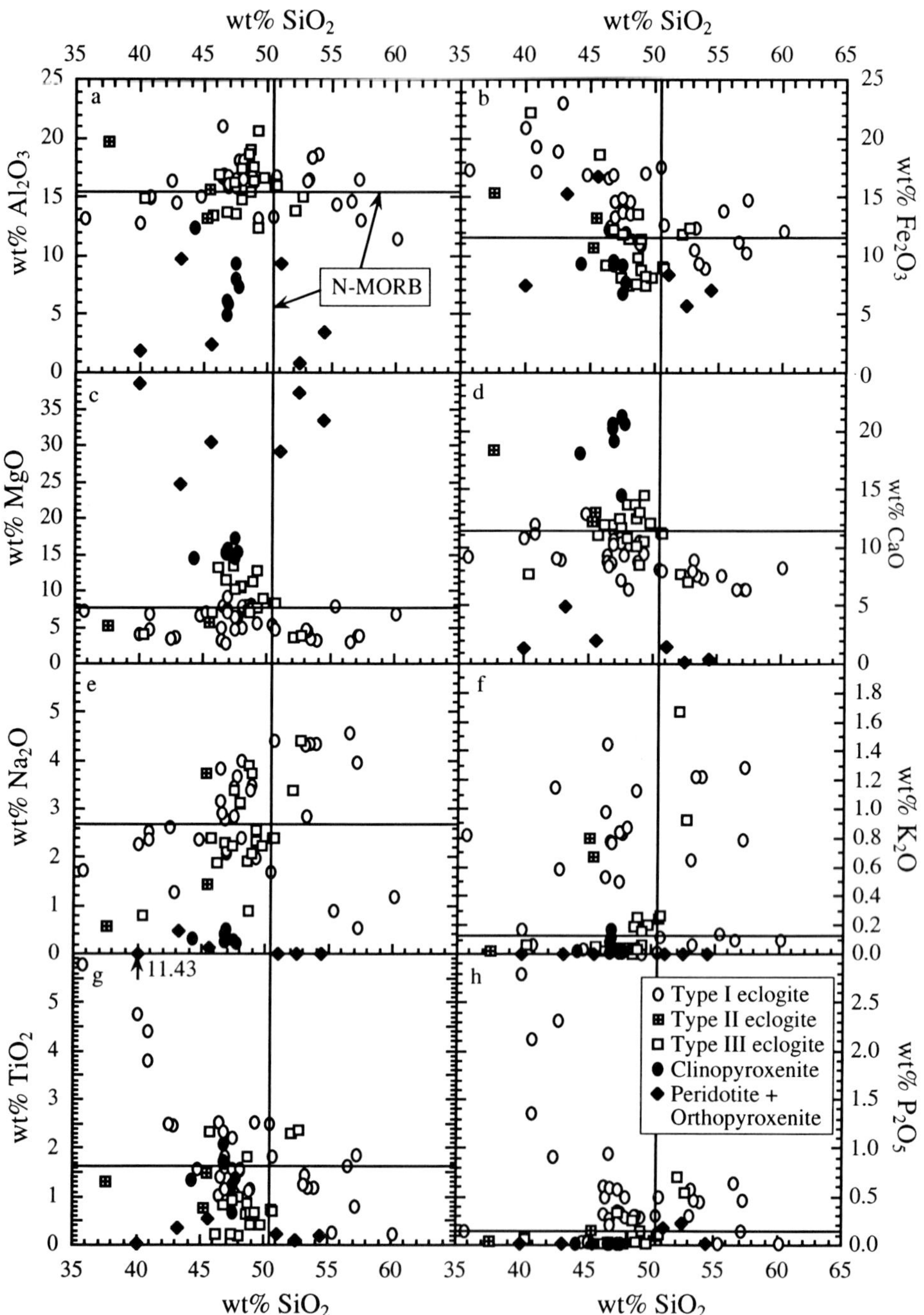

Figure 5. Major element variations of eclogites and associated ultramafic rocks from the Su–Lu and Dabie areas. The eclogites are generally basaltic and quartz-normative, with some showing the cumulate nature of their protoliths. No clear distinction is visible among the three types of eclogite.

probably derived from marls or siliceous carbonates, appear to have higher CaO than type I eclogites. Ultramafic rocks, peridotites and pyroxenites, are distinguishable from eclogites based on the abundances of Al_2O_3, MgO, CaO, Na_2O, and possibly K_2O.

Traditional classification based on SiO_2 vs. total alkalis for these rocks is misleading because alkalis like K_2O might have been mobilized during UHP metamorphism and subsequent exhumation. This is demonstrated by ubiquitous Rb depletion relative to Sr as deduced from the Rb-Sr isotopic systematics shown later. The large compositional variability observed in the UHP eclogites may be explained by the heterogeneity of their protoliths and possible metasomatism before, during or after eclogite-facies metamorphism. The protolithic heterogeneity includes the original variation in source compositions which gave rise to basaltic or gabbroic rocks, and the contrasting chemistry between cumulate rocks and residual liquids. Beard *et al.* (1992) suggested that variable accumulation of garnet and clinopyroxene may explain the wide scatter in major element variation of eclogites. However, garnet and clinopyroxene (omphacite) are metamorphic minerals, they cannot accumulate as in magma differentiation processes, but they could be concentrated via metamorphic differentiation processes. It is difficult to obtain a genuinely representative whole-rock sample for a coarse-grained rock with compositional banding produced by metamorphic segregation. Consequently, the compositional variability is due to both igneous and metamorphic processes, and in some cases, compounded by metasomatic effects.

Some eclogites of andesitic compositions, particularly those from Weihai (SiO_2 = 54–60%, Table 1), are demonstrably of metasomatic origin (Jahn *et al.*, 1996). Their protoliths were originally of basaltic composition as inferred from relatively high MgO (6.3–7.8%), Ni (73–200 ppm) and Cr (280–410 ppm) contents. Andesitic rocks of island arc or continental margin are renowned for their Ni (and Cr) depletion, but this is not the case for the Weihai eclogites. Moreover, the Weihai eclogites have been severely depleted in K, Rb, Sr and LREE, which is manifest by the extraordinarily high Sm/Nd ratio (1.8–2.3) produced during earlier metasomatism. Such high Sm/Nd ratios led to rapid growth of radiogenic Nd, producing world-record high $^{143}Nd/^{144}Nd$ ratios of about 0.522–0.526 at the present time ($\varepsilon_{Nd}(0)$ = +170 to +260; Jahn *et al.*, 1996).

A few samples from Dabie Shan have very high TiO_2 (4–11%), Fe_2O_3 (19–22%) and V (200–1190 ppm) contents, comparable with those observed in the ferrogabbros of some layered intrusions (*e.g.*, Kiglapait, (Morse, 1981); Cayman Trough, (Eggler *et al.*, 1973)). The high TiO_2 eclogites do not represent a melt composition, and are likely of cumulate origin. Two eclogites with high Ti and Fe contents (TiO_2 = 3.4–3.8%; ΣFeO = 21–20%) from the Bixiling mafic-ultramafic layered massif have also been reported (Zhang *et al.*, 1995).

Trace element abundances in eclogites show variation that exceeds the entire field of basaltic rocks and the total range of liquid compositions produced by magmatic differentiation. Such considerable variability has also been reported for the lower crust xenolith suites (eclogites and granulites) from Lesotho (Griffin *et al.*, 1979), southeastern Australia, and the southwestern Colorado Plateau (Arculus *et al.*, 1988). In other words, the eclogites are not isochemically metamorphosed basalts. Fig. 6 shows that eclogites from Dabie and Su–Lu have highly variable Ti/V ratios ranging from 5–240. This range encompasses all fields of basaltic rocks, from island arc tholeiites (IAT) through MORB and back-arc basin basalt (BABB) to ocean island basalts (OIB). Eclogites of very low Ti/V ratios (~5) are from Weihai where a metasomatic origin has been identified (Jahn *et al.*, 1996). By contrast, eclogites of very high Ti/V ratios (≥100) are exclusively from Wumiao (Dabie Shan) where microdiamonds have been reported (Xu *et al.*, 1992). No relation between the high Ti/V and diamond can be inferred. Overall, the majority of the data fall in the fields for MORB, BABB and continental flood basalts (CFB), with Ti/V in the range of 20–50.

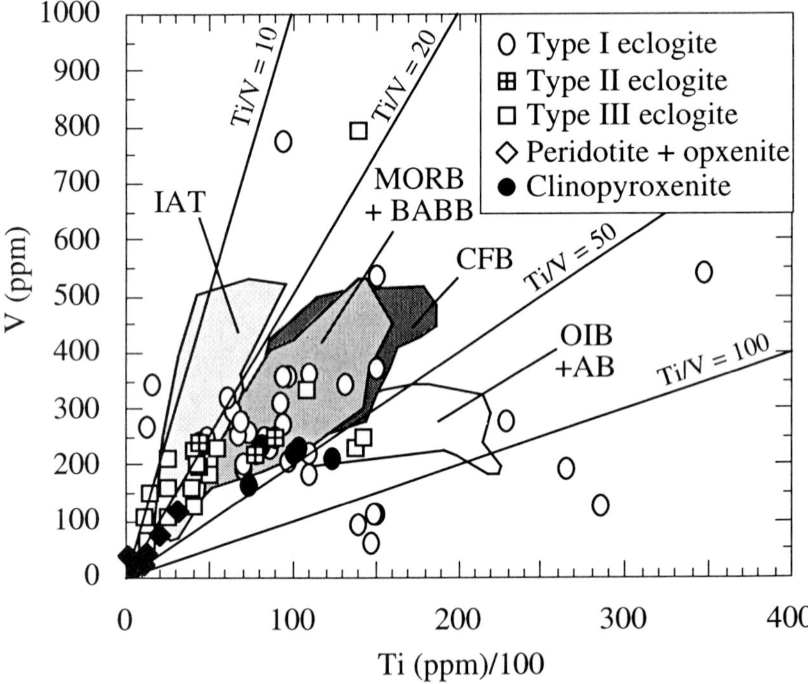

Figure 6. V vs. Ti shows that the UHP eclogites are widely scattered, exceeding the entire field of basaltic rocks. IAT = island arc tholeiite, MORB = mid-ocean ridge basalt, BABB = back-arc basin basalt, CFB = continental flood basalt, OIB = ocean island basalt, AB = alkaline basalt. Reference fields from (Shervais, 1982).

Ba/Nb ratios vary from ~1 to 250 and La/Nb ranges from 0.3–15. Ba/Nb ratios for most oceanic basalts (MORB, OIB) are ≤20, whereas island arc volcanic rocks average about 200. Most N-MORB, OIB and alkali basalts have La/Nb ratios ranging from 0.5–2.5. Many alkali basalts and associated rocks (nephelinites, lamprophyres, carbonatites, etc.) show positive Nb-Ta anomalies in "spidergrams" and low La/Nb ratios (Fitton and Upton, 1987). By contrast, island arc basalts and rocks of continental affinity (granitoids, granulites, sediments, etc.) are characterized by negative Nb-Ta anomalies and show much more dispersed but relatively high La/Nb ratios (1–15). Although the range of La/Nb ratios in the eclogites is high, it is nevertheless remarkably similar to that commonly observed in ancient amphibolites and basic granulites in Precambrian terranes (Jahn, 1990a; 1990b).

Ni and Cr contents in the eclogites are highly variable but roughly correlated, with Cr/Ni ~2–4. The high Ni (300–900 ppm) and Cr (1000–3600 ppm) concentrations in garnet pyroxenites of Rizhao (Hujialin) are consistent with a cumulate origin. Rocks produced in arcs are depleted in Ni and Cr, and the protoliths of eclogites with very low Ni and Cr contents may have formed in ancient subduction zones.

REE analyses by isotope dilution and ICP-MS are given in Table 3. Chondrite-normalized REE patterns for Type I eclogites from Su–Lu and Dabie (Fig. 7) are characterized by LREE enrichment, as commonly observed in continental basalts and amphibolites of Precambrian terranes (Hall and Hughes, 1990). LREE-depleted MORB-type patterns have not been identified in this study. However, there are a few LREE-depleted Type III eclogites from Junan (southern Su–Lu) that occur in association with serpentinites (Yang, 1991; Zhang *et al.*, 1994).

Spidergrams of Type I eclogites from Su–Lu and Dabie are shown in Fig. 8. Rock standard BCR-1, a continental flood basalt (CFB) from the Columbia River basalt province, was analyzed and shown for comparison. The conspicuous negative Nb and Zr anomalies are a classical signature of magmas emplaced in island arc or continental crust. Negative Ti anomalies are clear in Su–Lu but are not so pronounced in Dabie eclogites except for Wumiao (Fig. 8b). Again, none of the spidergrams mimic the abundance patterns of oceanic basalts.

In conclusion, the geochemical characteristics of the eclogites from the Dabie orogen can be summarized as follows:

1. They are mainly basaltic, mostly quartz-normative (not illustrated), and some are of cumulate origin. They show significant variation in major element compositions (Fig. 5): SiO_2 from ~40–60%, 7–20% total Fe_2O_3, 3–14% MgO, 6–15% CaO, 0.6–4.6% Na_2O, 0–1.7% K_2O, and 0.1–11% TiO_2.

2. The variation in immobile elemental ratios (Ti/V, La/Nb) exceeds the entire range of basaltic rocks from different tectonic settings. REE patterns of Type I

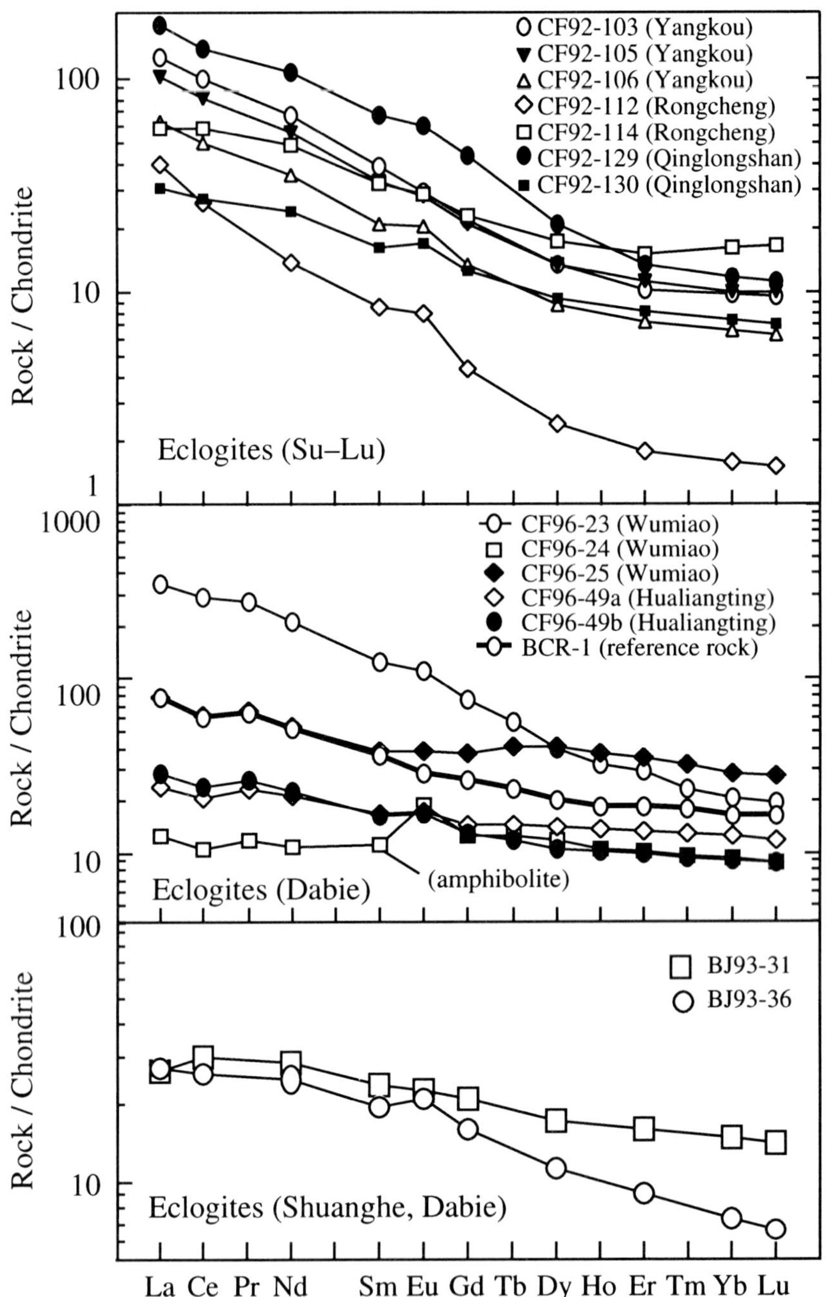

Figure 7. REE distribution patterns of eclogites. All are Type I except one Type III sample from Rongcheng (CF92–112). The common feature is the LREE enrichment which contrasts with N-MORB but resembles continental mafic rocks (basalt, amphibolite and basic granulite). Chondrite values are from Masuda *et al.* (1973) divided by 1.2.

Table 3. REE Abundances (in ppm) in Eclogites and Ultramafic Rocks from the Su–Lu and Dabie Areas.

Sample	Rock	Locality	La	Ce	Pr	Nd	Sm	Eu	Gd	Tb	Dy	Ho	Er	Tm	Yb	Lu
Type I: enclaves in granitic gneisses, Su–Lu area																
CF92-103	EC	Yangkou	39.92	81.48		39.84	7.50	2.12	5.61		4.35		2.18		2.05	0.308
CF92-105	EC	Yangkou	32.40	66.17		33.20	6.35	2.03	5.39		4.41		2.40		2.09	0.321
CF92-106	EC	Yangkou	19.80	40.45		21.21	3.98	1.46	3.51		2.82		1.54		1.37	0.203
CF92-107	EC	Wehai	0.13	0.26		0.59	1.35	0.76	3.55		5.14		3.53		3.55	0.551
CF92-108	EC	Wehai	0.04	0.12		0.56	1.02	0.60	2.89		4.60		3.34		3.30	0.511
CF92-109a	GG	Wehai	7.46	25.95		19.33	4.81	1.20	5.53		6.26		4.05		4.04	0.647
CF92-109b	GG	Wehai	21.57	73.59		52.94	11.53	2.49	10.62		9.51		5.08		4.35	0.656
CF92-114	EC	Rongcheng	18.64	47.27		29.03	6.21	2.09	5.98		5.66		3.20		3.40	0.536
CF92-129	EC	Qinglongshan	56.30	111.35		63.57	12.95	4.31	11.26		6.80		2.89		2.42	0.362
CF92-130	EC	Qinglongshan	9.74	22.24		14.08	3.16	1.22	3.26		3.05		1.73		1.52	0.227
Type II: enclaves in or interlayers with marbles, Su–Lu area																
CF92-116	GE	Rongcheng	0.37	1.23		2.67	3.13	1.53	5.80		5.96		3.33		2.81	0.410
Type III: enclaves in or interlayers with ultramafic rocks, Su–Lu area																
CF92-111	GL	Rongcheng	0.53	1.11		0.56	0.11	0.03	0.12		0.14		0.10		0.12	0.019
CF92-112	EC	Rongcheng	12.64	21.20		8.26	1.65	0.57	1.14		0.77		0.38		0.33	0.049
CF92-120	GX	Rizhao	1.46	5.57		6.27	2.00	0.70	2.25		1.70		0.66		0.42	0.056
CF92-122	GX	Rizhao	2.17	7.22		8.94	2.74	1.01	2.92		2.01		0.72		0.42	0.055
CF92-123	GX	Rizhao	2.21	7.61		8.57	2.62	0.81	2.79		1.91		0.72		0.41	0.055
CF92-124	GX	Rizhao	2.05	7.90		9.07	2.64	0.80	2.71		1.86		0.69		0.40	0.051
Type I: enclaves in granitic gneisses, Dabie area																
BJ93-31	EC	Shuanghe	8.32	24.24		16.89	4.57	1.62	5.38		5.61		3.36		3.05	0.457
BJ93-36	EC	Shuanghe	8.50	21.34		14.89	3.72	1.51	4.09		3.68		1.91		1.52	0.214
CF96-23	EC	Wumiao	108.67	233.49	31.56	126.85	24.07	8.00	19.29	2.80	13.00	2.34	6.23	0.71	4.27	0.63
CF96-24	AM	Wumiao	3.91	8.64	1.37	6.45	2.16	1.35	3.28	0.63	3.90	0.78	2.18	0.30	1.93	0.29
CF96-25	EC	Wumiao	24.13	50.23	7.49	31.36	7.31	2.78	9.68	2.04	13.23	2.72	7.51	0.99	6.02	0.89
CF96-49a	CE	Hualiangting	7.59	16.75	2.70	12.61	3.25	1.26	3.72	0.73	4.58	1.00	2.84	0.39	2.57	0.38
CF96-49b	CE	Hualiangting	9.11	19.52	3.03	13.45	3.16	1.20	3.35	0.59	3.47	0.74	2.13	0.29	1.88	0.29
BCR-1		basalt standard	24.64	48.36	7.17	30.20	6.93	2.09	6.76	1.16	6.61	1.33	3.93	0.54	3.43	0.52
Chondrite normalization values:			0.315	0.813	0.115	0.595	0.193	0.072	0.259	0.050	0.325	0.073	0.213	0.031	0.208	0.032

All analyses except CF96- and BCR-1 were done by isotope dilution in Rennes; uncertainties ± 2% for all elements except La and Lu (± 3%). The last five samples (CF96-23 to CF96-49b) and BCR-1 were analyzed by ICP-MS at Guangzhou Institute of Geochemistry; uncertainties = 3-5%.

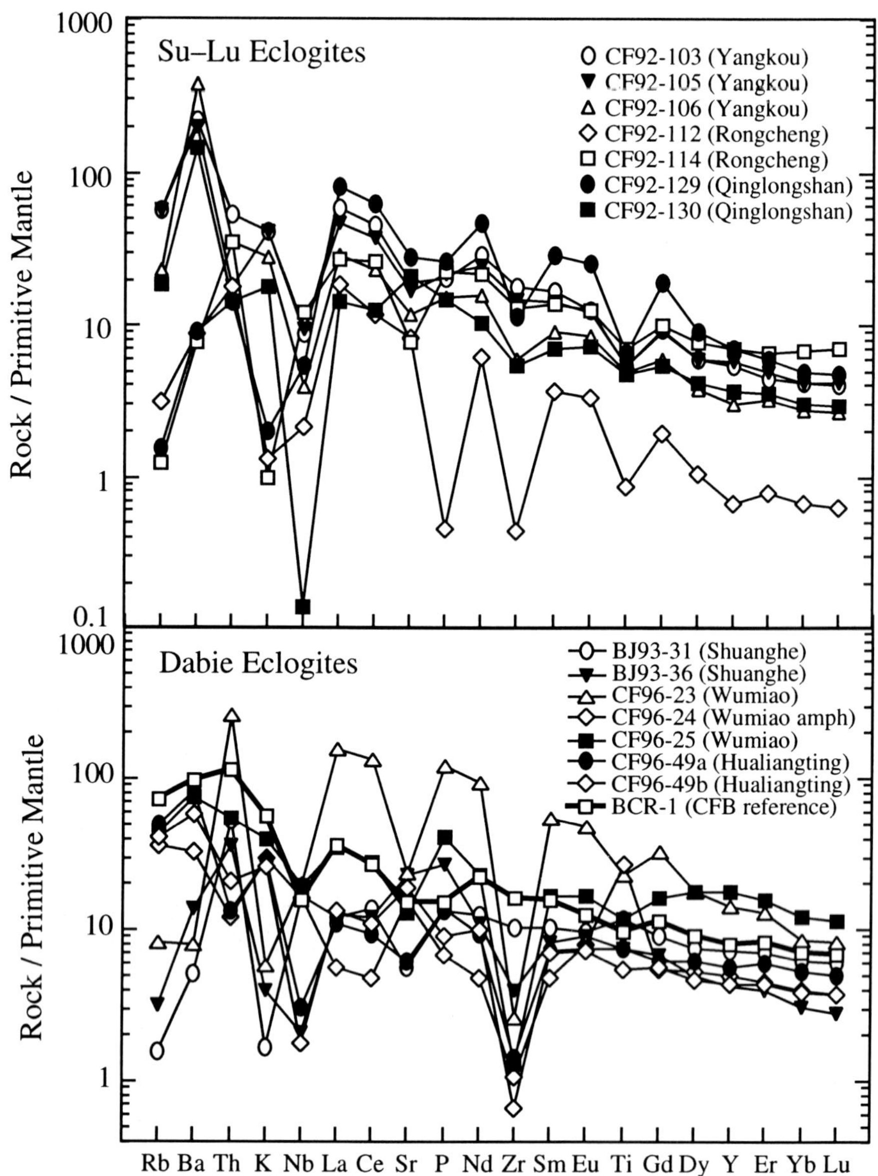

Figure 8. Primitive-mantle (PM) normalized spidergrams of eclogites from the Su–Lu and Dabie areas. Negative Nb and Zr anomalies are a classical signature of magmas emplaced in island arcs or continental crust. Negative Ti anomalies are clear in Su–Lu eclogites but are not so pronounced in Dabie eclogites except for Wumiao. PM values from Sun and McDonough (1989).

eclogites show LREE enrichment. In spidergrams, significant negative anomalies in Nb, Zr and Ti are observed. This indicates a continental affinity for most Type

I and II eclogite protoliths. Very large variation in mobile elements (*e.g.*, K, Rb) is also observed, but no unique interpretation is possible.

3. Typical normal MORB has not been identified as a protolith of the eclogites studied. The geochemical variability cannot be explained by magmatic differentiation alone. Multiple sources of protoliths are more probable and metasomatism has imprinted some eclogites (*e.g.*, Weihai). To some extent, the chemical variation may be ascribed to metamorphic segregation (*e.g.*, garnet-rich Type II eclogite from Rongcheng, CF92–116, Table 1). Nevertheless, the overall feature of chemical variability is reminiscent of that observed in the mafic components, such as amphibolites or basic granulites, of granitic-gneiss dominated Precambrian terranes (Hall and Hughes, 1990).

4.2 Nd–Sr Isotopic Characteristics

Rb-Sr and Sm-Nd isotopic data on whole rocks are presented in Tables 4 and 5. Analyses of individual phases are obviously vital to studying the ages of various events (peak metamorphism, cooling, reheating, etc.) and to understanding the isotopic evolution and petrogenesis of the eclogites, but this aspect is reported elsewhere (Chavagnac and Jahn, 1996; Jahn *et al.*, 1996; Jahn *et al.*, in preparation-a; in preparation-b) and is beyond the scope of this chapter. Petrogenetic interpretation based only on whole-rock analyses can be in error if post-magmatic and post-metamorphic modifications of chemical and isotopic compositions have taken place, however, whole-rock analyses provide the most direct means of protolith identification. Some authors have tried to use clinopyroxene and garnet analyses to reconstitute whole-rock compositions (Shervais *et al.*, 1988; Jerde *et al.*, 1993), but the validity of this approach hinges on the assumption that the bulk rocks contain only two phases and both phases are free of inclusions. This approach is not favored in the present study.

Figure 9 shows the Rb-Sr isotopic compositions of the eclogites and ultramafic rocks from Su–Lu and Dabie. The data for Type I and II eclogites are highly dispersed, suggesting that many (most?) Rb-Sr systems were not closed. A significant number of data show relatively high $^{87}Sr/^{86}Sr$ (0.705–0.710) but very low $^{87}Rb/^{86}Sr$ ($\leq$0.02) ratios. These high $^{87}Sr/^{86}Sr$ ratios are unsupported by their Rb/Sr ratios, indicating preferential Rb depletion. This point has also been made by Ames *et al.* (1996). Because of the highly mobile nature of Rb relative to Sr, no unique interpretation is possible for the scatter of these data.

Type III eclogites and associated ultramafic rocks of the Bixiling massif (Dabie) and garnet pyroxenites of the Rizhao area (Su–Lu) have the lowest $^{87}Sr/^{86}Sr$ ratios (0.703–0.7042). Although the intrusive age of the Bixiling layered massif is not precisely known, the very low Rb/Sr ratios in three samples (an eclogite and two garnet pyroxenites) readily suggest that the initial $^{87}Sr/^{86}Sr$ ratio

Table 4. Sr-Nd Isotopic Compositions of Eclogites and Ultramafic Rocks from the Su–Lu Area.

Sample	Rock	Locality	Rb ppm	Sr ppm	$^{87}Rb/^{86}Sr$	$^{87}Sr/^{86}Sr$	±2σ	I_{Sr} (220 Ma)	Sm ppm	Nd ppm	$^{147}Sm/^{144}Nd$	$^{143}Nd/^{144}Nd$	±2σ	ε_{Nd0}	$\varepsilon_{Nd(t)}$ (220 Ma)	$f_{Sm/Nd}$	T_{DM} Ga	Reference
Type I: enclaves in granitic gneisses																		
CF92-103	EC	Yangkou	36.11	395.95	0.264	0.709209	8	0.708383	7.501	39.84	0.1138	0.511673	5	-18.8	-16.5	-0.42	2.24	this study
CF92-105	EC	Yangkou	36.10	343.37	0.304	0.709316	8	0.708365	6.354	33.20	0.1157	0.511675	3	-18.8	-16.5	-0.41	2.28	this study
CF92-106	EC	Yangkou	13.52	238.89	0.164	0.708494	9	0.707981	3.983	21.21	0.1135	0.511764	4	-17.0	-14.7	-0.42	2.10	this study
CF92-107	EC	Wehai	7.40	15.98	1.340	0.711766	9	0.707573	1.347	0.592	1.3802	0.525950	5	260	226.5	6.02	1.67	Jahn et al. (1996)
CF92-108	EC	Wehai	5.24	11.34	1.338	0.712666	8	0.708480	1.015	0.562	1.0937	0.522480	5	192	166.8	4.56	1.61	Jahn et al. (1996)
HNG-1	EC	Wehai							0.845	0.462	1.1097	0.521550	4	174	148.2	4.64	1.43	Jahn et al. (1996)
HNG-2	EC	Wehai							0.934	0.497	1.1398	0.521619	5	175	148.7	4.79	1.39	Jahn et al. (1996)
HNG-3	EC	Wehai							1.022	0.509	1.2188	0.524207	5	226	197.0	5.20	1.67	Jahn et al. (1996)
CF92-109a	GG	Wehai	20.31	90.49	0.649	0.708334	9	0.706303	4.807	19.33	0.1503	0.511825	5	-15.9	-14.6	-0.24	3.16	Jahn et al. (1996)
CF92-109b	GG	Wehai	56.86	174.20	0.944	0.708640	9	0.705686	11.525	52.94	0.1316	0.511792	4	-16.5	-14.7	-0.33	2.51	Jahn et al. (1996)
CF92-114	EC	Rongcheng	1.22	158.10	0.022	0.705592	9	0.705523	6.214	29.03	0.1294	0.511885	4	-14.7	-12.8	-0.34	2.28	this study
CF92-115	EC	Rongcheng	1.09	262.75	0.012	0.705393	9	0.705355	1.997	5.065	0.2384	0.512328	4	-6.0	-7.2	0.21	-5.17	this study
CF92-129	EC	Qinglongshan	1.27	597.05	0.006	0.707738	8	0.707719	12.948	63.57	0.1231	0.511891	3	-14.6	-12.5	-0.37	2.11	this study
QL-1	EC	Qinglongshan	7.92	545.60	0.041	0.706556	11	0.706427	7.260	34.03	0.1290	0.511865	11	-15.1	-13.2	-0.34	2.30	Li et al. (1994)
ZB-4	EC	Zhubian	9.47	658.30	0.043	0.712901	16	0.712766	7.450	31.48	0.1431	0.511652	13	-19.2	-17.7	-0.27	3.21	Li et al. (1994)
Type II: enclaves in or interlayers with marbles																		
CF92-116	GT	Rongcheng	1.13	45.91	0.071	0.709350	8	0.709128	3.135	2.670	0.7098	0.512790	3	3.0	-11.4	2.61	-0.11	this study
Type III: enclaves in or interlayers with ultramafic rocks																		
CF92-111	GL	Rongcheng	0.28	18.71	0.043	0.708087	9	0.707952	0.111	0.555	0.1210	0.512275	4	-7.1	-5.0	-0.38	1.44	this study
CF92-112	EC	Rongcheng	1.04	152.55	0.020	0.705933	9	0.705870	1.650	8.26	0.1208	0.512321	3	-6.2	-4.1	-0.39	1.36	this study
CF92-120	GX	Rizhao	0.92	112.04	0.024	0.703890	8	0.703815	1.996	6.272	0.1924	0.512712	4	1.4	1.6	-0.02	3.12	this study
CF92-121	GX	Rizhao	0.60	80.28	0.021	0.703813	9	0.703747	2.132	6.519	0.1977	0.512790	5	3.0	2.9	0.01	3.41	this study
CF92-122	GX	Rizhao	2.05	127.46	0.047	0.704362	8	0.704215	2.741	8.941	0.1853	0.512704	5	1.3	1.6	-0.06	2.38	this study
CF92-123	GX	Rizhao	1.00	124.01	0.023	0.703980	5	0.703907	2.621	8.573	0.1848	0.512705	5	1.3	1.6	-0.06	2.34	this study
CF92-124	GX	Rizhao	1.29	145.69	0.026	0.704313	9	0.704232	2.641	9.072	0.1760	0.512669	4	0.6	1.2	-0.11	1.94	this study
JZA	GX	Junan	0.76	294.68	0.007	0.705993	19	0.705970	3.241	11.67	0.1679	0.512346	5	-5.7	-4.9	-0.15	2.66	Yang (1991)
JC313B	EC	Junan	38.57	96.47	1.137	0.706701	22	0.703143	1.299	2.41	0.3264	0.512566	6	-1.4	-5.0	0.66	-0.79	Yang (1991)
H06	GX	Rizhao	0.78	75.71	0.030	0.704390	9	0.704296	2.416	6.85	0.2134	0.512564	8	-1.4	-1.9	0.08	151.33	Yang (1991)
Type not defined																		
SL91-12	EC	Rongcheng	0.60	129.50	0.126	0.705550	20	0.705156	0.500	1.69	0.1789	0.512282	8	-6.9	-6.4	-0.09	3.76	Ames *et al.* (1996)
SL91-21	EC	Junan							12.570	54.33	0.1399	0.512708	18	1.4	3.0	-0.29	0.91	Ames *et al.* (1996)
SL91-23	EC	Junan	0.98	364.60	0.073	0.708470	10	0.708242	0.790	4.30	0.1111	0.511268	10	-26.7	-24.3	-0.44	2.78	Ames *et al.* (1996)

$^{143}Nd/^{144}Nd$ ratios have been corrected for mass fractionation relative to $^{146}Nd/^{144}Nd = 0.7219$ and are reported relative to the La Jolla Nd standard = 0.511860 or Ames Nd standard = 0.511966. $^{87}Sr/^{86}Sr$ ratios have been corrected for mass fractionation relative to $^{86}Sr/^{88}Sr = 0.1194$ and are reported relative to the NBS-987 Sr standard = 0.710250. CHUR (chondritic uniform reservoir): $^{147}Sm/^{144}Nd = 0.1967$; $^{143}Nd/^{144}Nd = 0.512638$. DM (depleted mantle): $^{147}Sm/^{144}Nd = 0.2137$; $^{143}Nd/^{144}Nd = 0.51315$.

Table 5. Sr–Nd Isotopic Compositions of Dabie Eclogites and Ultramafic Rocks.

Sample	Rock	Locality	Rb	Sr	$^{87}Rb/^{86}Sr$	$^{87}Sr/^{86}Sr$	± 2 σ	I_{Sr}	Sm	Nd	$^{147}Sm/^{144}Nd$	$^{143}Nd/^{144}Nd$	± 2 σ	ε_{Nd0}	$\varepsilon_{Nd(t)}$	$f_{Sm/Nd}$	T_{DM}	Reference
			ppm	ppm				(220 Ma)	ppm	ppm					(220 Ma)		Ga	
Type I: enclaves in granitic gneisses																		
BJ93–27a	EC	Wumiao	27.71	160.3	0.500	0.709973	7	0.70841	5.48	22.10	0.1499	0.512240	20	–7.8	–6.5	–0.24	2.17	this study
BJ93–27c	EC	Wumiao	21.30	397.5	0.155	0.708533	7	0.70805	12.27	57.06	0.1300	0.512242	7	–7.7	–5.9	–0.34	1.65	this study
BJ93–27d	EC	Wumiao	2.43	214.0	0.032	0.709016	7	0.70891	7.77	28.91	0.1625	0.512237	7	–7.8	–6.9	–0.17	2.70	this study
BJ93–31	EC	Shuanghe	0.19	112.1	0.005	0.710811	6	0.71080	4.57	16.89	0.1636	0.511695	7	–18.4	–17.5	–0.17	4.38	this study
BJ93–36	EC	Shuanghe	2.70	487.3	0.016	0.704919	7	0.70487	3.72	14.89	0.1510	0.512229	7	–8.0	–6.7	–0.23	2.23	this study
BJ95–11	EC	Shuanghe	1.15	182.7	0.018	0.713258	10	0.71320	5.62	21.08	0.1612	0.511729	10	–17.7	–16.7	–0.18	4.08	this study
BJ95–12	EC	Shuanghe	12.19	452.2	0.078	0.705634	7	0.70539	3.62	15.06	0.1451	0.512133	10	–9.9	–8.4	–0.26	2.25	this study
BJ95–13	EC	Shuanghe	0.66	492.6	0.004	0.704941	7	0.70493	3.15	12.47	0.1528	0.512226	6	–8.0	–6.8	–0.22	2.30	this study
BJ93–39	CE	Tanshuao	1.31	364.4	0.010	0.705744	7	0.70571	6.84	33.10	0.1249	0.512029	5	–11.9	–9.9	–0.36	1.92	this study
BJ95–20	CE	Zhujiachung	15.23	447.9	0.098	0.706378	7	0.70607	5.43	25.37	0.1293	0.512069	14	–11.1	–9.2	–0.34	1.95	this study
BJ95–21	CE	Zhujiachung	27.70	332.1	0.241	0.707029	7	0.70627	4.50	21.41	0.1269	0.512108	8	–10.3	–8.4	–0.35	1.82	this study
BJ95–22	CE	Huangzhen	2.34	299.1	0.023	0.705090	7	0.70502	5.69	25.28	0.1361	0.512198	8	–8.6	–6.9	–0.31	1.86	this study
252C	EC	Shuanghe	12.00	616.0	0.056	0.705220	10	0.70504	4.09	18.87	0.1310	0.512212	15	–8.3	–6.5	–0.33	1.73	Okay et al. (1993)
Type II: enclaves in or interlayers with marbles																		
BJ93–15	EC	Taolichung	19.35	165.1	0.339	0.711030	9	0.70997	9.26	24.91	0.2249	0.511656	4	–19.2	–19.9	0.14	–21.97	this study
230D	EC	S of Wumaio	18.00	347.0	0.150	0.710620	10	0.71015	5.43	31.87	0.1030	0.511524	15	–21.7	–19.1	–0.48	2.23	Okay *et al.* (1993)
Type III: enclaves in or interlayers with ultramafic rocks																		
BJ93–02	AM	Bixiling	38.08	172.6	0.638	0.708791	8	0.70679	6.84	25.82	0.1601	0.512457	8	–3.5	–2.5	–0.19	1.96	Chavagnac & Jahn (1996)
BJ93–03	EC	Bixiling	0.14	456.3	0.001	0.703785	5	0.70378	1.96	7.51	0.1578	0.512520	4	–2.3	–1.2	–0.20	1.71	Chavagnac & Jahn (1996)
BJ93–04	EC	Bixiling	5.81	232.3	0.072	0.704056	8	0.70383	1.63	5.55	0.1778	0.512608	6	–0.6	–0.1	–0.10	2.29	Chavagnac & Jahn (1996)
BJ93–05	EC	Bixiling	6.35	207.0	0.089	0.704093	9	0.70382	1.45	4.75	0.1845	0.512596	4	–0.8	–0.5	–0.06	2.88	Chavagnac & Jahn (1996)
BJ93–06	EC	Bixiling	4.40	168.4	0.076	0.704308	6	0.70407	1.03	3.37	0.1839	0.512590	4	–0.9	–0.6	–0.07	2.84	Chavagnac & Jahn (1996)
BJ93–07	EC	Bixiling	6.56	229.7	0.083	0.704063	8	0.70380	1.84	6.74	0.1651	0.512584	4	–1.1	–0.2	–0.16	1.77	Chavagnac & Jahn (1996)
BJ93–08	GP	Bixiling	0.09	177.6	0.002	0.703648	9	0.70364	0.77	2.85	0.1633	0.512500	5	–2.7	–1.8	–0.17	1.96	Chavagnac & Jahn (1996)
BJ93–09	GP	Bixiling	0.22	86.9	0.007	0.703711	9	0.70369	0.43	1.36	0.1919	0.512594	17	–0.9	–0.7	–0.02	3.85	Chavagnac & Jahn (1996)
BJ93–20	EC	Taolichung	1.81	21.1	0.248	0.707385	18	0.70661	1.46	3.44	0.2566	0.512378	7	–5.1	–6.8	0.30	–2.78	this study
BJ95–17	OX	Maowu	0.23	9.0	0.074	0.707603	34	0.70737	0.17	0.33	0.3092	0.512484	23	–3.0	–6.2	0.57	–1.07	this study
BJ95–18	OX	Maowu	0.13	26.2	0.015	0.707269	8	0.70722	0.06	0.23	0.1549	0.512174	13	–9.1	–7.9	–0.21	2.52	this study
BJ95–19	WE	Maowu	0.19	13.9	0.040	0.707243	13	0.70712	3.54	15.85	0.1352	0.512205	12	–8.4	–6.7	–0.31	1.83	this study
R–4	EC	Raobazhai							4.96	14.63	0.2049	0.512515	15	–2.4	–2.6	0.04	10.66	Li *et al.* (1994)
Type not defined																		
DB91–28	EC	Dabie	0.16	69.8	0.058	0.706410	50	0.70623	3.39	11.20	0.1830	0.512460	14	–3.5	–3.1	–0.07	3.40	Ames *et al.* (1996)
DB91–34	EC	Dabie	0.01	47.4	0.001	0.704940	20	0.70494	4.74	20.08	0.1427	0.512168	10	–9.2	–7.7	–0.27	2.10	Ames *et al.* (1996)
Wumiao	EC	Dabie	0.84	220.9	0.108	0.708260	50	0.70792	3.22	18.51	0.1052	0.512277	12	–7.0	–4.5	–0.47	1.23	Ames *et al.* (1996)
DB91–64	EC	Dabie							1.14	7.51	0.0918	0.512401	12	–4.6	–1.7	–0.53	0.94	Ames *et al.* (1996)
DB91–65	EC	Dabie	0.02	144.7	0.004	0.704330	30	0.70432	0.42	1.45	0.1751	0.512526	19	–2.2	–1.6	–0.11	2.45	Ames *et al.* (1996)
DB91–15a	EC	Dabie	0.05	118.4	0.011	0.705860	50	0.70583	3.52	14.10	0.1509	0.512553	9	–1.7	–0.4	–0.23	1.45	Ames *et al.* (1996)
DB91–61–2	EC	Dabie	0.19	138.6	0.004	0.704130	20	0.70412	4.89	38.20	0.0774	0.512397	10	–4.7	–1.4	–0.61	0.84	Ames *et al.* (1996)
Hu–105	EC	Hong'an	0.22	137.1	0.090	0.705240	60	0.70496	3.49	10.65	0.1981	0.512940	8	5.9	5.9	0.01	2.05	Ames *et al.* (1996)
Hu–117	EC	Hong'an	0.48	201.6	0.135	0.706760	30	0.70634	6.36	24.55	0.1566	0.512445	8	–3.8	–2.6	–0.20	1.88	Ames *et al.* (1996)
Hu–113	EC	Hong'an	0.50	251.3	0.116	0.705400	30	0.70504	7.45	27.00	0.1668	0.512531	7	–2.1	–1.2	–0.15	2.01	Ames *et al.* (1996)
Hu–115	EC	Hong'an	0.05	86.3	0.031	0.710630	90	0.71053	2.91	8.95	0.1966	0.513030	9	7.6	7.7	0.00	1.07	Ames *et al.* (1996)
Hu–121	EC	Hong'an	0.01	135.5	0.003	0.705610	70	0.70560	7.95	30.87	0.1557	0.512420	8	–4.3	–3.1	–0.21	1.91	Ames *et al.* (1996)

$^{143}Nd/^{144}Nd$ ratios have been corrected for mass fractionation relative to $^{146}Nd/^{144}Nd = 0.7219$ and are reported relative to the La Jolla Nd standard = 0.511860 or Ames Nd standard = 0.511966. $^{87}Sr/^{86}Sr$ ratios have been corrected for mass fractionation relative to $^{86}Sr/^{88}Sr = 0.1194$ and are reported relative to the NBS–987 Sr standard = 0.710250. CHUR (chondritic uniform reservoir): $^{147}Sm/^{144}Nd = 0.1967$; $^{143}Nd/^{144}Nd = 0.512638$. DM (depleted mantle): $^{147}Sm/^{144}Nd = 0.2137$; $^{143}Nd/^{144}Nd = 0.51315$.

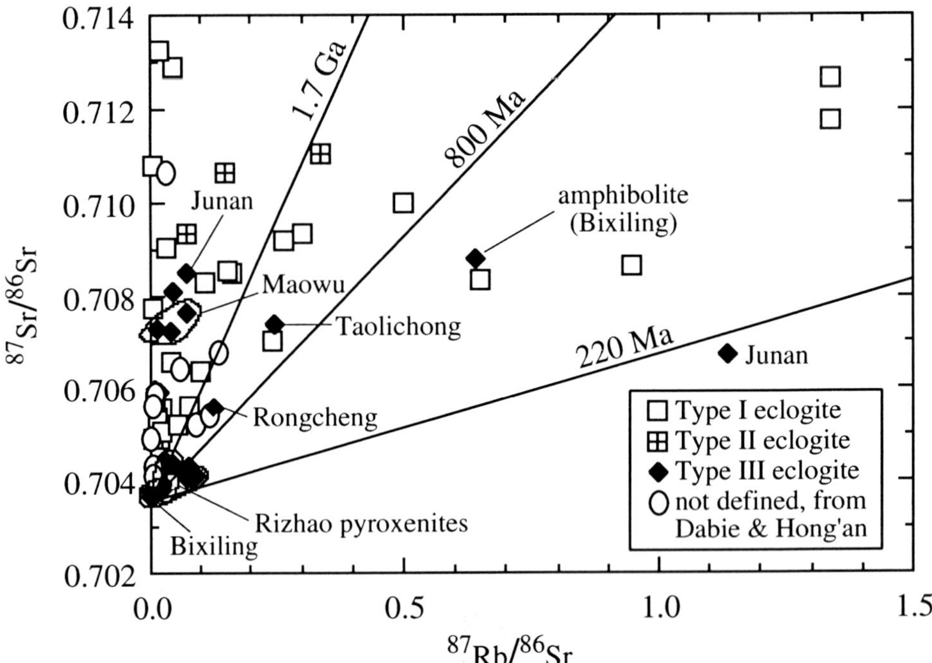

Figure 9. $^{87}Sr/^{86}Sr$ vs. $^{87}Rb/^{86}Sr$ diagram for UHP eclogites of Su–Lu and Dabie. Three reference isochrons based on the protolith age (1.7 Ga) of Weihai eclogites (Jahn *et al.*, 1996), granitic gneisses of the Su–Lu and Dabie complexes (~800 Ma), and the UHP metamorphic event (~220 Ma) are shown for assessment of Rb-Sr isotopic systems in different types of eclogites. Sources: this paper, and Ames *et al.* (1996) for those "not defined". The data are highly dispersed and many show "unsupported" Sr isotopic compositions (data left of the 1.7 Ga isochron).

of their formation must be close to 0.7036. For the garnet pyroxenites of Rizhao the initial $^{87}Sr/^{86}Sr$ ratio could be estimated at 0.7032–0.7036 regardless of age. Because the Bixiling layered massif represents a mantle-derived melt–cumulate association and the Rizhao garnet pyroxenites are probably of mantle origin in view of their intimate association with serpentinite and harzburgite, the $^{87}Sr/^{86}Sr$ ratio of about 0.7036 can be regarded as the Sr isotopic composition of the subcontinental lithosphere beneath the Dabie and Su–Lu regions. Other type III eclogites with unsupported high $^{87}Sr/^{86}Sr$ ratios, such as those of the Maowu layered complex, Junan eclogite and Rongcheng (Chijiadian) peridotite, could have been produced by crustal contamination.

REE are generally considered to be less mobile than Rb and Sr in metamorphic processes. Thus, Sm-Nd isotope tracers may provide more accurate information about the nature of the eclogite protoliths. Fig. 10 shows Sm-Nd isotopic data for eclogites from the European and Pan-African orogenic belts. Eclogites from NW Scotland and the Western Gneiss Region of Norway (a Caledonian belt) have negative $\varepsilon_{Nd}(t)$ values that are clearly different from the positive values for eclogites from the Hercynian and Alpine chains of western Europe and the Pan-African belt of North Africa. Eclogites from the Hercynian, Alpine and Pan-African chains were produced from subducted oceanic crusts;

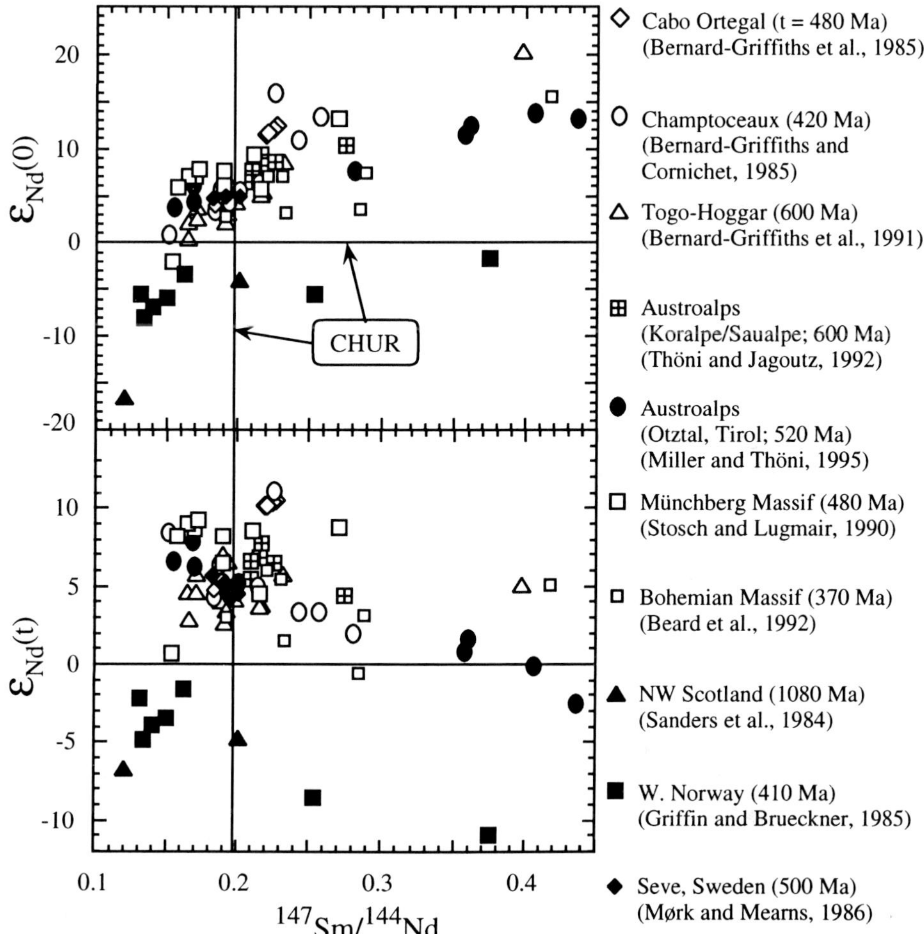

Figure 10. $\varepsilon_{Nd}(0)$ vs. $^{147}Sm/^{144}Nd$ and $\varepsilon_{Nd}(t)$ vs. $^{147}Sm/^{144}Nd$ for eclogites from the Alpine (Austrian Alps), Hercynian (Cabo Ortegal, Champtoceaux, Münchberg Massif, Bohemian Massif), Caledonian (NW Scotland, W. Norway, and Seve, Sweden), and Pan-African (Togo-Hoggar) orogenic belts. The Glenelg eclogite occurs within the Caledonian belt, but was metamorphosed in Grenvillian time (~1000 Ma, Sanders et al. 1984). Rocks with very high $^{147}Sm/^{144}Nd$ ratios (≥0.25) could be garnet-rich eclogites produced by metamorphic differentiation, and hence are less likely to represent "whole-rock" eclogites. If only data with "ordinary" $^{147}Sm/^{144}Nd$ ratios (≤0.25) are considered, most eclogites from the Alpine and Hercynian belts have positive $\varepsilon_{Nd}(t)$ values, suggesting protoliths of oceanic crust or ocean islands (or depleted mantle sources). Most eclogites from the Caledonian belt (in black) have contrasting negative $\varepsilon_{Nd}(t)$ values (–2 to –7), suggesting their affinity with ancient continental crust. However, the Nd isotopic compositions of the Seve eclogites (Sweden) suggest their affinity with subducted oceanic crust.

whereas those from Caledonian Norway and Proterozoic NW Scotland have a continental affinity.

Figures 11 and 12 illustrate the Sm-Nd and Nd-Sr isotopic plots for eclogites from the Su–Lu and Dabie areas. Except for Weihai, Type I and Type II eclogites from both areas have low $\varepsilon_{Nd}(t)$ values (–6 to –20) at the time of peak metamorphism or continental collision (~220 Ma). These values are among the lowest known for eclogite and are significantly different from eclogites in the Hercynian and Alpine chains in western Europe (Fig. 10). To some extent, only eclogites from the Caledonian belt (*e.g.*, NW Scotland and Western Norway) have Nd isotopic compositions similar to those of the Dabie orogen. The isotopic data and geochemical characteristics unambiguously indicate that the eclogite protoliths of the Dabie orogen cannot be subducted Tethyan oceanic slab. The very low $\varepsilon_{Nd}(t)$ values require that the protoliths of these eclogites formed long before the Triassic collision. The Tethys ocean, regardless of its petrological and geochemical nature, was simply too young to have brought about such isotopic evolution.

Type III eclogites also show variable Nd isotopic compositions, but are generally similar to their associated ultramafic rocks. Some of them are close to the mantle isotopic value as exemplified by garnet pyroxenites from the Rizhao area (Fig. 11b). Such a feature is consistent with the idea of their being "mantle xenoliths" of tectonic origin. Both the Sr and Nd isotopic compositions may be characteristic of the lithospheric mantle in the Su–Lu and Dabie areas. The Bixiling eclogites and ultramafic rocks (garnet peridotite and garnet pyroxenite) have a narrow range of $\varepsilon_{Nd}(t)$ values (–0.1 to –2.5) and $^{147}Sm/^{144}Nd$ ratios (0.16–0.19). In the ε_{Nd} vs. $^{87}Sr/^{86}Sr$ diagram (Fig. 12b) these rocks lie just beneath the mantle array, which suggests an involvement of depleted lower crustal material in the petrogenesis and evolution of this layered complex (Chavagnac and Jahn, 1996).

In the Su–Lu region, eclogite and associated garnet lherzolite from Rongcheng, and eclogite and associated garnet pyroxenite from Junan, have $\varepsilon_{Nd}(t)$ values of about –5 (Fig. 11b). In the Dabie Shan, eclogite and associated ultramafic rocks from Maowu have $\varepsilon_{Nd}(t)$ values of –6 to –7 (Fig. 12b). These rocks were probably derived from long-term enriched mantle segments or were subjected to crustal contamination. If the latter, it is not clear whether such contamination was effected in the mantle sources, during magma differentiation (Maowu), or during tectonic transport of mantle slices.

Eclogite samples from the Weihai area show the most spectacular Nd isotopic compositions (off scale in Fig. 11). Their chemistry has been severely modified by metasomatism and they preserve world record whole-rock $\varepsilon_{Nd}(0)$ values ranging from +260 to +174 with extreme Sm/Nd fractionation ($f_{Sm/Nd}$ = +6.1 and +4.6). The Sm-Nd isotopic compositions were used to calculate precise T_{DM} ages

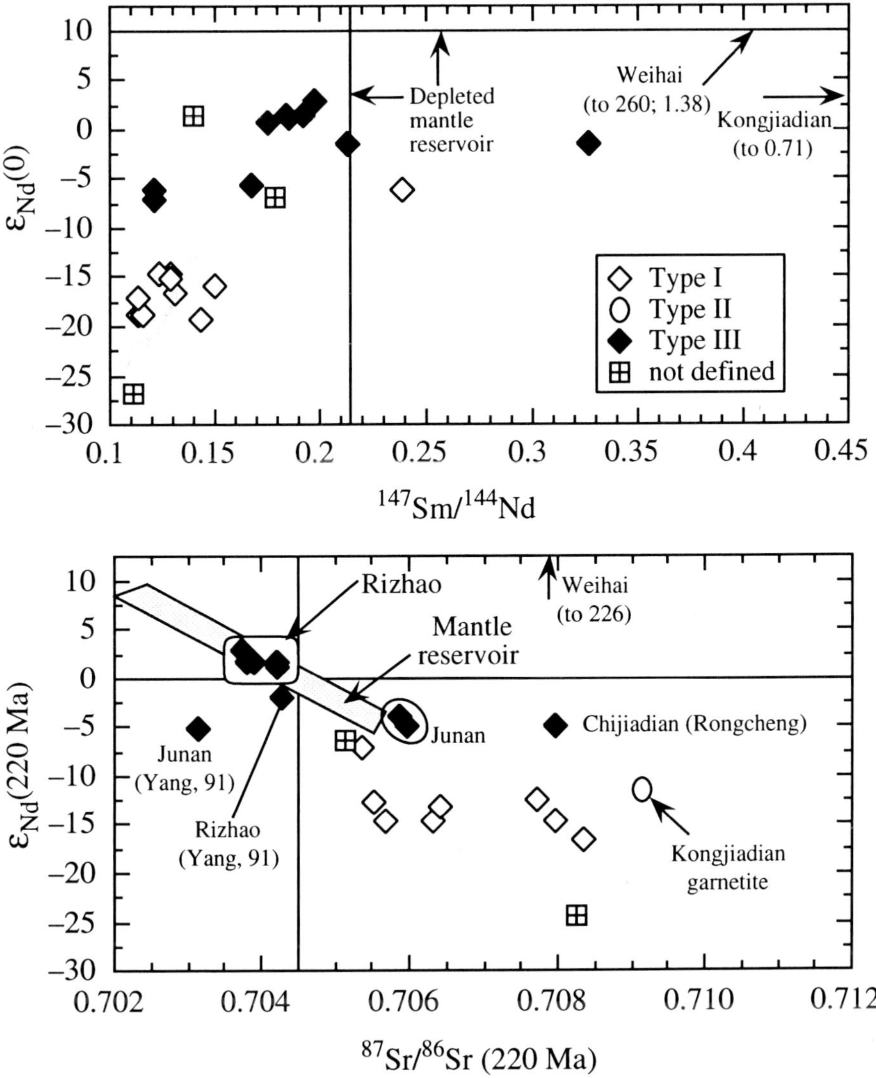

Figure 11. $\varepsilon_{Nd}(0)$ vs. $^{147}Sm/^{144}Nd$ and $\varepsilon_{Nd}(t)$ vs. I_{Sr} (= initial $^{87}Sr/^{86}Sr$) for Su–Lu eclogites. t is assumed to be 220 Ma but does not represent the emplacement time of the protoliths. Eclogites of Types I and II have low $\varepsilon_{Nd}(t)$ values (–7 to –20) and relatively high I_{Sr} ratios (0.705–0.708). Because of the mobility of Rb and possibly Sr during alteration and metamorphism, identification of the protolith using Sr isotopic signature is more difficult. However, the highly negative $\varepsilon_{Nd}(t)$ values suggest that the eclogite protoliths are old (Proterozoic?) mafic rocks emplaced into the continental crust of the Yangtze or Sino-Korean cratons. The Rizhao garnet pyroxenites have Nd-Sr isotopic compositions falling in the "mantle array" (stippled band), pointing to their mantle origin. Type III eclogites have $\varepsilon_{Nd}(t)$ between 0 and –5. This may be explained by a small amount of crustal contamination during the tectonic emplacement of these rocks. Sources: this study, Yang (1991), Li *et al.* (1993b), Ames *et al.* (1996).

of 1.67 and 1.61 Ga, which constrain the minimum time for protolith formation to the middle Proterozoic (~1.7 Ga, Jahn *et al.*, 1996). The metasomatism was not related to the UHP metamorphism, because the small time interval of ~220 Ma is not sufficient to develop such large isotopic growth (270 ε_{Nd} units).

5. DISCUSSION

5.1 Nature and Age of Protoliths of Type I and II Eclogites

The geochemical analyses presented here indicate that the protoliths of eclogites from the Dabie orogen have diverse origins. They were derived from heterogeneous sources and further complicated by fractional crystallization. In some cases, chemical compositions were modified by metamorphic recrystallization and fluid metasomatism, as exemplified by the Weihai eclogites. In others, eclogites in association with marbles and metasediments (Type II) could have been derived from marl or calcareous silicate layers. Overall, chemical (particularly REE) evidence suggests that the Type I eclogites represent the mafic components (amphibolites and basic granulites) of Precambrian gneiss terranes. Further, the generally low ε_{Nd}(220 Ma) values (–6 to –20) indicate that most eclogite protoliths are of Proterozoic age. This is supported by U-Pb zircon upper intercept ages (Ames *et al.*, 1996) and Sm-Nd model ages (Fig. 13). In other words, the protoliths of both Type I and Type II eclogites are integral parts of ancient continental crust. They do not resemble typical ocean floor basalts (MORB-type) nor ophiolites, thus are not part of the subducted Tethyan ocean.
Figure 13 shows $^{147}Sm/^{144}Nd$ vs. T_{DM} diagrams for the Su–Lu and Dabie eclogites. Sm/Nd ratios are a measure of REE fractionation that yields information about the chemical type of eclogite. T_{DM} calculated with $^{147}Sm/^{144}Nd$ ratios close to chondritic or depleted mantle values (shaded area) have large uncertainties or aberrant ages, hence they are excluded from the discussion. Only T_{DM} calculated with $^{147}Sm/^{144}Nd$ ratios ≤0.15 or ≥0.30 may provide a certain degree of age significance. The model ages (T_{DM}) of ~1.7 Ga or the isochron age of about 1.8 Ga for the Weihai eclogites set the minimum age for the emplacement of their protoliths at about 1.7–1.8 Ga (Jahn *et al.*, 1996). It is unlikely that all protoliths of eclogites from different localities formed in the same period, but the $\varepsilon_{Nd}(t)$ values and T_{DM} ages are consistent with their formation in late Archean to Proterozoic times. At least four rocks with Archean T_{DM} ages (2.5–3.2 Ga) are recognized in the Su–Lu area but none in the Dabie Shan (Fig. 13). U-Pb analyses of zircons from Type I eclogites and host granitic gneisses frequently indicate upper intercept ages between 700 and 800 Ma (Li *et al.*, 1993a; Ames *et al.*, 1996; Rowley *et al.*, 1997). This age range was considered by these authors as the approximate protolith ages for some eclogites

and gneisses from the Dabie and Su–Lu complexes. From a compilation of extensive Sm-Nd isotope data and using a model age approach, Chen and Jahn (1998) have concluded that the crustal evolution of the bulk of southeastern China (Yangtze and Cathaysia included) is essentially limited to post-Archean time except for the Kongling gneiss terrane.

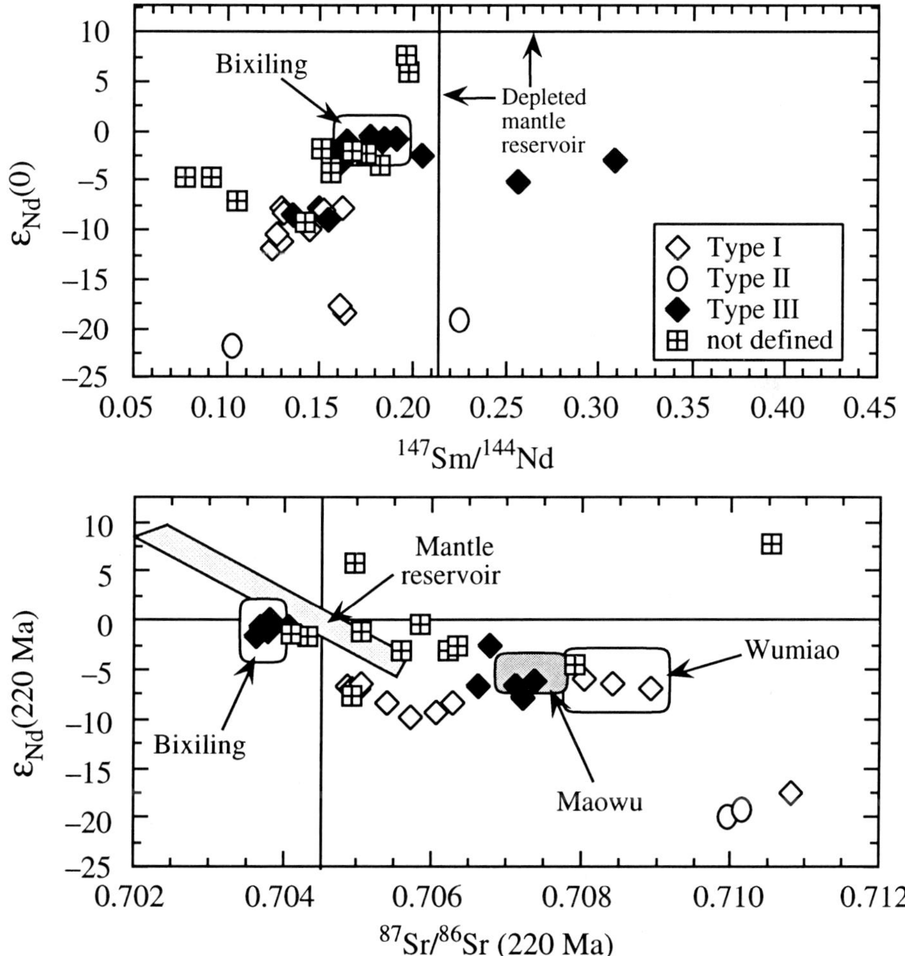

Figure 12. $\varepsilon_{Nd}(0)$ vs. $^{147}Sm/^{144}Nd$ and $\varepsilon_{Nd}(t)$ vs. I_{Sr} for Dabie eclogites. Interpretation is similar to Su–Lu eclogites. The Bixiling complex has probably been contaminated by small amounts of lower crust (Chavagnac and Jahn, 1996); whereas the Maowu complex has seen significant upper crustal contamination. Sources: this study, Okay *et al.* (1993), Ames *et al.* (1996), Chavagnac and Jahn (1996), Jahn *et al.* (in preparation-a).

5.2 Nature and Age of Protoliths of Type III Eclogite–Ultramafic Suites

Type III eclogite-ultramafic suites in the Su–Lu and Dabie areas could have two different origins. Either the suites, such as the Bixiling and Maowu complexes, represent layered intrusions initially emplaced at crustal levels and then subjected

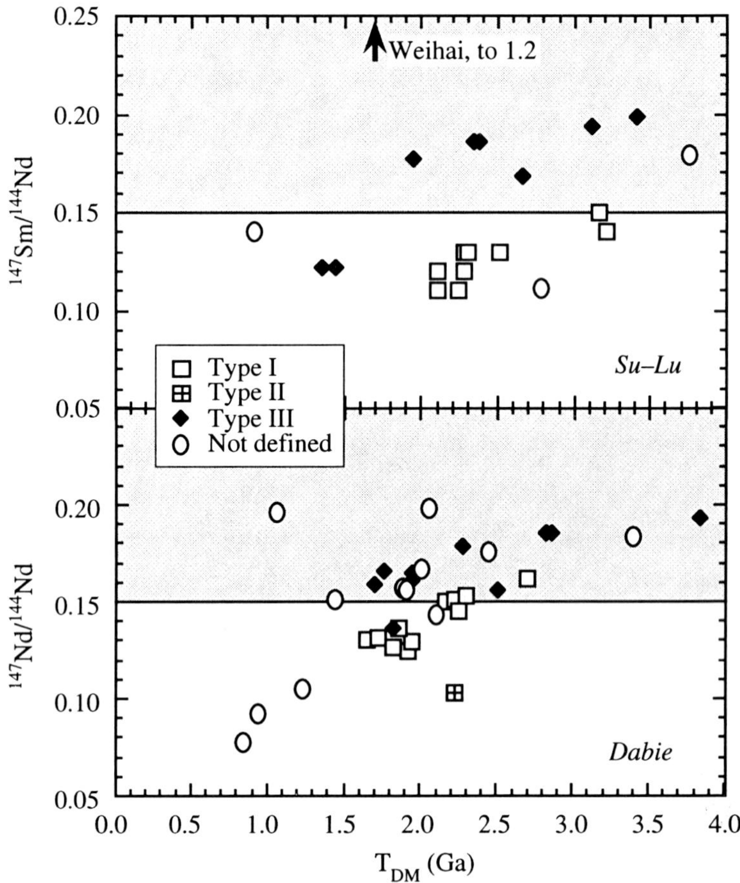

Figure 13. $^{147}Sm/^{144}Nd$ vs. T_{DM} (model age) for Su–Lu and Dabie eclogites. $^{147}Sm/^{144}Nd$ ratios are a measure of REE fractionation, which yields information about the chemical type of eclogite. T_{DM} calculated with $^{147}Sm/^{144}Nd$ ratios close to chondritic (0.1967) or depleted mantle values (0.2137) have large uncertainties or aberrant ages with no rigorous meaning. This is shown by some Type I and most Type III eclogites (in shaded area). Only T_{DM} calculated with $^{147}Sm/^{144}Nd$ ratios ≤0.15 may provide a certain degree of age significance. Except for a few Su–Lu eclogites with Archean T_{DM} (up to 3.2 Ga), most eclogites have Proterozoic model ages (2.3–0.8 Ga).

to UHP metamorphism as a result of continental subduction (Okay, 1993; Zhang *et al.*, 1995; Chavagnac and Jahn, 1996; Fan *et al.*, 1996; Liou *et al.*, 1997), or the suites are mantle rocks exhumed together with crustal UHP metamorphic rocks. In this case, they are considered here to be tectonic enclaves of mantle origin within granitic gneisses. A probable example is the garnet pyroxenites of Rizhao. Even though these pyroxenites represent igneous cumulate rocks from a mantle-derived liquid, the magma differentiation probably occurred within the mantle.

The time of intrusion for the Bixiling complex has not been precisely determined. However, based on garnet Sm-Nd and phengite Rb-Sr mineral isotopic systematics, Chavagnac and Jahn (1996) argued that the Bixiling complex was intruded ≤300 Ma before the UHP metamorphism (~220 Ma). A recent zircon U-Pb dating of Type III eclogite from Maowu gave an upper intercept age of ~450 Ma, which was interpreted as the time of protolith crystallization (Rowley *et al.*, 1997). It remains to be determined whether all the Type III eclogites in the Dabie orogen have similarly "young" protolith ages. The isotopic systematics of the Maowu layered intrusion ($\varepsilon_{Nd}(t)$ –5 to –6, I_{Sr} ~0.707–0.708) are different from those of the Bixiling complex (Fig. 12b). Upper crustal contamination during magma chamber processes is likely to have occurred in the Maowu intrusion (Jahn *et al.*, in preparation-a). However, the timing of this process cannot be easily determined. Sapphirine, clinochlore, and enstatite inclusion minerals in garnet led Okay (1994) to conclude that the Maowu intrusion underwent granulite facies metamorphism (730°C, 0.4 GPa) in Precambrian time, followed by UHP metamorphism (740°C, >4 GPa) during Triassic continental collision. Because most Type III rocks possess $^{147}Sm/^{144}Nd$ ratios higher than 0.15 (Fig. 13), the calculated T_{DM} ages have no rigorous meaning. The Maowu rocks show an extreme variation in T_{DM}, from –6.2 Ga for garnet orthopyroxenite to +5.8 Ga for eclogite (Jahn *et al.*, in preparation-a)

The interpretation of Rizhao and Rongcheng garnet peridotites as mantle slices is not ironclad. It is still possible that a pre-UHP low-pressure event might be identified for these ultramafic rocks, which would imply their earlier emplacement in the crust, as demonstrated for garnet peridotites from Donghai (Yang and Jahn, in preparation). In any case, the temperature/pressure ratios recorded in the mineral assemblages of garnet pyroxenites and garnet lherzolites are significantly lower than the upper mantle (Yang *et al.*, 1993; Zhang *et al.*, 1994; Hiramatsu *et al.*, 1995). Consequently, these ultramafic rocks have implicitly undergone UHP metamorphism as the associated eclogites.

5.3 Age of UHP Metamorphism and Continental Collision

The above geochemical constraints indicates a tectonic scenario in which a gneiss terrane of ancient continental crust containing layers or enclaves of amphibolite or basic granulite was subducted to mantle depths. These basic layers or enclaves were transformed into eclogites at UHP. With the exhumation of crustal blocks containing these eclogites, some mantle slices or fragments were also transported and emplaced as tectonic enclaves within the UHP metamorphic complex.

Dating of UHP metamorphic rocks of the Dabie orogen is potentially a difficult task because: i) UHP terranes are often affected by multiple geologic events—the Dabie Shan witnessed an important Cretaceous thermal event with massive granitic intrusions about 100 m.y. after the Yangtze/Sino-Korean collision; ii) the duration of elevated temperature and rate of cooling is spatially variable; iii) the effect of fluids on isotopic systems has not been precisely understood; and iv) the behavior of elements and isotopes at UHP is not sufficiently known. Any or all of the above factors can lead to chemical and isotopic disequilibrium.

Nevertheless, the available data obtained from different geochronological studies indicate that the age of UHP metamorphism and the time of continental collision is most probably ~220 Ma in the Dabie Shan (Ames *et al.*, 1993; Li *et al.*, 1993b; 1994; Hacker and Wang, 1995; Ames *et al.*, 1996; Chavagnac and Jahn, 1996; Rowley *et al.*, 1997). The age for the Su–Lu region could be slightly older if the collision took place first in the east as argued from paleomagnetic data (Zhao and Coe, 1987). Although isotopic disequilibrium cannot be discounted, the range of Sm-Nd ages is in agreement with U-Pb zircon ages of coesite-bearing eclogites (Ames *et al.*, 1993; Hacker *et al.*, 1997; Rowley *et al.*, 1997). $^{40}Ar/^{39}Ar$ phengite dates often yielded inconsistent results due to the presence of excess ^{40}Ar (Li *et al.*, 1994; Hacker and Wang, 1995). However, some phengites from the schists and gneisses of the Hong'an Block seem to have recorded "correct" $^{40}Ar/^{39}Ar$ ages (Eide *et al.*, 1994). Hacker and Wang (1995) also determined eight internally concordant $^{40}Ar/^{39}Ar$ ages, ranging from 179–227 Ma, on phengite, biotite and hornblende from amphibolitized eclogite and UHP gneiss from the Dabie Mountains.

A Late Triassic collision is also supported by sedimentological and paleomagnetic data. As outlined by Hsü *et al.* (1987), Upper Triassic flysch in the Tongbai Mountains marks the first influx of continental material into the pre-collisional Qinling ocean; and Upper Triassic and younger sediments are less deformed than, and unconformably overlie, older deformed marine strata in the Qinling Mountains. Recent paleomagnetic data suggest that the Sino-Korean and Yangtze cratons moved farther apart during the Middle Permian to Middle–Late Triassic, but then approached each other in Late Triassic–Early Jurassic time. Yang *et al.* (1992) concluded that the South China Block underwent final accretion to the North China Block in the Middle Jurassic (Yang *et al.*, 1991).

6. CONCLUSIONS

The present geochemical and Nd-Sr isotopic analyses lead to the following conclusions:

1) The eclogites are generally of basaltic composition, but show a wide range of major and trace element abundances, suggesting multiple origins and heterogeneous protoliths due to igneous differentiation and metamorphic segregation. Type II eclogites may have been derived from marls or siliceous carbonates.

2) Types I and II eclogites show light-rare-earth enriched REE patterns and highly negative $\varepsilon_{Nd}(t)$ values (–6 to –20). The protoliths of the eclogites resemble continental mafic rocks of Precambrian gneiss terranes, and constitute an integral component of continental cratons. They are not oceanic crust from the Tethyan Ocean. Sm-Nd model age data suggest that the eclogite protoliths were emplaced at different times but mainly in the Proterozoic. Depletion and enrichment of certain elements via metasomatism can be identified in some protoliths (*e.g.*, Weihai). This metasomatism took place in the Proterozoic (~1.7 Ga) and was unrelated to the Triassic UHP metamorphic event.

3) Type III eclogites are genetically related to their associated ultramafic rocks (lherzolite, harzburgite, or pyroxenites). Their Sr-Nd isotopic signatures suggest that they were derived from differentiation of basaltic magmas in layered intrusions at crustal levels (*e.g.*, Bixiling, Maowu), within the mantle (*e.g.*, Rizhao garnet clinopyroxenite), or represent mantle slices that were emplaced as tectonic enclaves during exhumation (Rongcheng and Donghai garnet peridotites). Minor contamination by lower crustal materials in the Bixiling layered intrusion has been identified, whereas upper crustal contamination during magma chamber processes is likely to have occurred in the Maowu intrusion. Alternatively, the Maowu intrusion could have been derived from a mantle source enriched by upper crustal components. The isotopic signatures of the tectonic enclaves indicate that some mantle segments evolved in long-term enriched sources (Rongcheng) whereas others are from slightly depleted reservoirs (Rizhao). The UHP parageneses of layered intrusions imply continental subduction, whereas those of other ultramafic rocks do not direct imply continental subduction.

4) The UHP metamorphism took place at ~220 Ma based on a variety of reliable age data (U-Pb, Sm-Nd, $^{40}Ar/^{39}Ar$) obtained in different laboratories. Other "aberrant" ages can be due to lack of isotope equilibrium at UHP (Sm-Nd, U-Pb systems), or to excess ^{40}Ar. No ages can be determined from the Rb-Sr systems because most eclogites have experienced Rb depletion relative to Sr, leading to unsupported high $^{87}Sr/^{86}Sr$ ratios. However, the lowest $^{87}Sr/^{86}Sr$ (~0.7036) for the Bixiling complex and Rizhao pyroxenites may represent the Sr isotopic composition of the subcontinental lithosphere of the Dabie orogen.

ACKNOWLEDGMENTS

I am indebted to B.L. Cong and his associates of the Chinese Academy of Sciences (1993–1995), and D.Y. Liu and J.J. Yang of the Chinese Academy of Geological Sciences (1996) for their co-operation and kind assistance in the field. Laboratory analyses in Rennes were ably assisted by Jean Cornichet, Joël Macé, Odile Henin, Nicole Morin and Martine Le Coz-Bouhnik. I thank the editors Brad Hacker and J.G. Liou for inviting me to write this chapter. This manuscript has benefited by the constructive reviews of the editors in addition to J.M. Mattinson, Bolin Cong, Kai Ye, J.D. Walker and an anonymous reviewer. Many of my vague viewpoints in petrology have been clarified through frequent conversation with Jianjun Yang during his visit in Rennes. My research on UHP metamorphic terranes of China has been supported by Actions Spécifiques DSPT-3/MST 1995 (Ministère de l'Education Nationale), DBT II of INSU (ATP Dynamique et Bilans de la Terre 1996), and "Intérieur de la Terre" of INSU–1997.

REFERENCES

Ames, L., Tilton, G.R., and Zhou, G. (1993) Timing of collision of the Sino-Korean and Yangtse Cratons: U-Pb zircon dating of coesite-bearing eclogites, *Geology* **21**, 339–342.

Ames, L., Zhou, G., and Xiong, B. (1996) Geochronology and geochemistry of ultrahigh-pressure metamorphism with implications for collision of the Sino-Korean and Yangtze cratons, central China, *Tectonics* **15**, 472–489.

Arculus, R.J., Ferguson, J., Chappel, B.W., Smith, D., McCulloch, M.T., Jackson, I., Hensel, H.D., Taylor, S.R., Knutson, J., and Gust, D.A. (1988) Trace element and isotopic characteristics of eclogites and other xenoliths derived from the lower continental crust of southeastern Australia and southwestern Colorado Plateau, U.S.A., in D.C. Smith (ed.), *Eclogites and eclogite-facies rocks*, Elsevier, Amsterdam, pp. 335–386.

Baker, J., Matthews, A., Mattey, D., Rowley, D., and Xue, F. (1997) Fluid-rock interactions during ultra-high pressure metamorphism, Dabie Shan, China, *Geochimica Cosmochimica et Acta* **61**, 1685–1696.

Beard, B.L., Medaris, L.G., Johnson, C.M., Brueckner, H.K., and Z., M. (1992) Petrogenesis of Variscan high-temperature Group A eclogites from the Moldanubian zone of the Bohemian Massif, Czechoslovakia, *Contributions to Mineralogy and Petrology* **111**, 468–483.

Bernard-Griffiths, J. and Cornichet, J. (1985) Origin of eclogites from South Brittany, France: A Sm-Nd isotopic and REE study, *Chemical Geology* **52**, 185–201.

Bernard-Griffiths, J., Peucat, J.J., Cornichet, J., Ponce de Leon, M.I., and Ibarguchi, J.I.G. (1985) U-Pb, Nd isotopic and REE geochemistry in eclogites from the Cabo Ortegal Complex, Galicia, Spain: an example of REE immobility conserving MORB-like patterns during high-grade metamorphism, *Chemical Geology* **52**, 217–225.

Bernard-Griffiths, J., Peucat, J.J., and Menot, R.P. (1991) Isotopic (Rb-Sr, U-Pb, and Sm-Nd) and trace element geochemistry of eclogites from the pan-African Belt: a case study of REE fractionation during high-grade metamorphism, *Lithos* **27**, 43–57.

Carswell, D.A., Wilson, R.N., and Zhai, M. (1997) Ultra-high pressure aluminous titanites in carbonate-bearing eclogites at Shuanghe in Dabie Shan, central China, *Mineralogical Magazine* **60**, 461-471.

Chavagnac, V. and Jahn, B.-m. (1996) Coesite-bearing eclogites from the Bixiling Complex, Dabie Mountains, China; Sm-Nd ages, geochemical characteristics and tectonic implications, *Chemical Geology* **133**, 29–51.

Chen, J.F. and Jahn, B.M. (1998) Crustal evolution of southeastern China: Nd and Sr isotopic evidence, *Tectonophysics*.

Chopin, C. (1984) Coesite and pure pyrope in high-grade blueschists of the western Alps: a first record and some consequences, *Contributions to Mineralogy and Petrology* **86**, 107–118.

Cong, B. (1996) *Ultrahigh-Pressure Metamorphic Rocks in the Dabie Shan-Sulu Region of China*, Science Press, Beijing.

Eggler, D.H., Fahlquist, D.A., Pequegnat, W.E., and Herndon, J.M. (1973) Ultrabasic rocks from the Cayman Trough, Caribbean Sea, *Geological Society of America Bulletin* **84**, 2133–2138.

Eide, L., McWilliams, M.O., and Liou, J.G. (1994) 40Ar/39Ar geochronologic constraints on the exhumation of HP-UHP metamorphic rocks in east-central China, *Geology* **22**, 601-604.

Enami, M. and Zang, Q. (1990) Quartz pseudomorph after coesite in eclogites from Shandong province, east China, *American Mineralogist* **75**, 381-386.

Ernst, W.G. and Liou, J.G. (1995) Contrasting plate-tectonic styles of the Qinling-Dabie-Sulu and Franciscan metamorphic belts, *Geology* **23**, 353–356.

Fan, Q.C., Liu, R.X., Ma, B.L., Zhao, D.S., and Zhang, Q. (1996) The protolith and ultrahigh-pressure metamorphism of Maowu mafic-ultramafic rock block in Dabie Shan Mountains, *Acta Petrologica Sinica* **12**, 29–47.

Fitton, J.G. and Upton, B.G.J. (1987) Introduction, in J.G. Fitton and B.G.J. Upton (eds.), *Alkaline Igneous Rocks*, 30, Geological Society of London, London, pp. IX-XIV.

Griffin, W.L. and Brueckner, H.K. (1985) REE, Rb-Sr and Sm-Nd studies of Norwegian eclogites, *Chemical Geology* **52**, 249–271.

Griffin, W.L., Carswell, D.A., and Nixon, P.H. (1979) Lower crustal granulites and eclogites from Lesotho, Southern Africa, in F.R. Boyd and H.O. Meyer (eds.), *The mantle sample: Inclusions in kimberlites and other volcanics*, **2**, American Geophysical Union, Washington, D.C., pp. 59–86.

Hacker, B.R., Ratschbacher, L., Webb, L., Ireland, T., Walker, D., Calvert, A., and Dong, S.W. (1997) Exhumation of ultrahigh-pressure rocks, Dabie-Hong'an-Tongbai Shan, China, *Geological Society of America Abstracts with Programs* **28**.

Hacker, B.R., Ratschbacher, L., Webb, L., Ireland, T., Walker, D., and Dong, S. (in press) U/Pb zircon ages constrain the architecture of the ultrahigh-pressure Qinling-Dabie Orogen, China, *Earth and Planetary Science Letters*.

Hacker, B.R. and Wang, Q.C. (1995) Ar/Ar geochronology of ultrahigh-pressure metamorphism in central China, *Tectonics* **14**, 994–1006.

Hacker, B.R., Wang, X., Eide, E.A., and Ratschbacher, L. (1996) Qinling-Dabie ultrahigh-pressure collisional orogen, in A. Yin and T.M. Harrison (eds.), *The Tectonic Evolution of Asia*, Cambridge University Press, Cambridge, United Kingdom.

Hall, R.P. and Hughes, D.J. (1990) *Early Precambrian basic magmatism*, Blackie, Glasgow and London.

Hirajima, T., Wallis, S.R., Zhai, M., and Ye, K. (1993) Eclogitized metagranitoid from the Su-Lu ultrahigh-pressure (UHP) province, eastern China, *Proceedings Of The Japan Academy. Series B* **69**, 249–254.

Hiramatsu, N., Banno, S., Hirajima, T., and Cong, B. (1995) Ultrahigh-pressure garnet lherzolite from Chijiadian, Rongcheng County, in the Su-Lu region of eastern China, *The Island Arc* **4**, 324–333.

Hiramatsu, N. and Hirajima, T. (1995) Petrology of the Hjialin garnet clinopyroxenite in the Su-Lu ultrahigh-pressure province, eastern China, *The Island Arc* **4**, 310–323.

Hsü, K.J., Wang, Q., Li, J., Zhou, D., and Sun, S. (1987) Tectonic evolution of the Qinling Mountains, China, *Eclogae Geologicae Helvetiae* **80**, 735–752.

Ishizaka, K., Hirajima, T., and Zheng, X.S. (1994) Rb-Sr dating for the Jiaodong gneiss of the Su-Lu ultrahigh pressure province, eastern China, *The Island Arc* **3**, 232–241.

Jahn, B.-M., Cornichet, J., Cong, B., and Yui, T.-F. (1996) Ultrahigh-eNd eclogites from an ultrahigh-pressure metamorphic terrane of China, *Chemical Geology* **127**, 61-79.

Jahn, B.M. (1990a) Early Precambrian basic rocks of China, in R.P. Hall and D.J. Hughes (eds.), *Early Precambrian Basic Magmatism*, Blackie, Glasgow, pp. 294–316.

Jahn, B.M. (1990b) Origin of granulites: geochemical constraints from Archean granulite facies rocks of the Sino-Korean Craton, China, in D. Vielzeuf and P. Vidal (eds.), *Granulites and Crustal Evolution*, Kluwer, Dordrecht, pp. 471-492.

Jahn, B.M., Cornichet, J., and Cong, B.L. (1995) Crustal evolution of the Qinling-Dabie Orogen: isotopic and geochemical constraints from coesite-bearing eclogites of the Su-Lu and Dabie terranes, China, *Chinese Science Bulletin* **40**, 116–119.

Jahn, B.M., Cornichet, J., Henin, O., Le Coz-Bouhnik, M., and Cong, B.L. (1994) Geochemical and isotopic investigation of ultrahigh pressure (UHP) metamorphic terranes in China: Su-Lu and Dabie complexes, *Stanford Workshop on Ultrahigh-P metamorphism and tectonics*, A71-A74.

Jahn, B.M., Fan, Q.C., Yang, J.J., and Henin, O. (in preparation-a) Petrogenesis of the UHPM complex of Maowu, Dabie Shan, China.

Jahn, B.M., Wu, F.Y., and Lo, C.H. (in preparation-b) Mafic-ultramafic intrusions of the Northern Dabie Complex, central China: Geochemical and isotopic evidence for post-collisional crust-mantle interaction.

Jerde, E.A., Taylor, L.A., Crozaz, G., Sobolev, N.V., and Sobolev, V.N. (1993) Diamondiferous eclogites from Yakutia, Siberia: evidence for a diversity of protoliths, *Contributions to Mineralogy and Petrology* **114**, 189–202.

Li, S., Chen, Y.Z., Ge, N.J., Liu, D.L., Zhang, Z.M., Zhang, Q.D., and Zhao, D.M. (1993a) U-Pb zircon ages of eclogite and gneiss from Jiaonan Group in Qingdao area, *Chinese Science Bulletin* **38**, 1773–1777.

Li, S., Wang, S., Chen, Y., Liu, D., Qiu, J., Zhou, H., and Zhang, Z. (1994) Excess argon in phengite from eclogite: Evidence from the dating of eclogite minerals by the Sm-Nd, Rb-Sr and 40Ar/39Ar methods, *Chemical Geology* **112**, 343.

Li, S., Xiao, Y., Liou, D., Chen, Y., Ge, N., Zhang, Z., Sun, S.-S., Cong, B., Zhang, R., Hart, S.R., and Wang, S. (1993b) Collision of the North China and Yangtze blocks and formation of coesite-bearing eclogites: Timing and processes, *Chemical Geology* **109**, 89–111.

Liou, J.G. and Zhang, R.Y. (1995) Occurrence of ultrahigh-P talc-bearing eclogite assemblages and their tectonic significance, *Mineralogical Magazine* **59**, 93–102.

Liou, J.G., Zhang, R.Y., Eide, E.A., Maruyama, S., Wang, X., and Ernst, W.G. (1996) Metamorphism and tectonics of high-P and ultrahigh-P belts in Dabie-Sulu Regions, eastern central China, in A. Yin and T.M. Harrison (eds.), *The Tectonic Evolution of Asia*, Rubey Volume IX, Cambridge University Press, Cambridge, United Kingdom, pp. 300–343.

Liou, J.G., Zhang, R.Y., and Jahn, B.M. (1997) Petrology, geochemistry and isotope data on a ultrahigh-pressure jadeite quartzite from Shuanghe, Dabie Mountains, east-central China, *Lithos* **41**, 59–78.

Maruyama, S., Liou, J.G., and Zhang, R. (1994) Tectonic evolution of the ultrahigh-pressure (UHP) and high-pressure (HP) metamorphic belts from central China, *The Island Arc* **3**, 112–121.

Masuda, A., Nakamura, N., and Tanaka, T. (1973) Fine structures of mutually normalized rare-earth patterns of chondrites, *Geochimica Cosmochimica et Acta* **37**, 239–244.

Mattauer, M., Matte, P., Malavieille, J., Tapponnier, P., Maluski, H., Xu, Z.Q., Lu, Y.L., and Tang, Y.Q. (1985) Tectonics of the Qinling belt: Build-up and evolution of eastern Asia, *Nature* **317**, 496–500.

Miller, C. and Thöni, M. (1995) Origin of eclogites from the Austroalpine Otztal basement (Tirol, Austria): geochemistry and Sm-Nd vs Rb-Sr isotope systematics, *Chemical Geology* **122**, 199–225.

Mørk, M.B.E., Kullerud, K., and Stabel, A. (1988) Sm-Nd dating of Seve eclogites, Norrbotten, Sweden - Evidence for early Caledonian (505Ma) subduction, *Contributions to Mineralogy and Petrology* **99**, 344–351.

Mørk, M.B.E. and Mearns, E.W. (1986) Sm-Nd isotopic systematics of a gabbro-eclogite transition, *Lithos* **19**, 255–267.

Morse, S.A. (1981) Kiglapait geochemistry IV: the major elements, *Geochimica Cosmochimica et Acta* **45**, 461-479.

Okay, A.I. (1993) Petrology of a diamond and coesite-bearing metamorphic terrain: Dabie Shan, China, *European Journal of Mineralogy* **5**, 659–675.

Okay, A.I. (1994) Sapphirine and Ti-clinohumite in ultra-high-pressure garnet-pyroxenite and eclogite from Dabie Shan, China, *Contributions to Mineralogy and Petrology* **116**, 145–155.

Okay, A.I. and Sengör, A.M.C. (1992) Evidence for intracontinental thrust-related exhumation of the ultrahigh-pressure rocks in China, *Geology* **20**, 411-414.

Okay, A.I., Sengör, A.M.C., and Satir, M. (1993) Tectonics of an ultrahigh-pressure metamorphic terrane: the Dabie Shan/Tongbai Shan orogen, China, *Tectonics* **12**, 1320–1334.

Peucat, J.J., Bernard-Griffiths, J., Ibarguchi, J.I.G., Dallmeyer, R.D., Menot, R.P., Cornichet, J., and Ponce de Leon, M.I. (1990) Geochemical and geochronological cross section of the deep Variscan crust: The Cabo Ortegal high-pressure nappe (northwestern Spain), *Tectonophysics* **177**, 263–292.

R.G.S. Anhui. (1987) *Regional Geology of Anhui Province*, Geological Publishing House, Beijing.

Rowley, D.B., Xue, F., Tucker, R.D., Peng, Z.X., Baker, J., and Davis, A. (1997) Ages of ultrahigh pressure metamorphism and protolith orthogneisses from the eastern Dabie Shan: U/Pb zircon geochronology, *Earth and Planetary Science Letters* **151**, 191-203.

Rumble, D. (this volume) Stable Isotope Geochemistry Of Ultrahigh-Pressure Rocks, in B.R. Hacker and J.G. Liou (eds.), *When Continents Collide: Geodynamics and Geochemistry of Ultrahigh-Pressure Rocks*, Kluwer Academic Publishers, Dordrecht.

Sanders, I.S., van Calsteren, P.W.C., and Hawkesworth, C.J. (1984) A Grenville Sm-Nd age for the Glenelg eclogite in northwest Scotland, *Nature* **312**, 439–440.

Schertl, H.-P. and Okay, A.I. (1994) A coesite inclusion in dolomite in Dabie Shan, China: petrological and rheological significance, *European Journal of Mineralogy* **6**, 995–1000.

Shervais, J.W. (1982) Ti-V plots and the petrogenesis of modern and ophiolitic lavas, *Earth and Planetary Science Letters* **57**, 101-118.

Shervais, J.W., Taylor, L.A., Lugmair, G.W., Clayton, R.N., Mayeda, T.K., and Korotev, R.L. (1988) Early Proterozoic oceanic crust and the evolution of subcontinental mantle: Eclogites and related rocks from southern Africa, *Geological Society of America Bulletin* **100**, 411-423.

Stosch, H.G. and Lugmair, G.W. (1990) Geochemistry and evolution of MORB-type eclogites from the Münchberg Massif, southern Germany, *Earth and Planetary Science Letters* **99**, 230–249.

Sun, S.S. and McDonough, W.F. (1989) Chemical and isotopic systematics of oceanic basalts: implications for mantle composition and processes, in A.D. Saunders and M.J. Norry (eds.), *Magmatism in the Ocean Basins, Special Publications*, **42**, Geological Society of London, London, pp. 313–345.

Thöni, M. and Jagoutz, E. (1992) Some new aspects of dating eclogites in orogenic belts : Sm-Nd, Rb-Sr, and Pb-Pb isotopic results from the Austroalpine Saualpe and Koralpe type-locality (Carinthia/Styria, southeastern Austria), *Geochimica Cosmochimica et Acta* **56**, 347–368.

Tilton, G.R., Schreyer, W., and Schertl, H.-P. (1991) Pb-Sr-Nd isotopic behavior of deeply subducted crustal rocks from the Dora Maira massif, western Alps, Italy-II: what is the age of the ultrahigh pressure metamorphism, *Contributions to Mineralogy and Petrology* **108**, 22–33.

Tilton, G.R., Schreyer, W., and Schertl, H.P. (1989) Pb-Sr-Nd behavior of deeply subducted crustal rocks from the Dora Maira Massif, western Alps, Italy, *Geochimica Cosmochimica et Acta* **53**, 1391-1400.

Wang, Q.C., Ishiwatari, A., Zhao, Z., Hirajima, T., Enami, M., Zhai, M., Li, J., and Cong, B. (1993) Coesite-bearing granulite retrograded from eclogite in Weihai, eastern china: a preliminary study, *European Journal of Mineralogy* **5**, 141-152.

Wang, Q.C., Zhai, M.G., and Cong, B.L. (1996) Regional geology, in B.L. Cong (ed.), *Ultrahigh-Pressure Metamorphic Rocks in the Dabie Shan-Sulu Region of China*, Science Press, Beijing, pp. 8–26.

Wang, X. (1991) Petrology of coesite-bearing eclogites and ultramafic rocks from ultrahigh-pressure metamorphic terrane of the Dabie Mountains and implications to regional tectonics in central China. Stanford University, p. 213.

Wang, X., Jing, Y., Liou, J.G., Pan, G., Liang, W., Xia, M., and Maruyama, S. (1990) Field occurrences and petrology of eclogites from the Dabie Mountains, Anhui, central China, *Lithos* **25**, 119–131.

Wang, X. and Liou, J.G. (1991) Regional ultrahigh-pressure coesite-bearing eclogitic terrane in central China: evidence from country rocks, gneiss, marble, and metapelite, *Geology* **19**, 933–936.

Wang, X. and Liou, J.G. (1993) Ultrahigh-pressure metamorphism of carbonate rocks in the Dabie Mountains, central China, *Journal of Metamorphic Geology* **11**, 575–588.

Xu, S., Liu, Y.C., Jiang, L.L., Su, W., and Ji, S.Y. (1994) *Tectonic regime and evolution of Dabie Mountains*, Science Press, Beijing.

Xu, S., Okay, A.I., Ji, S., Sengör, A.M.C., Su, W., Liu, Y., and Jiang, L. (1992) Diamond from the Dabie Shan metamorphic rocks and its implication for the tectonic setting, *Science* **256**, 80–82.

Xue, F., Rowley, D.B., Tucker, R.D., and Peng, Z.X. (1997) U-Pb zircon ages of granitoid rocks in the north Dabie complex, eastern Dabie Shan, China, *Journal of Geology* **105**, 744–753.

Yang, J. (1991) *Eclogite, garnet pyroxenites and related ultrabasics in Shandong and north Jiangsu of east China*, Geological Publishing House, Beijing.

Yang, J., Godard, G., Kienast, J.-R., Lu, Y., and Sun, J. (1993) Ultrahigh-pressure (60 kbar) magnesite-bearing garnet peridotites from northeastern Jiangsu, China, *Journal of Geology* **101**, 541-554.

Yang, J. and Jahn, B.M. (in preparation) Metamorphism and metasomatism of garnet peridotites from northern Jiangsu of the Su-Lu UHP terrane, China.

Yang, J. and Smith, D.C. (1989) Evidence for a former sanidine-coesite eclogite at Lanshantou, eastern China, and the recognition of the Chinese "Su-Lu coesite eclogite province", East China, *Terra Abstracts* **1**, 26.

Yang, Z., Courtillot, V., Besse, J., Ma, X., Xing, L., Xu, S., and Zhang, J. (1992) Jurassic paleomagnetic constraints on the collision of the North and South China Blocks, *Geophysical Research Letters* **19**, 577–580.

Yang, Z., Ma, X., Besse, J., Courtillot, V., Xing, L., Zhang, J., and Xu, S. (1991) Paleomagnetic results from the Triassic sections in the Ordos basin, North China, *Earth and Planetary Science Letters* **104**, 258–277.

Yin, A. and Nie, S. (1993) An indentation model for the North and South China collision and the development of the Tanlu and Honam fault systems, eastern Asia, *Tectonics* **12**, 801-813.

Yui, T.F., Rumble, D., and Lo, C.H. (1995) Unusually low 18O ultrahigh-pressure metamorphic rocks from the Sulu terrain, eastern China, *Geochimica Cosmochimica et Acta* **59**, 2859–2864.

Zhai, M., Cong, B., Zhang, Q., and Wang, Q. (1994) The northern Dabie Shan terrain: A possible Andean-type arc, *International Geology Review* **36**, 867–883.

Zhai, M.G. and Cong, B.L. (1996) Major and trace element geochemistry of eclogites and related rocks, in B.L. Cong (ed.), *Ultrahigh-Pressure Metamorphic Rocks in the Dabie Shan-Sulu Region of China*, Science Press, Beijing, pp. 69–89.

Zhang, R.Y., J.G., L., and B., C. (1995) Talc-, magnesite- and Ti-clinohumite-bearing ultrahigh-pressure meta-mafic and ultramafic complex in the Dabie Mountains, *Journal of Petrology* **36**, 1011-1037.

Zhang, R.Y., Liou, J.G., and Cong, B. (1994) Petrogenesis of garnet-bearing ultramafic rocks and associated eclogites in the Su-Lu ultrahigh-P metamorphic terrane, eastern China, *Journal of Metamorphic Geology* **12**, 169–186.

Zhang, R.Y., Liou, J.G., and Tsai, C.H. (1996a) Petrogenesis of a high-temperature metamorphic terrane: a new tectonic interpretation for the northern Dabieshan, central China, *Journal of Metamorphic Geology* **14**, 319–333.

Zhang, R.Y., Liou, J.G., and Ye, K. (1996b) Mineralogy of UHPM rocks, in B.L. Cong (ed.), *Ultrahigh-Pressure Metamorphic Rocks in the Dabie Shan-Sulu Region of China*, Science Press, Beijing, pp. 106–127.

Zhang, R.Y., Liou, J.G., and Ye, K. (1996c) Petrography of UHPM rocks and their country rock gneisses, in B.L. Cong (ed.), *Ultrahigh-Pressure Metamorphic Rocks in the Dabie Shan-Sulu Region of China*, Science Press, Beijing, pp. 49–68.

Zhao, X. and Coe, R.S. (1987) Palaeomagnetic constraints on the collision and rotation of North and South China, *Nature* **327**, 141-144.

Zhao, Z., Wang, Q., and Cong, B. (1992) Coesite-bearing ultrahigh-pressure metamorphic rocks from Donghai, northern Jiangsu Province, eastern China: "foreign" or "in-situ"?, *Scientia Geologica Sinica* **1**, 43–58.

Zheng, Y.F., Fu, B., Gong, B., and Li, S. (1996) Extreme 18O depletion in eclogite from the Su-Lu terrane in East China, *European Journal of Mineralogy* **8**, 317–323.

Zhou, T.X., Chen, J.F., Li, X.M., and Foland, K.A. (1992) $^{40}Ar/^{39}Ar$ isotopic dating of intrusions from Huoshan-Shucheng syenite zone, Anhui Province, *Geology of Anhui* **2**, 4–11.

Chapter 9

Stable Isotope Geochemistry of Ultrahigh-Pressure Rocks

Douglas Rumble
Geophysical Laboratory, 5251 Broad Branch Rd., N. W., Washington, D. C. 20015, USA, rumble@gl.ciw.edu

Abstract: Stable isotope geochemistry gives unequivocal evidence that the protoliths of ultrahigh-pressure (UHP) metamorphic rocks were at Earth's surface prior to continental collision and subduction. Unusually low values of $\delta^{18}O$ and δD in eclogite and other Chinese UHP rocks record a Neoproterozoic geothermal area in which cold-climate meteoric waters altered basalts, granites, and associated sediments. The presence of a pre-metamorphic, distinctive isotopic signature in both eclogites and their host rocks supports the hypothesis of *in situ* metamorphism for coesite-bearing eclogites. The length scale of structural coherence in these Chinese UHP terrains is at least 100 km, for that is the distance over which subsequent deformation failed to obliterate artifacts of surface geothermal activity. Stable isotope evidence denies the existence of a pervasive fluid free to infiltrate across lithologic contacts during UHP metamorphism. Data from UHP terrains in China, Italy, and Norway show that dissimilar rocks failed to equilibrate isotopically on a centimeter to meter scale. Furthermore, the preservation of high-temperature oxygen isotope fractionations among minerals argues against the presence of free fluid after peak metamorphism, during exhumation and cooling. Residence in the upper mantle had no discernible metasomatic effect on the stable isotope composition of crustal rocks subducted during continental collision.

1. INTRODUCTION

The ramifications and implications of the discovery of ultradeep or ultrahigh-pressure (UHP) metamorphism extend beyond petrology to understanding the physical and chemical processes at convergent margins of colliding continental plates. Among the disciplines employed in the study of the phenomena are mineralogy, metamorphic petrology, experimental petrology, geochronology,

B. R. Hacker and J. G. Liou (eds.),
When Continents Collide: Geodynamics and Geochemistry of Ultrahigh-Pressure Rocks, 241–259.
© 1998 *Kluwer Academic Publishers. Printed in the Netherlands.*

structural geology, and geodynamic modelling. These studies have succeeded in establishing a basic set of facts as well as a logical framework of problems and questions so that efficient progress can be made toward understanding (see Coleman and Wang, 1995b; Harley and Carswell, 1995; Schreyer, 1995; Cong, 1996; Schreyer and Stockhert, 1997). Stable isotope geochemistry is a latecomer to UHP investigations. Recent results show, however, that measurements of oxygen, hydrogen, and carbon isotope ratios in coexisting minerals lead to definitive conclusions regarding three significant questions of UHP metamorphism: 1) Were the protoliths of UHP rocks ever at or near Earth's surface? 2) To what extent were UHP rocks subducted and exhumed as coherent structural units? Or, are UHP terrains a mélange of dismembered tectonic fragments, not all of which experienced ultradeep metamorphism? 3) What is the nature and extent of interaction between fluids of subducted crustal rocks and the upper mantle? This chapter considers these questions in the light of recent discoveries by stable isotope geochemists.

That rocks from Earth's surface survived a trip down into the upper mantle and back is, in itself, remarkable. But UHP rocks offer more than astonishment. They record a complete geodynamic pathway and are ground truth for testing hypotheses of continental collision. The coherence or dismemberment of UHP terrains reflects their strength or weakness during subduction and exhumation. Deformation structures exposed in outcrop, be they brittle *vs.* ductile faults, normal *vs.* reverse faults, fold vergence, strain markers, etc., display the physical processes of continental collision. Finally, investigating the interaction between crustal and mantle rocks offers revealing comparisons between continental collision and oceanic subduction zones. Subducted oceanic crust remains identifiable seismically to depths of hundreds of kilometers. Continental crust, as exemplified in UHP terrains, likewise, retains distinctive mineralogical, chemical, and isotopic properties to depths of at least 100 km.

2. NOTATION

Isotope data are reported as parts per thousand differences from a reference standard. The δ notation is defined as

$$\delta_x = 1000 \times (R_x - R_{STD})/R_{STD}$$

where R_x and R_{STD} are the isotope ratios $^{18}O/^{16}O$ or D/H ($^2H/^1H$) of the sample (x) and standard (STD), respectively. Data for oxygen and hydrogen are reported relative to V_{SMOW} (Vienna Standard Mean Ocean Water) and for carbon relative to V_{PDB} (Vienna Peedee Belemnite) (Coplen, 1995).

3. WERE PROTOLITHS OF UHP ROCKS EVER AT EARTH'S SURFACE?

Previous workers recognized that in their chemistry and stratigraphy, UHP rocks resemble sedimentary and igneous rocks from shallow crustal depths. Concordant and conformable layers of eclogite in quartzite, marble, and biotite gneiss suggest basaltic sills or lava flows. Interlayered quartzite, marble, and mica schist are similar to sedimentary sequences of sandstone, limestone, and shale found along continental margins (cf. Coleman and Wang, 1995a).

The oxygen and hydrogen isotope compositions of surface waters are sufficiently distinct from the crust and mantle as to afford a sensitive monitor of the residence of such rocks at Earth's surface. Crustal rocks range widely in composition but are generally greater than zero in $\delta^{18}O$. Mantle rocks are nearly homogeneous at +5 to +6‰ $\delta^{18}O$. Ocean water is zero in both $\delta^{18}O$ and δD. Meteoric waters (rain, groundwater, snow, ice) have negative $\delta^{18}O$ and δD values. At temperatures above 80°C, groundwater begins to exchange isotopes with its host rocks in an effort to achieve equilibrium. Waters become enriched and rocks, concomitantly, depleted in ^{18}O. Deuterium–hydrogen ratios in waters are little changed under these conditions, owing to the much greater amount of H in water than in rocks. Rocks themselves, however, become depleted in D. Thus, ^{18}O and D depletion in rocks in contact with heated groundwater provide a definitive record of surface proximity. Geothermal areas, where heat is provided by shallow igneous intrusions, are the focus of intense ^{18}O and D depletion in rocks (Criss and Taylor, 1986; Sheppard, 1986; Taylor and Sheppard, 1986; Gat, 1996).

The discovery of $\delta^{18}O$ values of garnet and omphacite as low as –10‰ and quartz at –7‰ (Fig. 1; Fig. 2) in coesite-bearing eclogites from Qinglongshan in the Su–Lu UHP terrain of eastern China was quite surprising because the values are so much lower than those previously measured (Yui *et al.*, 1994; 1995; Zheng *et al.*, 1996). Also surprising were reports of eclogite garnets with $\delta^{18}O$ of –5 to –7 (the equilibrium value for quartz at these conditions would be –2 to –4‰) (Baker *et al.*, 1997; Yui *et al.*, 1997). A comprehensive review of published $\delta^{18}O$ values of quartz from metamorphic rocks world-wide shows the extreme lower limit to be +7‰ (Fig. 7 of Sharp *et al.*, 1993). Considering that the entire natural range of variation in metamorphic quartz was known to extend only from +7 to +30‰ , the discoveries were not merely surprising but shocking! Unusually low δD values of –113 to –114‰ (V_{SMOW}) have been found in phengite from eclogite and quartzite at Qinglongshan (T.F. Yui, personal communication). The δD values are not as surprising as the low $\delta^{18}O$, but they are at the low end of the total range of natural variation in metamorphic micas (Fig. 8 of Sharp *et al.*, 1993).

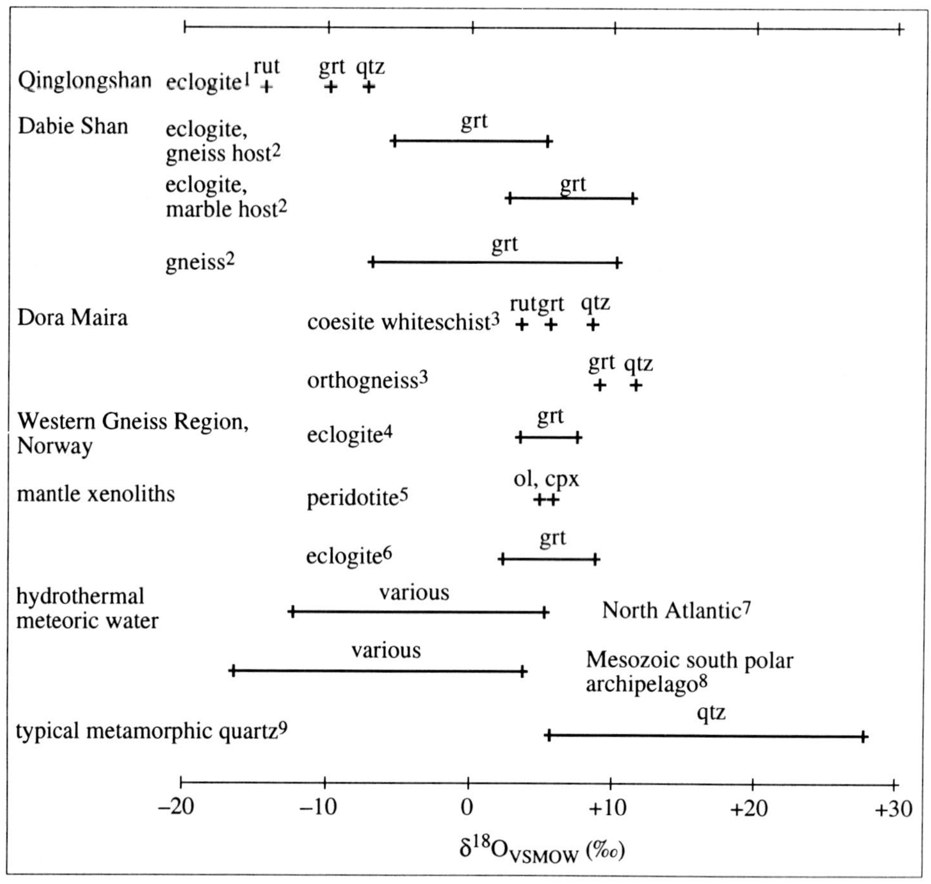

Figure 1. Oxygen isotope compositions of UHP minerals from Qinglongshan and Dabie Shan in China, Dora Maira in the western Alps of Italy, and the Western Gneiss region, Norway. Also shown are the compositions of minerals from mantle peridotite and eclogite xenoliths. Note that the only reservoirs overlapping the Qinglongshan and Dabie Shan values are rocks hydrothermally altered by meteoric water from cold climates. cpx, clinopyroxene; grt, garnet; ol, olivine; qtz, quartz; rut, rutile; various, variety of hydrous minerals and altered rocks. Sources: 1, Yui, *et al.* (1995); Zheng *et al.* (1996); 2, Baker *et al.* (1997); 3, Sharp *et al.* (1993); 4, Agrinier *et al.* (1985); 5, Mattey *et al.* (Mattey *et al.*, 1994b); Wiechert *et al.* (1997); 6, MacGregor and Manton (1986); Ongley, *et al.* (1987); Shervais *et al.* (1988); Deines, *et al.* (1991); 7, Taylor and Forester (1979); Hattori and Muehlenbachs (1982); Fehlhaber and Bird (1991); Brandriss *et al.* (1995); 8, Blattner *et al.* (1997); and 9, Sharp *et al.* (1993).

The question arose immediately, however, as to the time at which the Qinglongshan rocks acquired a negative ^{18}O signature. Analysis of individual minerals separated from coesite eclogite and interlayered quartzite revealed that oxygen isotope exchange equilibrium had been closely approached among coexisting adjacent mineral grains. Geothermometric estimates of isotopic exchange equilibrium temperatures range from 655 to 765°C (Yui *et al.*, 1995; Zheng *et al.*, 1996). The estimates overlap conditions calculated from mineral geothermometers applied to UHP parageneses, *e.g.* 700–850°C and 3.0 GPa (Zhang *et al.*, 1995a). From these relationships, it is apparent that a negative ^{18}O signature was acquired prior to UHP metamorphism (cf. Baker *et al.*, 1997). As for the age of UHP metamorphism, itself, Sm-Nd and Rb-Sr mineral isochrons (whole rock-phengite-omphacite-garnet) give ages of 226 ± 4 Ma and 220 ± 0.5 Ma, respectively (Li *et al.*, 1994). Zircons from eclogite have metamorphic ages of ~219 Ma (lower intercept of zircon discordia) with Neoproterozoic protolith

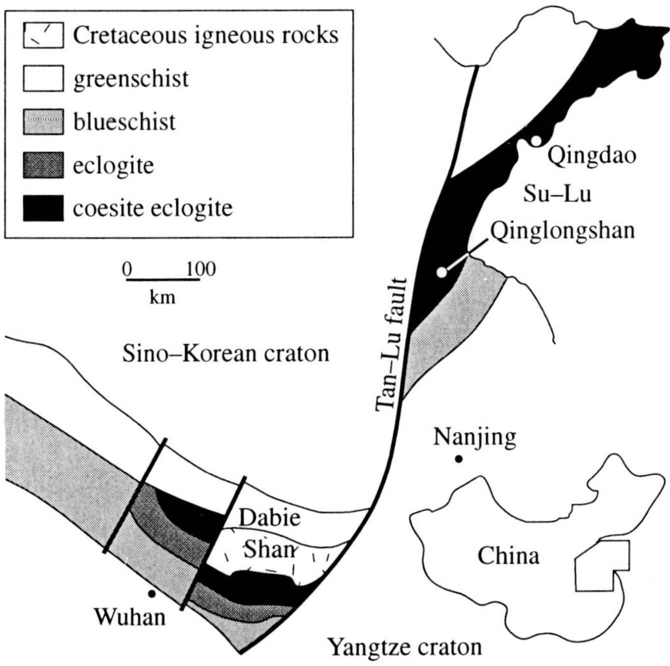

Figure 2. Location of Dabie Shan, Qinglongshan, Tan–Lu Fault and metamorphic facies belts of eastern China (map simplified after Wang *et al.*, 1995).

age of ~750 Ma (upper intercept) (Ames *et al.*, 1996; Rowley *et al.*, 1997; Hacker *et al.*, in press). The foregoing data establish that a negative $\delta^{18}O$ signature was acquired by basaltic and quartzose rocks prior to UHP metamorphism at 220 Ma. Given that meteoric waters are the most prominent reservoir of negative $\delta^{18}O$ and δD values, and given the unusually low values of both $\delta^{18}O$ and δD in Qinglongshan rocks, the O and H isotopic compositions must have been established when the rocks resided at Earth's surface (Yui *et al.*, 1995; Zheng *et al.*, 1996; Baker *et al.*, 1997; Yui *et al.*, 1997).

Stable isotope geochemistry provides a definitive positive answer to the question: Did UHP protoliths reside at Earth's surface? Not only did the rocks originate near the surface but they were participants in a characteristic surface process: hydrothermal alteration in an active geothermal system. The heat source may have been one of the pre-metamorphic granitic intrusions near Qinglongshan which show unusually low $\delta^{18}O$ values (Rumble, unpublished data). A number of authors have called attention to the paleoclimatological implications of low ^{18}O meteoric water alteration in geothermal systems (Gregory *et al.*, 1989; Nevle *et al.*, 1994; Brandriss *et al.*, 1995). Blattner *et al.* (1997) suggested that ancient examples of ^{18}O depletion record epochs during which the Earth had cold poles, a warm equator, and variations in $\delta^{18}O$ and δD with both latitude and altitude that were similar to today's climate. Application of this hypothesis to the Qinglongshan data suggests the former existence of a geothermal area in a cold climate of the Neoproterozoic Earth.

4. WERE UHP ROCKS SUBDUCTED AND EXHUMED AS COHERENT STRUCTURAL UNITS?

The question of whether UHP terrains were subducted and exhumed as coherent structural units or whether they are a jumble of dismembered tectonic fragments has been, and continues to be, a topic of spirited debate (Coleman and Wang, 1995a; Harley and Carswell, 1995; Krogh and Carswell, 1995; Schreyer, 1995; Smith, 1995; Cong and Wang, 1996; Ernst and Peacock, 1996; Liou *et al.*, 1996). The common outcrop occurrence of eclogite pods with sheared borders enclosed in gneiss or marble suggests the possibility that the pods are allochthonous. The scarcity of UHP minerals in enclosing gneisses reinforces arguments for tectonic emplacement (Massonne, 1995). In contrast, however, detailed mapping of well-exposed areas shows structurally coherent successions of UHP eclogite, UHP quartzite, metamorphosed granites, country-rock gneisses, and marbles that crop out over more than 100 km^2 (Compagnoni *et al.*, 1995; Michard *et al.*, 1995; Xue *et al.*, 1996). The mapped sequences are contained within thrust-bounded, crystalline nappes and may be no more than a few km thick. Relict UHP minerals and mineral assemblages compatible with UHP eclogites are known from country rocks (Wang *et al.*, 1995; Tabata *et al.*, this volume). U/Pb dating of zircons from eclogites and enclosing gneisses give identical ages of metamorphism, within error, of 220 Ma (Ames *et al.*, 1996; Hacker *et al.*, in press).

There is convincing evidence for and against the *in situ* and allochthonous models of UHP metamorphism. To a certain extent, the prevalence of "pro" or "contra" evidence in separated regions may reflect authentic differences in the tectonic history of different UHP terrains. But, the disparities may also exemplify differences in quality of exposure or differences in the perspectives of field geologists.

In an attempt to reconcile conflicting viewpoints it may be observed that there is little doubt of the existence of pre-, syn-, and post-metamorphic faulting in UHP terrains. But, there is also strong evidence that some UHP terrains retain the same level of coherence long recognized in Barrovian metamorphic terrains of mid-crustal orogenic belts world-wide. Perhaps it would be fruitful to recast the debate of allochthonous *vs. in situ* origin and ask: "What are the largest exposures of coherent UHP lithologic successions that can be identified by geologic, petrologic, and geochemical mapping?" The pre-eminent significance of lithologic mapping and structural analysis is undisputed (cf. Compagnoni *et al.*, 1995; Michard *et al.*, 1995; Xue *et al.*, 1996). But, other techniques such as stable isotope geochemistry may provide independent corroboration.

Regional mapping of stable isotope compositions of minerals in UHP terrains offers a useful technique for measuring the minimum lateral extent of coherent lithologic successions. Consider the discovery of unusually low $\delta^{18}O$ and δD rocks in the UHP terrains of China. The low ^{18}O values were acquired prior to subduction, prior to UHP metamorphism, and prior to exhumation during a geothermal episode when the protoliths were at Earth's surface (Yui *et al.*, 1995; Zheng *et al.*, 1996; Baker *et al.*, 1997). The existence of a laterally extensive, pre-subduction geochemical signature provides a benchmark by which subsequent tectonic disruption can be measured. It is important to recognize the limitations of the technique from the outset: geothermal areas are known to be of limited geographic extent, smaller than the 1000+ km length of most orogenic belts (Criss and Taylor, 1986). Thus, the technique of stable isotope mapping may yield useful estimates of the minimum size of coherent terrains but is unlikely to span their maximum dimensions. Furthermore, the use of the technique requires the adventitious prior existence in UHP protoliths of an ancient geothermal system in a cold climate.

The UHP terrain of Dabie Shan, Anhui province, China, has negative or zero $\delta^{18}O$ values in a majority of some 70 rocks analyzed by Baker *et al.* (1997). The lowest values in eclogite are –5.5‰ (cf. Table 2 of Yui *et al.*, 1997). Surprisingly, country rocks, including biotite gneiss, hornblende gneiss, and amphibolite, also show negative values as low as –6.8‰ (Fig. 1; Fig. 2). That both eclogites and their country rocks share unusually low ^{18}O supports the *in situ* model of UHP metamorphism. Equally significant is the finding that sample localities with negative values are distributed over an area measuring 25 x 25 km (Fig. 2 of Baker *et al.*, 1997). These dimensions establish a minimum length scale over which deformation did not obliterate traces of an ancient geothermal system.

The length scale based on the data of Baker *et al.* (1997), however, does not do full justice to estimates of structural coherence. Additional published and unpublished data suggest evidence of an ancient geothermal system persists over as much as 75–100 km of the Chinese UHP terrain. The UHP terrain of Dabie Shan is separated from a correlative UHP belt on the Jiangsu–Shandong Peninsula, termed the Su–Lu belt, by 600 km of left-lateral offset on the Tan–Lu strike-slip fault (Xu, 1993; Wang *et al.*, 1995). If left-lateral displacement along the Tan–Lu fault is restored, the Dabie Shan and Su–Lu UHP terrains are juxtaposed (Fig. 2; Fig. 3), and the ^{18}O depleted rocks of Baker *et al.* (1997) are directly opposite those in Su–Lu (Fig. 3). The ^{18}O depleted rocks at Qinglongshan reported by Yui *et al.* (1995) and Zheng *et al.* (1996) are located approximately 50 km east of the Tan–Lu fault. New data show that a zone of low ^{18}O eclogites and country rocks extends 30 km NNE–SSW and 15 km E–W from Qinglongshan (Rumble and Yui, in press). Adding together the distances over which evidence of ancient geothermal activity is preserved, 25 km in Dabie Shan, 50 km from the Tan–Lu fault to Qinglongshan, and 25 km beyond, one obtains a length of 100 km.

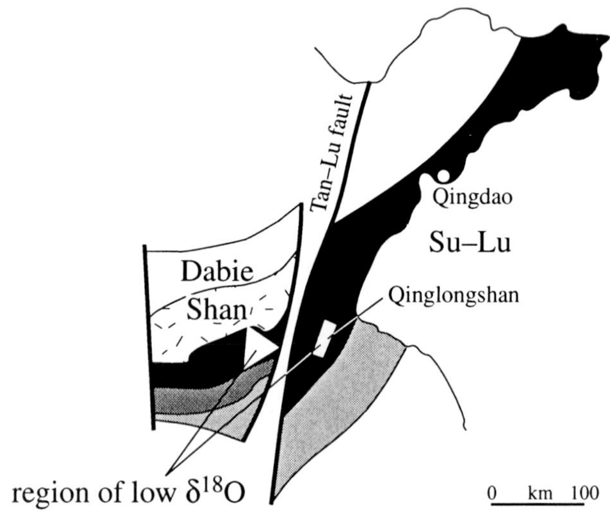

Figure 3.. Schematic restoration of 600 km left-lateral offset on the Tan–Lu Fault. Note juxtaposition of low $\delta^{18}O$ regions in Dabie Shan and Su–Lu.

5. WHAT ARE THE NATURE AND EXTENT OF INTERACTION BETWEEN FLUIDS OF SUBDUCTED CRUSTAL ROCKS AND THE UPPER MANTLE?

The question of interaction between fluids from subducted crust and mantle rocks is one of the most fascinating in Earth science. Fluids of subducted oceanic crust are believed to play a crucial role in the occurrence of intermediate depth earthquakes and in the generation of arc magmas (Bebout, 1996; Hacker, 1996; Kirby *et al.*, 1996; Peacock, 1996). Dehydration reactions lead to densification, pulling the slab downward. Fluids released from the downgoing slab are thought to have a profound effect on phase relations in the overlying mantle wedge, leading to melting, diapirism, and eruptions at Earth's surface that build volcanic arcs. Continental collision, the locus of UHP metamorphism, has associated intermediate depth earthquakes, but magmatic arcs are not present. Of course, most collision zones are ancient and extinct. Modern seismicity is not to be expected and ancient arcs may have been eroded. Nevertheless, UHP terrains, as examples of familiar crustal rocks subjected to upper mantle conditions, offer unique insights into fluid–rock processes under extreme pressure and temperature. In the following section, the topic of fluid behavior during UHP metamorphism will be addressed first, followed by a discussion of fluid interaction between UHP rocks and the upper mantle. The problem of partial melting of supracrustal rocks during UHP metamorphism has been addressed by a number of authors (Philippot, 1993; Sharp *et al.*, 1993; Patiño Douce and McCarthy, this volume). Evidence of melting lies primarily in mineral assemblages and phase equilibria. The available data of stable isotope geochemistry do not provide definitive testing of UHP melting hypotheses, therefore, the subject will not be discussed.

Petrologic studies have determined the first-order behavior of fluids during UHP metamorphism. The occurrence of relict UHP minerals in outcrops at Earth's surface implies the absence of even small amounts of free water during exhumation and cooling. Had any water remained along grain boundaries following the culmination of recrystallization, its catalytic effect would have converted coesite to quartz during exhumation (Rubie, 1990; Hacker and Peacock, 1994; Liou and Zhang, 1996; Liou *et al.*, 1997; Mosenfelder and Bohlen, 1997) Minerals susceptible to retrogression such as omphacite would be represented only by aggregates of hydrated-equivalent mineral assemblages in surface outcrops. But, UHP rocks are not strictly anhydrous. The hydrous minerals phengite, zoisite–epidote, talc, and amphibole occur as minor phases throughout the eclogite facies (Carswell, 1990). Phengite is considered to be a primary phase stable in coesite-bearing eclogites and coesite pseudomorphs have been found as inclusions in epidote (Liou *et al.*, 1995). Rare hydrous UHP minerals include ellenbergerite, bearthite, Mg-staurolite, talc, nyböite, and Ti-clinohumite (Hirajima *et al.*, 1992; Chopin and Sobolev, 1995; Zhang *et al.*, 1995b; Liou and Zhang, 1996). The importance of water is recorded by the formation of stable eclogite-facies mineral assemblages from metastable, dry, granulite-facies minerals driven by an influx of water, as observed in outcrops of eclogitized shear zones in the Western Gneiss region, Norway (Austrheim, 1987; 1990; Jamtveit *et al.*, 1990; Mattey *et al.*, 1994a). Thus, water, the most effective catalytic agent known to petrologists of the shallow, middle, and deep crust, continues to play its indispensable role under UHP conditions. In the absence of water, mineralogical reactions are greatly retarded, despite overstepping equilibrium pressure-temperature boundaries. Yet, because of its low density and because of reaction- and deformation-enhanced permeability, free water escapes to continue its work at shallower crustal levels.

Because surface water reservoirs have distinctive isotope signatures and because stable isotopes are readily exchanged between fluids and rocks, their geochemistry is a sensitive record of fluid–rock interaction not only at Earth's surface, but also during diagenesis, hydrothermal alteration, and metamorphism. Stable isotope evidence of fluid behavior is, in a sense, indirect: what is actually recorded in outcrops is fluid-mediated isotopic exchange. To measure the extent of isotopic exchange, or whether any exchange took place at all, pre-metamorphic rock compositions must be inferred. The difficulty in defining an unambiguous datum by which to measure isotopic exchange is, in many cases, the chief limitation of such studies. A common practice is to sample dissimilar rock types in mutual contact at closely spaced intervals (typically cm-scale). Samples are taken with the expectation that pre-metamorphic differences in isotopic composition persist to some degree throughout the metamorphic cycle and, if persistent, will supply a datum to measure isotopic exchange. Analysis of individual minerals from different lithologies gives data on the extent of isotopic equilibration among rock types and the length scale of material transport. Estimates of the relative efficacy of diffusive *vs*. advective transport may be made by examining the shape of composition profiles and the position of profiles in relation to inert markers such as lithologic contacts or vein walls. (Baumgartner and Rumble, 1988; Blattner and Lassey, 1989; Bickle and Baker, 1990).

The persistence of recognizable pre-metamorphic isotopic compositions throughout a metamorphic cycle denies pervasive infiltration of an exotic fluid strongly out of isotopic equilibrium with protoliths. If such infiltration had occurred, protolith isotopic compositions would have been changed (Ganor *et al.*, 1996; Matthews *et al.*, 1996). The prevalence of a pre-subduction, pre-metamorphic, high-latitude, meteoric water isotopic signature in both the Dabie Shan and Su–Lu UHP terrains of China has been extensively documented (Yui *et al.*, 1995; Zheng *et al.*, 1996; Baker *et al.*, 1997; Yui *et al.*, 1997; Zheng, 1997; Rumble and Yui, in press). Evidence of pre-metamorphic carbon isotope compositions is found in marbles of Dabie Shan, China. The marbles contain eclogite pods with relict coesite inclusions in garnets (Wang *et al.*, 1995). Both Yui *et al.* (1997) and Baker *et al.* (1997) report $\delta^{13}C$ (PDB) values for calcite in the range +4 to +6‰. The high values probably represent secular variations in $\delta^{13}C$ at the time of protolith deposition in the Neoproterozoic (cf. Wickham and

Peters, 1993). Regional scale, pervasive infiltration of exotic fluid across lithologic contacts did not occur during UHP metamorphism of these rocks.

Evidence of fluid behavior on a local, outcrop scale is provided by measurements of the extent to which dissimilar rock types in mutual contact have equilibrated isotopically. The coesite-bearing, pyrope quartzite of Dora Maira failed to achieve oxygen isotope equilibrium with the enclosing orthogneiss. The minerals of each rock type are homogeneous but differ by 3‰ in $\delta^{18}O$ between the two lithologies (Sharp *et al.*, 1993). Similar relationships are seen in the Chinese UHP terrains. Differences in $\delta^{18}O$ of as much as 7‰ were observed between garnet from marble-hosted eclogite pods and garnet from eclogite enclosed in biotite gneiss from Shima, Dabie Shan, China (Baker *et al.*, 1997). The same authors, however, cite opposite examples in which oxygen isotope equilibrium has been achieved. Garnets from small eclogite pods and garnets from adjacent host gneiss are identical in $\delta^{18}O$; but, garnets from the largest eclogite body in the area (Bixiling, 1 km in diameter) differ from those of local host gneisses (Baker *et al.*, 1997). Similarly, garnets from small eclogite pods embedded in impure marble are as high as 9.5–11‰ $\delta^{18}O$ whereas those from a 3–meter thick boudin are 2.3‰. It is thus concluded that fluid-mediated isotope exchange crossing lithologic contacts was limited to length scales less than 3 m but greater than 10 cm (Baker *et al.*, 1997). Observations at Qinglongshan in the Su–Lu terrain support these conclusions: a 10–15–cm thick quartzite layer has the same $\delta^{18}O$ values for quartz, garnet, and epidote as eclogites in the same outcrop (Yui *et al.*, 1994; 1995; Zheng *et al.*, 1996; Rumble and Yui, in press). These observations deny fluid flow across lithologic layers. The existence of a static fluid along grain boundaries cannot be disproven; witness the equilibration of oxygen isotopes on a scale of 10–15 cm. But, whether an intergranular film had the properties of free water or consisted of polar water molecules weakly bound to mineral surfaces is not readily resolved with available isotopic data.

The preservation of high-temperature, oxygen isotope fractionations among minerals signifies an absence of fluid following peak metamorphism that would otherwise, if present, have facilitated retrograde isotope exchange at lower temperatures during exhumation. (See Farquhar *et al.* (1996) and references therein for discussion of "dry" *vs.* "wet" retrograde isotope exchange.) The attainment of high-temperature isotope fractionations probably signifies the presence of a grain-boundary film of water to facilitate isotope exchange between minerals. Eclogites from the UHP Western Gneiss region, Norway, have high-temperature intermineral fractionations among quartz, pyroxene, garnet, rutile, and phengite with calculated temperatures of 595–875°C (Agrinier *et al.*, 1985). The minerals quartz, garnet, and rutile of the coesite-bearing pyrope quartzite at Dora Maira, and its enclosing orthogneiss have intermineral fractionations of 700–750°C (Sharp *et al.*, 1993). High-temperature oxygen isotope partitioning has been reported from both the Dabie Shan and Su–Lu UHP terrains of China.

Quartz, kyanite, epidote, omphacite, and garnet in eclogite and quartzite from Qinglongshan record temperatures of 655–765°C (Yui *et al.*, 1995; Zheng *et al.*, 1996). Garnet–epidote fractionations from three different rock types, eclogite, gneiss, and amphibolite in Dabie Shan are remarkably consistent and virtually zero in magnitude. Nominal temperatures are 800°C, but precise geothermometry is not possible because partitioning in this mineral pair is small and depends only weakly on temperature at high temperature (Baker *et al.*, 1997).

The discussion of fluid behavior has emphasized, thus far, characteristic features obtained during peak pressure-temperature conditions. Rocks from UHP terrains are hardly free from retrograde mineralogical effects, however (Rubie, 1990). Commonly observed retrograde reactions are i) conversion of eclogite to amphibolite; ii) replacement of omphacite by clinopyroxene + sodic plagioclase symplectite; iii) coronas of pargasite and epidote around garnet; iv) phengite replaced by biotite, itself replaced by chlorite; and v) kyanite engulfed by muscovite or margarite + quartz (Wang *et al.*, 1995). Fluid inclusions have been found in retrograde metamorphic minerals but not in UHP minerals (J.G. Liou, personal communication). Nor are UHP rocks free from retrograde stable isotope effects. Non-equilibrium fractionation of $^{18}O/^{16}O$ in garnet–omphacite, garnet–phengite, and garnet–amphibole pairs has been found by Yui *et al.* (1997). Disequilibrium partitioning correlates with replacement of omphacite by clinopyroxene + sodic plagioclase symplectite, with retrograde growth of amphibole, and with the replacement of phengite by biotite. Garnet was found to be resistant to retrograde oxygen isotope exchange (Yui *et al.*, 1997). These results show that once free water gains access to UHP minerals, it leaves a clearly recognizable record of its presence.

Published studies of three UHP terrains, the Western Gneiss region, Norway, Dora Maira, Italy, and Dabie–Su–Lu, China, are in accord concerning fluid behavior during UHP metamorphism. Water may have been present along grain boundaries but it was not free to infiltrate across lithologic layers. Studies of lower pressure eclogites lacking coesite inclusions are in agreement with this. Pre-metamorphic differences in $\delta^{18}O$ between rocks in mutual contact are preserved throughout the entire range of eclogite-facies conditions (Nadeau *et al.*, 1993; Getty and Selverstone, 1994; Barnicoat and Cartwright, 1997). A consistent picture of fluids in UHP metamorphism emerges. All free water is expelled mechanically near the surface during compaction. Water structurally bound in clays, zeolites, and other low-temperature hydrous minerals is carried down by subducted crustal rocks. Free fluids released by dehydration reactions appear episodically along grain boundaries as the successive thermal stabilities of various hydrous minerals are exceeded. Fluids do not remain in place but are expelled from their release site by continued compression and buoyancy, facilitated by transient, reaction- and deformation-enhanced permeability. The mobile fluids are potent agents of change. They may act locally within subducted continental crust to promote melting and metasomatism; they may be entrained in propagating fractures for return to the surface via the suture; or, they may penetrate the hanging wall and there activate hydration, metasomatism and partial melting in the mantle wedge.

The answer to the final question as to the nature and extent of interaction between subducted continental crust and mantle is already clear. Rocks of the continental crust retain their identity just as stubbornly as those of the oceanic crust during subduction. There is no evidence that crustal rocks subducted into the upper mantle suffered stable isotope metasomatism. Comparison of mantle xenoliths erupted in Neogene volcanics along the Tan–Lu fault reveals a profound oxygen isotopic disequilibrium. Clinopyroxene from xenoliths has $\delta^{18}O$ of +5.5 to +5.8‰ (Xu *et al.*, 1996), in comparison to omphacite values of –9 to –10‰ in eclogites from Qinglongshan (Yui *et al.*, 1995; Zheng *et al.*, 1996). Perhaps the most crucial question is to what extent did subducted crustal rocks influence the adjacent upper mantle. Unfortunately, there are no data from the mantle, itself, to answer the question.

6. CONCLUSIONS

The stable isotope geochemistry of UHP rocks retains a record of conditions of protolith formation despite subduction, metamorphism, and exhumation. The Chinese UHP rocks record protolith conditions so faithfully that it is possible to reconstruct their depositional environment. Extrusive basalts, volcaniclastics, and associated sediments were heated by shallow granitic intrusions in a cold climate. An active geothermal system ensued in which the rocks acquired strongly depleted $\delta^{18}O$ and δD values. These events took place in the Neoproterozoic as shown by the upper intercepts of zircon discordia. The depositional age is corroborated by the unusually high $\delta^{13}C$ values of UHP marbles that correlate stratigraphically with world-wide secular variations in $\delta^{13}C$ (Kaufman and Knoll, 1995). It is tempting to correlate the cold climate indicated by $\delta^{18}O$ and δD depletion with Neoproterozoic glaciation but UHP marbles could not have been deposited then because the glaciation was a time of lowered $\delta^{13}C$ in limestones. Thus, the cold climate indicated by oxygen and hydrogen isotopes may have anticipated or followed glaciation but did not coincide with it.

But finally, one is forced to admit that UHP rocks, so unique in their mineralogy, tectonic history and P-T conditions are not so special from the perspective of fluid–rock interactions. The same processes seen in UHP rocks are found in mid-crustal metamorphic terrains. Vast tracts of these terrains show little evidence isotopically of pervasive fluid infiltration across lithologic contacts. There are many examples of dissimilar mid-crustal rocks in mutual contact that failed to equilibrate during metamorphism.

There appears to be a dearth of evidence of channelized fluid flow in UHP rocks whether it is in fractures, shears, or metamorphic aquifers (see, however, Austrheim, 1987; 1990). But this may be due to incomplete study. The isotopic studies of sample profiles discussed above are sensitive to layer-perpendicular flow and diffusion but insensitive to flow parallel to lithologic contacts. The present understanding of fluid–rock interaction during UHP metamorphism therefore remains incomplete. Layer-parallel flow can be estimated by measuring the progress of mineral–fluid and isotope exchange reactions driven by infiltration of reactive fluid (Ferry, 1994; Bickle *et al.*, 1997). Acquiring an improved understanding of fluid flow during UHP metamorphism will require a lot of hard work including: i) systematic rock sampling parallel to the strike of lithologic contacts and across metamorphic isograds; ii) painstaking modal analysis of mineral abundances and their chemical analysis followed by estimation of reaction stoichiometry and progress; and iii) *in situ* chemical and isotopic analysis of mineral grains at a spatial scale small enough to resolve intra-mineral heterogeneity in order to distinguish successive metamorphic episodes. Such hard work will be rewarded by a new and deeper understanding of the mechanisms, magnitude, and spatial distribution of fluid flow during UHP metamorphism.

ACKNOWLEDGMENTS

Preparation of this paper was supported by the National Science Foundation, EAR-9526700. The writer is very grateful to R.W. Birnie, C. Blum, C.P. Chamberlain, S. Fullerton, R. and G. Morse, and R.C. Reynolds for hospitality at Dartmouth College. Reviews of the manuscript by G.E. Bebout, W.G. Ernst, J.M. Ferry, B.R. Hacker, J.G. Liou, A. Matthews, and T.Z. Yui were most helpful. The writer is especially grateful to B. Cong, J.G. Liou, Q.C. Wang, and R.Y. Zhang for introducing him to the wonders of UHP metamorphism.

REFERENCES

Agrinier, P., Javoy, M., Smith, D.C., and Pineau, F. (1985) Carbon and oxygen isotopes in eclogites, amphibolites, veins, and marbles from Western Gneiss region, Norway, *Chemical Geology* **52**, 145–162.

Ames, L., Zhou, G., and Xiong, B. (1996) Geochronology and geochemistry of ultrahigh-pressure metamorphism with implications for collision of the Sino–Korean and Yangtze cratons, central China, *Tectonics* **15**, 472–489.

Austrheim, H. (1987) Eclogitization of lower crustal granulites by fluid migration through shear zones, *Earth and Planetary Science Letters* **81**, 221–232.

Austrheim, H. (1990) The granulite–eclogite facies transition: a comparison of experimental work and a natural occurrence in the Bergen Arcs, western Norway, *Lithos* **25**, 163–169.

Baker, J., Matthews, A., Mattey, D., Rowley, D., and Xue, F. (1997) Fluid-rock interactions during ultra-high pressure metamorphism, Dabie Shan, China, *Geochimica Cosmochimica et Acta* **61**, 1685–1696.

Barnicoat, A.C. and Cartwright, I. (1997) Focused fluid flow during subduction: oxygen isotope data from high pressure ophiolites of the western Alps, *Earth and Planetary Science Letters* **132**, 53–61.

Baumgartner, L.P. and Rumble, D. (1988) Transport of stable isotopes: I: Development of a kinetic continuum theory for stable isotope transport, *Contributions to Mineralogy and Petrology* **98**, 417–430.

Bebout, G.E. (1996) Volatile transfer and recycling at convergent margins: mass-balance and insights from high-P/T metamorphic rocks, in G.E. Bebout, D.W. Scholl, S.H. Kirby, and J.P. Platt (eds.), *Subduction top to bottom, Geophysical Monograph*, **96**, American Geophysical Union, Washington, pp. 179–193.

Bickle, M.J. and Baker, J. (1990) Advective-diffusive transport of isotopic fronts: an example from Naxos, Greece, *Earth and Planetary Science Letters* **97**, 78–93.

Bickle, M.J., Chapman, H.J.F., Ferry, J.M., Rumble III, D., and Fallick, A.E. (1997) Fluid flow and diffusion in the Waterville Limestone, South-Central Maine: constraints from strontium, oxygen, and carbon isotope profiles, *Journal of Petrology* **38**, 1489–1512.

Blattner, P., Grindley, G.W., and Adams, C.J. (1997) Low ^{18}O terranes tracking Mesozoic polar climates in the South Pacific, *Geochimica et Cosmochimica Acta* **61**, 569–576.

Blattner, P. and Lassey, K.R. (1989) Stable-isotope exchange fronts, Damkohler numbers, and fluid to rock ratios, *Chemical Geology* **78**, 381–392.

Brandriss, M.E., Nevle, R.J., Bird, D.K., and O'Neil, J.R. (1995) Imprint of meteoric water on the stable isotope compositions of igneous and secondary minerals, Kap Edvard Holm Complex, East Greenland, *Contributions to Mineralogy and Petrology* **121**, 74–86.

Carswell, D.A. (1990) Eclogites and the eclogite facies: definitions and classifications, in D.A. Carswell (ed.), *Eclogite Facies Rocks*, Chapman and Hall, New York, pp. 1–13.

Chopin, C. and Sobolev, N.V. (1995) Principal mineralogic indicators of UHP in crustal rocks, in R.G. Coleman and X. Wang (eds.), *Ultrahigh Pressure Metamorphism*, Cambridge University Press, Cambridge, pp. 96–131.

Coleman, R.G. and Wang, X. (1995a) Overview of the geology and tectonics of UHPM, in R.G. Coleman and X. Wang (eds.), *Ultrahigh Pressure Metamorphism*, Cambridge University Press, New York, pp. 1–31.

Coleman, R.G. and Wang, X. (1995b) *Ultrahigh Pressure Metamorphism*, Cambridge University Press, New York.

Compagnoni, R., Hirajima, T., and Chopin, C. (1995) Ultra-high-pressure metamorphic rocks in the Western Alps, in R.G. Coleman and X. Wang (eds.), *Ultrahigh Pressure Metamorphism*, Cambridge University Press, Stanford, pp. 206–243.

Cong, B. (1996) *Ultrahigh-Pressure Metamorphic Rocks in the Dabie Shan-Sulu Region of China*, Science Press, Beijing.

Cong, B. and Wang, Q. (1996) A review on researches of UHPM rocks in the Dabieshan-Su–Lu region, in B. Cong (ed.), *Ultrahigh-Pressure Metamorphic Rocks in the Dabieshan-Su–Lu region of China*, Science Press, Beijing, pp. 1–7.

Coplen, T.B. (1995) Reporting of stable carbon, hydrogen, and oxygen isotopic abundances, *Anonymous Reference and Intercomparison Materials for Stable Isotopes of Light Elements*, International Atomic Energy Agency, Vienna, pp. 31–34.

Criss, R.E. and Taylor, H.P. (1986) Meteoric-hydrothermal systems, in J.W. Valley, H.P. Taylor, and J.R. O'Neil (eds.), *Stable Isotopes in High Temperature Geological Processes*, Mineralogical Society of America, Washington, D.C., pp. 373–424.

Deines, P., Harris, J.W., Robinson, D.N., Gurney, J.J., and Shee, S.R. (1991) Carbon and oxygen isotope variations in diamond and graphite eclogites from Orapa, Botswana, and the nitrogen content of their diamonds, *Geochimica et Cosmochimica Acta* **55**, 515–524.

Ernst, W.G. and Peacock, S.M. (1996) A thermotectonic model for preservation of ultrahigh-pressure phases in metamorphosed continental crust, in G.E. Bebout, D.W. Scholl, S.H. Kirby, and J.P. Platt (eds.), *Subduction Top to Bottom*, American Geophysical Union, Washington, D.C., pp. 171–178.

Farquhar, J., Chacko, T., and Ellis, D.J. (1996) Preservation of oxygen isotope compositions in granulites from Northwestern Canada and Enderby Land, Antarctica: implications for high-temperature isotopic thermometry, *Contributions to Mineralogy and Petrology* **125**, 213–224.

Fehlhaber, K. and Bird, D.K. (1991) Oxygen-isotope exchange and mineral alteration in gabbros of the lower layered series, Kap Edvard Holm complex, East Greenland, *Geology* **19**, 819–822.

Ferry, J.M. (1994) A historical review of metamorphic fluid flow, *Journal of Geophysical Research* **99**, 15487–15498.

Ganor, J., Matthews, A., Schliestedt, M., and Garfunkel, Z. (1996) Oxygen isotope heterogeneities of metamorphic rocks: an original tectonostratigraphic signature, or an imprint of exotic fluids? A case study of Sifnos and Tinos islands (Greece), *European Journal of Mineralogy* **8**, 719–732.

Gat, J.R. (1996) Oxygen and hydrogen isotopes in the hydrologic cycle, *Annual Reviews Earth and Planetary Sciences* **24**, 225–262.

Getty, S.R. and Selverstone, J. (1994) Stable isotope and trace element evidence for restricted fluid migration in 2 GPa eclogites, *Journal of Metamorphic Geology* **12**, 747–760.

Gregory, R.T., Douthitt, C.B., Duddy, I.R., and Rich, P.V. (1989) Oxygen isotopic composition of carbonate concretions from the Lower Cretaceous of Victoria, Australia: Implications for the evolution of meteoric waters of the Australian continent in a paleopolar environment, *Earth and Planetary Science Letters* **92**, 27–42.

Hacker, B.R. (1996) Eclogite formation and the rheology, buoyancy, seismicity, and H2O content of oceanic crust, in G.E. Bebout, Scholl, D., Kirby, S.H., Platt, J.P. (ed.), *Dynamics of Subduction*, Monograph, American Geophysical Union, Washington, D.C., pp. 337–246.

Hacker, B.R. and Peacock, S.M. (1994) Creation, preservation, and exhumation of coesite-bearing, ultrahigh-pressure metamorphic rocks, in R.G. Coleman and X. Wang (eds.), *Ultrahigh Pressure Metamorphism*, Cambridge University Press, Cambridge, United Kingdom

Hacker, B.R., Ratschbacher, L., Webb, L., Ireland, T., Walker, D., and Dong, S. (in press) U/Pb zircon ages constrain the architecture of the ultrahigh-pressure Qinling–Dabie Orogen, China, *Earth and Planetary Science Letters*.

Harley, S.L. and Carswell, D.A. (1995) Ultradeep crustal metamorphism: A prospective view, *Journal of Geophysical Research* **100**, 8367–8380.

Hattori, K. and Muehlenbachs, K. (1982) Oxygen isotope ratios in the Icelandic crust, *Journal of Geophysical Research* **87**, 6559–6565.

Hirajima, T., Zhang, R., Li, J., and Cong, B. (1992) Petrology of the nyboite-bearing eclogite in the Donghai area, Jiangsu Province, eastern China, *Mineralogical Magazine* **56**, 37–46.

Jamtveit, B., Bucher-Nurminen, K., and Austrheim, H. (1990) Fluid controlled eclogitization of granulites in deep crustal shear zones, Bergen arcs, western Norway, *Contributions to Mineralogy and Petrology* **104**, 184–193.

Kaufman, A.J. and Knoll, A.H. (1995) Neoproterozoic variations in the C-isotopic composition of seawater: stratigraphic and biogeochemical implications, *Precambrian Research* **73**, 27–49.

Kirby, S.H., Engdahl, E.R., and Denlinger, R. (1996) Intermediate-depth intraslab earthquakes and arc volcanism as physical expressions of crustal and uppermost mantle metamorphism in subducting slabs, in G.E. Bebout, D. Scholl, and S. Kirby (eds.), *Subduction top to bottom, Geophysical Monograph*, **96**, AGU, Washington, D.C., pp. 195–214.

Krogh, E.J. and Carswell, D.A. (1995) HP and UHP eclogites and garnet peridotites in the Scandinavian Caledonides, in R.G. Coleman and X. Wang (eds.), *Ultrahigh Pressure Metamorphism*, Cambridge University Press, Stanford, pp. 244–298.

Li, S., Wang, S., Chen, Y., Liu, D., Qiu, J., Zhou, H., and Zhang, Z. (1994) Excess argon in phengite from eclogite: Evidence from the dating of eclogite minerals by the Sm-Nd, Rb-Sr and 40Ar/39Ar methods, *Chemical Geology* **112**, 343.

Liou, J.G. and Zhang, R.Y. (1996) Occurrence of intergranular coesite in ultrahigh-P rocks from the Sulu region, eastern China: Implications for lack of fluid during exhumation, *American Mineralogist* **81**, 1217–1221.

Liou, J.G., Zhang, R.Y., Eide, E.A., Maruyama, S., Wang, X., and Ernst, W.G. (1996) Metamorphism and tectonics of high-P and ultrahigh-P belts in Dabie-Sulu Regions, eastern central China, in A. Yin and T.M. Harrison (eds.), *The Tectonic Evolution of Asia*, Rubey IX, Cambridge University Press, Cambridge, United Kingdom, pp. 300–343.

Liou, J.G., Zhang, R.Y., and Ernst, W.G. (1995) Occurrences of hydrous and carbonate phases in ultrahigh-pressure rocks from east-central China: Implications for the role of volatiles deep in cold subduction zones, *The Island Arc* **4**, 362–375.

Liou, J.G., Zhang, R.Y., and Jahn, B.M. (1997) Petrology, geochemistry and isotope data on a ultrahigh-pressure jadeite quartzite from Shuanghe, Dabie Mountains, east-central China, *Lithos* **41**, 59–78.

MacGregor, I. and Manton, W.I. (1986) Roberts Victor eclogites: ancient oceanic crust, *Journal of Geophysical Research* **91**, 14063–14079.

Massonne, H.J. (1995) Is the concept of 'in situ' metamorphism applicable to deeply buried continental crust with lenses of eclogites and garnet peridotites?, *Chinese Science Bulletin* **40** (supplement), 145–147.

Mattey, D., Jackson, D.H., Harris, N.B.W., and Kelley, S. (1994a) Isotopic constraints on fluid infiltration from an eclogite shear zone, Holsenøy, Norway, *Journal of Metamorphic Geology* **12**, 311–325.

Mattey, D., Lowry, D., and Macpherson, C. (1994b) Oxygen isotope composition of mantle peridotite, *Earth and Planetary Science Letters* **128**, 231–241.

Matthews, A., Liata, A., Mposkos, E., and Skarpelis, N. (1996) Oxygen isotope geochemistry of the Rhodope polymetamorphic terrain in northern Greece: evidence for preservation of pre-metamorphic isotopic compositions, *European Journal of Mineralogy* **8**, 1139–1152.

Michard, A., Henry, C., and Chopin, C. (1995) Structures in UHPM rocks: A case study from the Alps, in R.G. Coleman and X. Wang (eds.), *Ultrahigh Pressure Metamorphism*, 132–158, Cambridge University Press, Stanford

Mosenfelder, J.L. and Bohlen, S.R. (1997) Kinetics of the cœsite → quartz transformation, *Earth and Planetary Science Letters* **153**, 133–147.

Nadeau, S., Philippot, P., and Pineau, F. (1993) Fluid inclusion and mineral isotopic compositions (H-C-O) in eclogitic rocks as tracers of local fluid migration during high-pressure metamorphism, *Earth and Planetary Science Letters* **114**, 431–448.

Nevle, R.J., Brandriss, M.E., Bird, D.K., McWilliams, M.O., and O'Neil, J.R. (1994) Tertiary plutons monitor climate change in East Greenland, *Geology* **22**, 775–778.

Ongley, J.S., Basu, A.R., and Kyser, T.K. (1987) Oxygen isotopes in coexisting garnets, clinopyroxenes, and phlogopites of Roberts Victor eclogites; implications for petrogenesis and mantle metasomatism, *Earth and Planetary Science Letters* **83**, 80–84.

Patiño Douce, A.E. and McCarthy, T.C. (this volume) Melting of crustal rocks during continental collision and subduction

Peacock, S.M. (1996) Thermal and petrologic structure of subduction zones, in G.E. Bebout, D.W. Scholl, S.H. Kirby, and J.P. Platt (eds.), *Subduction Top to Bottom*, American Geophysical Union, Washington, D.C., pp. 119–134.

Philippot, P. (1993) Fluid-melt-rock interaction in mafic eclogites and coesite-bearing metasediments; constraints on volatile recycling during subduction, *Chemical Geology* **108**, 93–112.

Rowley, D.B., Xue, F., Tucker, R.D., Peng, Z.X., Baker, J., and Davis, A. (1997) Ages of ultrahigh pressure metamorphism and protolith orthogneisses from the eastern Dabie Shan: U/Pb zircon geochronology, *Earth and Planetary Science Letters* **151**, 191–203.

Rubie, D.C. (1990) Role of kinetics in the formation and preservation of eclogites, in D.A. Carswell (ed.), *Eclogite Facies Rocks*, Blackie, Glasgow, pp. 111–140.

Schreyer, W. (1995) Ultradeep metamorphic rocks: The retrospective viewpoint, *Journal of Geophysical Research* **100**, 8353–8366.

Schreyer, W. and Stockhert, B. (1997) High-pressure metamorphism in nature and experiment, *Lithos* **41**, 1–4.

Sharp, Z.D., Essene, E.J., and Hunziker, J.C. (1993) Stable isotope geochemistry and phase equilibria of coesite-bearing whiteschists, Dora Maira Massif, western Alps, *Contributions to Mineralogy and Petrology* **114**, 1–12.

Sheppard, S.M.F. (1986) Characterization and isotopic variations in natural waters, in J.W. Valley, H.P. Taylor, and J.R. O'Neil (eds.), *Stable Isotopes in High Temperature Geological Processes*, 165–183, Mineralogical Society of America, Washington, D.C.

Shervais, J.W., Taylor, L.A., Lugmair, G.W., Clayton, R.N., Mayeda, T.K., and Korotev, R.L. (1988) Early Proterozoic oceanic crust and the evolution of subcontinental mantle: Eclogites and related rocks from southern Africa, *Geological Society of America Bulletin* **100**, 411–423.

Smith, D.C. (1995) Microcoesites and microdiamonds in Norway, in R.G. Coleman and X. Wang (eds.), *Ultrahigh Pressure Metamorphism*, 299–355, Cambridge University Press, Stanford

Tabata, H., Yamauchi, K., Maruyama, S., and Liou, J.G. (this volume) Tracing the extent of a UHP metamorphic terrane: A mineral-inclusion study of zircons in gneisses from the Dabie Shan

Taylor, H.P. and Forester, R.W. (1979) An oxygen and hydrogen isotope study of the Skaergaard intrusion and its country rocks: a description of a 55–m.y. old fossil hydrothermal system, *Journal of Petrology* **20**, 355–419.

Taylor, H.P. and Sheppard, S.M.F. (1986) Igneous rocks: I. Processes of isotopic fractionation and isotope systematics, in J.W. Valley, H.P. Taylor, and J.R. O'Neil (eds.), *Stable Isotopes in High Temperature Geological Processes*, Mineralogical Society of America, Washington, D.C., pp. 227–271.

Wang, X., Zhang, R.Y., and Liou, J.G. (1995) UHPM terrane in east central China, in R.G. Coleman and X. Wang (eds.), *Ultrahigh Pressure Metamorphism*, Cambridge University Press, Stanford, pp. 356–390.

Wickham, S.M. and Peters, M.T. (1993) High $\delta^{13}C$ Neoproterozoic carbonate rocks in western North America, *Geology* **21**, 165–168.

Wiechert, U., Ionov, D.A., and Wedepohl, K.H. (1997) Spinel peridotite xenoliths from the Atsagin-Dush volcano, Dariganga lava plateau, Mongolia: a record of partial melting and cryptic metasomatism in the upper mantle, *Contributions to Mineralogy and Petrology* **126**, 345–364.

Xu, J. (1993) Basic characteristics and tectonic evolution of the Tanchung-Lujiang fault zone, in J. Xu (ed.), *The Tancheng-Lujiang Wrench Fault System*, John Wiley and Sons, New York, pp. 17–50.

Xu, Y.G., Menzies, M.A., Mattey, D., Lowry, D., Harte, B., and Hinton, R.W. (1996) The nature of the lithospheric mantle near the Tancheng-Lujiang fault, China: an integration of texture, chemistry, and oxygen isotopes, *Chemical Geology* **134**, 67–82.

Xue, F., Rowley, D.B., and Baker, J. (1996) Refolded syn-ultrahigh-pressure thrust sheets in the south Dabie Complex, China; field evidence and tectonic implications, *Geology* **24**, 455–458.

Yui, T.F., Rumble, D., and Lo, C.H. (1995) Unusually low ^{18}O ultrahigh-pressure metamorphic rocks from the Sulu terrain, eastern China, *Geochimica Cosmochimica et Acta* **59**, 2859–2864.

Yui, T.Z., Rumble, D., Chen, C.H., and Lo, C.H. (1997) Stable isotope characteristics of eclogites from the ultra-high-pressure metamorphic terrain, east-central China, *Chemical Geology* **137**, 135–147.

Yui, T.Z., Rumble, D., and Lo, C.H. (1994) Stable-isotope characteristics of the ultra-high pressure (UHP) metamorphic rocks of China, in J.G. Liou, W.G. Ernst, and R.G. Coleman (eds.), *Workshop on Ultrahigh-Pressure Metamorphism*, Stanford University, Stanford, pp. A76–A77.

Zhang, R., Hirajima, T., Banno, S., Cong, B., and Liou, J.G. (1995a) Petrology of ultrahigh-pressure rocks from the southern Su-Lu region, eastern China, *Journal of Metamorphic Geology* **13**, 659–675.

Zhang, R.Y., J.G., L., and B., C. (1995b) Talc-, magnesite- and Ti-clinohumite-bearing ultrahigh-pressure meta-mafic and ultramafic complex in the Dabie Mountains, *Journal of Petrology* **36**, 1011–1037.

Zheng, Y.F. (1997) Oxygen and carbon isotope anomalies in the ultrahigh pressure metamorphic rocks of the Dabie-Sulu terranes: implications for geodynamics, *Episodes* **20**, 104–108.

Zheng, Y.F., Fu, B., Gong, B., and Li, S. (1996) Extreme ^{18}O depletion in eclogite from the Su–Lu terrane in East China, *European Journal of Mineralogy* **8**, 317–323.

Chapter 10

Tracing the Extent of a UHP Metamorphic Terrane: Mineral-Inclusion Study of Zircons in Gneisses from the Dabie Shan

Hirokazu Tabata, Kazuhiro Yamauchi and Shigenori Maruyama
Department of Earth and Planetary Sciences, Tokyo Institute of Technology, Ookayama 2–12–1, Meguro, Tokyo, 152, Japan, tabata, yama, and smaruyam@geo.titech.ac.jp,

Juhn G. Liou
Department of Geological and Environmental Sciences, Stanford University, CA, 94305, USA, liou@geo.stanford.edu

Abstract: Coesite inclusions in zircon from the country-rock orthogneiss of the southern Dabie ultrahigh-pressure terrane, central China were identified by Raman spectroscopy and electron microprobe microanalysis. All the coesite-bearing samples analyzed show amphibolite-facies mineral assemblages in the matrix, indicating that these gneissic rocks experienced ultrahigh-pressure metamorphism and then retrogression. Quartz inclusions are also present in the same samples, but some of them are accompanied by fractures and/or fluid inclusion arrays. Coesite-bearing orthogneisses occur far from previously recognized UHP eclogites, indicating that coesite-bearing lithologies occur all over the southern Dabie unit, requiring subduction of the whole area to mantle depths.

1. INTRODUCTION

Studies of mineral inclusions in metamorphic minerals often provide qualitative and quantitative information about the reaction and growth history of a particular metamorphic rock or terrane (e.g., Liou *et al.*, 1997a). Numerous petrologic investigations have shown that porphyroblastic minerals (*e.g.*, garnet) commonly contain mineral inclusions not in equilibrium with the matrix assemblage. The relict inclusion assemblage records an earlier stage of the metamorphic history,

B. R. Hacker and J. G. Liou (eds.),
When Continents Collide: Geodynamics and Geochemistry of Ultrahigh-Pressure Rocks, 261–273.
© 1998 *Kluwer Academic Publishers. Printed in the Netherlands.*

and is preserved by virtue of being surrounded by the porphyroblast, whereas similar minerals in the rock matrix may have broken down or reacted with other phases. Geothermobarometry and phase equilibria considerations can help to determine P-T conditions prior to and during growth of the porphyroblasts, and aid in delineating a metamorphic P-T path for the rock (e.g., Spear, 1993).

For the study of ultrahigh-pressure (UHP) metamorphic terranes, armored relict inclusions within host minerals play a particularly important role in deciphering the metamorphic record prior to the formation of the matrix assemblage. UHP rocks, which generally occur as discrete blocks surrounded by greenschist- to amphibolite-facies country rock gneisses, have commonly been recognized by the presence of characteristic index minerals such as coesite and microdiamond. Except for rare intergranular occurrences (Liou and Zhang, 1996; Banno *et al.*, 1997), coesite is present as inclusions within garnet or other host minerals (Chopin, 1984; Smith, 1984; Okay *et al.*, 1989; Sobolev *et al.*, 1990; Xu *et al.*, 1992). Although thermobarometry for matrix assemblages sometimes gives pressure estimates consistent with UHP, most UHP rock matrixes are retrogressed and do not contain UHP index minerals. In addition, the country rocks to most eclogites almost always retain no vestige of primary UHP conditions. The pressure gap between UHP and country rocks has long caused controversy regarding the extent of UHP metamorphism (e.g., Harley and Carswell, 1995). This controversy as well as divergent tectonic models for the Dabie Shan (e.g., Okay *et al.*, 1993; Maruyama *et al.*, 1994; Hacker *et al.*, 1995) is essentially due to a lack of direct evidence of UHP relics in country rocks.

This chapter is a case study of micro-Raman spectroscopy to identify UHP minerals in zircons from amphibolite-facies gneisses in the Dabie Shan of central China, to determine the extent to which supracrustal materials have been subducted and metamorphosed at mantle depths. Zircon, despite its small grain size, is a promising container of diagnostic UHP mineral inclusions (Sobolev *et al.*, 1992; 1994; Schertl and Schreyer, 1996). It occurs as an accessory but ubiquitous mineral in most country-rock gneisses, and often remains as a stable matrix mineral, preserving its mineral inclusions, in host rocks that have experienced subsequent retrogression. Phases included in zircon tend to be too small to be easily identified with optical methods or electron microprobe analysis, however, micro-Raman spectroscopy (Roberts and Beattie, 1995) enables the identification of fine-grained inclusions of coesite and microdiamond (Smith, 1984; Boyer *et al.*, 1985; Sobolev *et al.*, 1990).

2. EXAMPLE OF MICRO-INCLUSIONS IN ZIRCON, THE SOUTHERN DABIE UHP UNIT

2.1 Geologic Setting and Problem of the Regional Extent of the Southern Dabie UHP Terrane

The Dabie UHP terrane is situated in the 2000 km-long suture zone between the Sino-Korean and Yangtze cratons, which collided in Triassic time (e.g., Wang *et al.*, 1995; Hacker *et al.*, 1996; Liou *et al.*, 1996). The Dabie block, bounded by several major faults, is divided into two units: the southern Dabie unit and the northern Dabie unit (Fig. 1). The northern Dabie unit consists chiefly of Cretaceous orthogneiss and plutonic rocks intruding an older basement (Hacker *et al.*, 1996; Zhang *et al.*, 1996). This unit may never have been subjected to high- or ultrahigh-pressure metamorphism.

The southern Dabie unit is composed chiefly of amphibolite-facies quartzofeldspathic gneisses including boudinaged layers or lenses of eclogite, garnet peridotite, amphibolite and marble. Mineral assemblages in metabasites generally show a northward increase in metamorphic grade: from greenschist to blueschist to eclogite facies. The eclogite-facies rocks are subdivided into a southern quartz-bearing eclogite or "cold eclogite" zone and northern coesite- and diamond-bearing eclogite or "hot eclogite: zone, based mainly on the distribution of relict coesite inclusion in eclogites (Okay, 1993). These zones are separated from each other by a southward-dipping normal fault (Okay *et al.*, 1993). Besides a difference in silica polymorphs, eclogites in the coesite eclogite zone are also distinguishable from those in the quartz eclogite zone by their garnet zonation and inferred metamorphic temperatures (Wang *et al.*, 1992; Okay, 1993; Carswell *et al.*, 1997).

Coesite and its pseudomorphs, which characterise the "hot" eclogite zone, occur as inclusions within garnet, omphacite, kyanite, epidote/zoisite, dolomite and zircon in eclogite boudins and in adjacent metapelites and marbles (Okay *et al.*, 1989; Wang *et al.*, 1989; Schertl and Okay, 1994; Sobolev *et al.*, 1994; Zhang and Liou, 1996). Most individual eclogite boudins consist of UHP minerals, including garnet, omphacite and coesite, in their cores which grade out into amphibolite-facies assemblages (hornblende + clinozoisite + plagioclase + quartz) toward the contact with the host gneisses. This suggests that, following peak UHP metamorphism, the eclogite boudins experienced amphibolite-facies overprinting at conditions similar to the surrounding gneisses.

Researchers have proposed two different interpretations for the origin of the UHP rocks and the extent of UHP metamorphism of the southern Dabie unit: an *in situ* model and a foreign source model. The *in situ* model holds that UHP metamorphism of regional extent affected a continental crust composed of

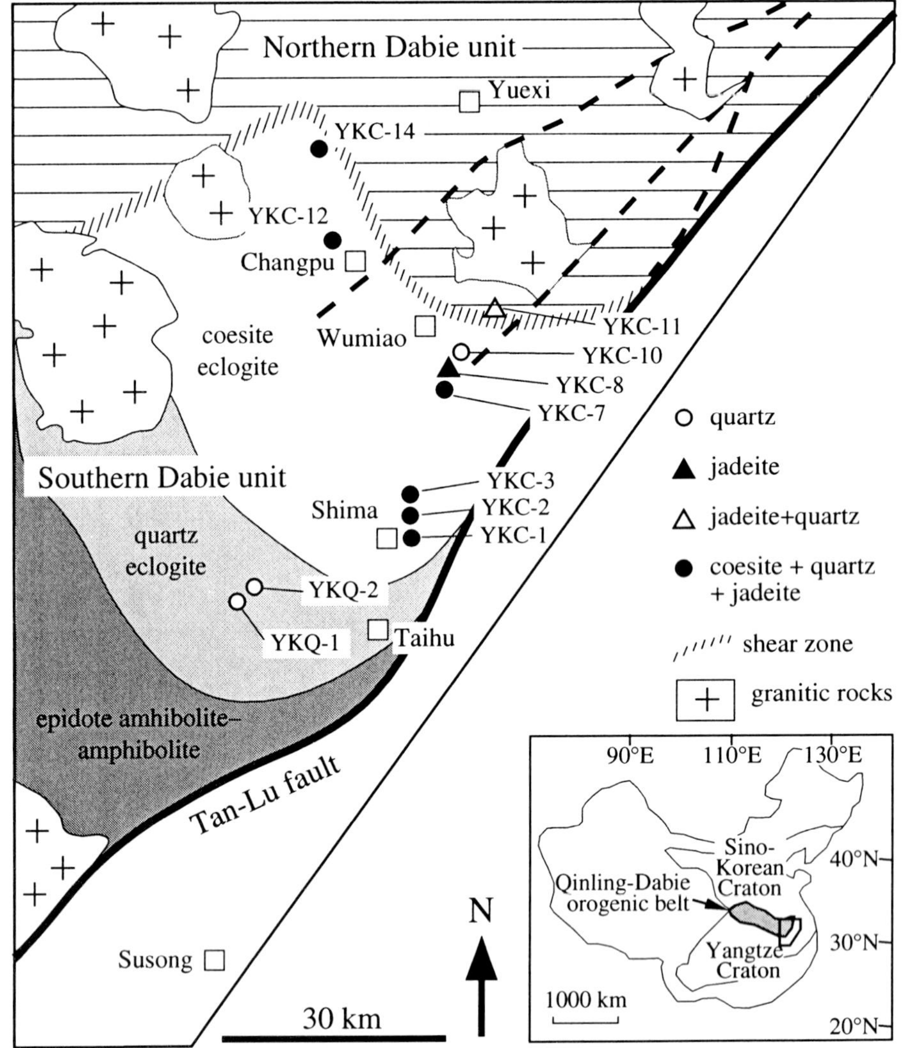

Figure 1. Geologic map of the study area, showing localities of analyzed samples containing silica and/or jadeite inclusions within zircon.

eclogite blocks and enclosing host rock gneisses. This model is favored because: i) metamorphic grade increases systematically northward in the eclogite zone, suggesting a coherent metamorphic unit (Wang and Liou, 1991; Wang *et al.*, 1992); ii) the coesite-bearing eclogites are structurally concordant with the surrounding gneisses (Xue *et al.*, 1996); and iii) marble, paragneiss and jadeite quartzite adjacent to eclogite blocks contain coesite and/or quartz pseudomorphs after coesite as inclusion minerals (Wang and Liou, 1991; Sobolev *et al.*, 1994; Zhang and Liou, 1996; Liou *et al.*, 1997b).

The foreign source model postulates that UHP tectonic blocks composed of coesite-bearing eclogite, coherent gneisses, and marble, were juxtaposed with country-rock orthogneisses after UHP metamorphism. Proponents of this model cite: i) an apparent pressure gap between eclogites and surrounding gneisses; and (2) coesite-bearing eclogite and adjacent paragneiss and marble occur as tectonic blocks discordant to the country-rock gneisses (Fig. 2 of Cong *et al.*, 1995).

One of the conclusive pieces of evidence in favor of the *in situ* model is the occurrence of coesite inclusions in surrounding paragneisses. If the eclogites and enclosing gneisses were contiguous during UHP metamorphism, the gneisses should contain coesite inclusions (e.g., Sobolev *et al.*, 1994). Furthermore, the systematic increase in metamorphic grade in quartz- and coesite-eclogite zones may be questionable (Carswell *et al.*, 1997). In particular, Wang *et al.* (1992) concluded on the basis of geothermometry of eclogitic mineral assemblages that peak metamorphic temperatures increase northward from 640–700°C at Shima to 740–780°C at Wumiao. However, Tabata *et al.* (in press) demonstrated no correlation between calculated peak temperatures and a north-south traverse across the coesite eclogite zone. In fact, a high temperature of 790°C was obtained from the Shima area. The idea of a gradient in peak temperature within the coesite unit as evidence supporting the *in situ* model should be disregarded.

2.2 Diamond in Dabie Eclogite

Microdiamond of UHP origin was first discovered as inclusions in zircon and garnet of the Kokchetav Massif by Sobolev and Shatsky (1990), and subsequently reported in garnet from eclogite, garnet-pyroxenite and jadeitite in the Dabie Shan by Xu *et al.* (1992), and in residues separated from Western Gneiss Region gneisses of Norway (Dobrzhinetskaya *et al.*, 1995). Microdiamonds from Norway are 20–50 μm in diameter and are round anhedral crystals. The Kokchetav microdiamonds are restricted to metasediments, have cubo-octahedral faces, and average 12 μm in diameter; some grains have been replaced by graphite. On the other hand, more than 20 diamond crystals have been identified mainly in eclogites of the Dabie Mountains (Xu *et al.*, 1992); they range in grain size from 150 to 700 microns, much coarser than those in the Kokchetav massif.

Since the discovery of microdiamond in garnet from eclogite by Xu *et al.* (1992), other occurrences in both the Dabie Mountains and the Su-Lu region have been reported (Hu and Liu, 1992; Okay, 1993; Xu and Su, 1997). However, the identification of microdiamond in the Chinese occurrences has been questioned. The Raman spectrum of diamond reported by Xu *et al.* (1992) has a peak intensity at 1331 cm^{-1} too low to be diagnostic of diamond (but cf. Xu and Su, 1997). Moreover, the original optical identification of diamond by Okay

(1993) proved to be erroneous (A. Okay, personal communication to C. Chopin, 1997).

2.3 Sampling and Analytical Techniques

Samples of typical quartzofeldspathic gneisses were collected from a regional north–south traverse across the southern Dabie unit; nine samples are from the coesite-eclogite zone, and two from the quartz-eclogite zone. These gneisses contain a common mineral assemblage of muscovite + plagioclase + quartz + apatite and a small amount of zircon ± epidote ± Ca-amphibole ± garnet ± biotite ± calcite. Each sample was collected tens to hundreds of meters away from the nearest eclogite outcrop. Sample YKC-11 is a fine-grained micaschist from a broad mylonitic zone between the southern and northern Dabie units. YKC-14 lies at the northern limit of the southern Dabie unit (Table 1; Fig. 1).

Table 1. Mineral inclusions in zircons.

Sample	Location	coe	qtz	jd	mus	rut	ttn
Quartz eclogite zone							
YKQ-1	30°37.3'N, 116°08.8'E		+				
YKQ-2	30°37,4'N, 116°08.9'E						
Coesite eclogite zone							
YKC-1	30°30.5'N, 116°17.8'E	+	+	+		+	
YKC-2	30°30.9'N, 116°18.1'E	+	+	+		+	
YKC-3	30°31.2'N, 116°18.1'E	+	+	+	+	+	
YKC-7	30°37.0'N, 116°20.0'E	+		+		+	
YKC-8	30°37.5'N, 116°20.3'E			+			
YKC-10	30°38.8'N, 116°21.4'E		+			+	+
YKC-11	30°42.1'N, 116°23.5'E		+	+			
YKC-12	30°44.4'N, 116°13.9'E	+	+	+			
YKC-14	30°49.3'N, 116°12.5'E	+	+	+	+		

Abbreviations : coe, coesite; qtz, quartz; jd, jadeite; rut, rutile; ttn, titanite.
All the samples contain apatite inclusions.

After magnetic and heavy liquid separation of minerals from a fist-sized hand specimen, 30–80 grains of zircon approximately 100–300 μm in diameter were handpicked from each residue. These zircon grains were then mounted in epoxy, polished and observed using a scanning electron microscope with a cathodoluminescence detector. Mineral inclusions were analyzed by laser-Raman microsampling spectroscopy, using a JASCO NRS-2000 instrument with the 514.5 nm line of an Ar-ion laser. In some samples, wavelength-dispersive microprobe analyses were also carried out to confirm identifications made via Raman spectroscopy.

2.4 Results

In cathodoluminesence images, most individual zircon grains exhibit distinct zoning patterns (Fig. 2b). Minerals included within gneiss zircons are listed in Table 1. Apatite is an ubiquitous additional phase in all analyzed samples. Muscovite and titanite are subordinate inclusion phases. Titanite inclusions from sample YKC-10 contain substantial amount of Al_2O_3, up to 4.6 wt%. Coesite inclusions were found in six samples, and jadeite inclusions in all but one sample of the coesite-eclogite zone, whereas neither phase was identified from the quartz-eclogite zone. Coesite inclusions are distributed in the cores and rims of zircon grains (Fig. 2a and b), and there seems to be no spatial correlation. Quartz inclusions are present in all samples except for one from the quartz-eclogite zone and one from the coesite-eclogite zone. Quartz inclusions from the coesite-eclogite zone are sometimes accompanied by cracks emanating from the inclusions, and by two-phase (liquid + vapor) fluid inclusions constituting inclusion arrays (Fig. 2c). Although every sample in which coesite inclusions were identified also contains quartz-bearing zircons, both coesite and *pure* quartz inclusions have not been found in the same grain. The occurrence of these phases as well as jadeite, however, is not common. For instance, among the 82 grains of separated zircons from YKC-3 only two contain coesite inclusions. Neither diamond nor graphite was found.

Representative Raman spectra of SiO_2 inclusions and host zircon are shown in Fig. 3. Coesite was identified by a characteristic band at 523 ± 2 cm^{-1} and a weak band at 271 ± 1 cm^{-1}, whereas quartz was recognized by a line at 468 ± 2 cm^{-1} (cf. Boyer *et al.*, 1985). In a few samples, Raman spectra from single inclusions contain coesite bands along with a peak characteristic of quartz. For example, the Raman spectrum of a silica inclusion shown in Fig.3a includes a weak but distinguishable quartz band (470 cm^{-1}) along with bands from coesite and host zircon, whereas Fig. 3b shows a strong quartz band along with weak coesite bands. These observations suggest substantial transformation of coesite to quartz even within rigid zircon grains.

The distribution of silica polymorphs and jadeite inclusions in zircons is given in Fig. 1. Samples YKC-1, 2, 3 and YKC-7 with coesite inclusions are from the Shima and the Wumiao areas, where coesite and/or its pseudomorph have been reported from eclogites (Okay, 1993; 1995). Our northernmost sample (YKC-14) occurs in the northern extreme of the southern Dabie unit, where UHP signatures and eclogite have not been identified.

The lack of diamond or graphite inclusions in zircons in the investigated gneissic rocks from Shima and Wumiao is disappointing as these are the localities from which the occurrence of diamond inclusions in eclogitic garnet has been reported (Xu *et al.*, 1992; Xu and Su, 1997). Petrological studies of phengite-bearing eclogites from the southern Dabie terrane suggest that the UHP metamorphism occurred, at least locally, within the diamond stability field (e.g., Carswell *et al.*, 1997).

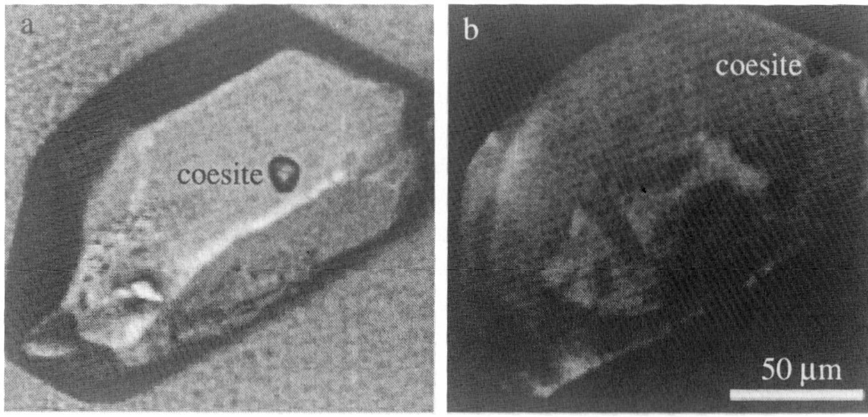

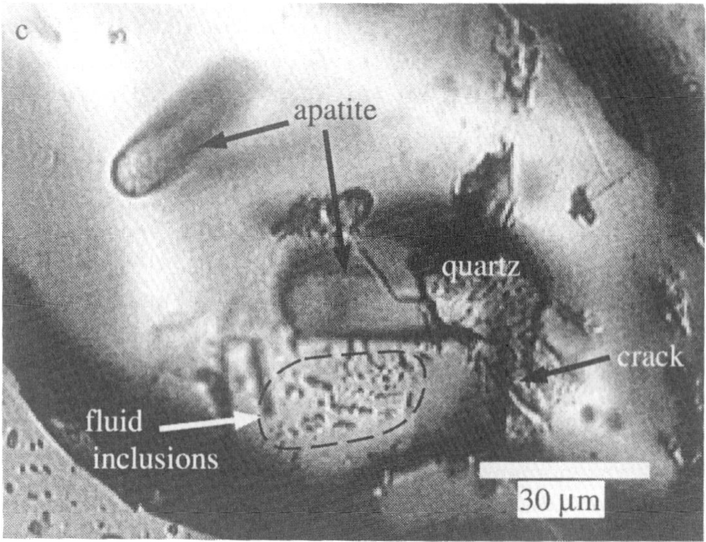

Figure 2. a) Photomicrograph (transmitted, plane polarized light) of coesite inclusion and host zircon from orthogneiss YKC-14. The inclusion is not exposed at the polished surface. b) Cathodoluminescence image of the same grain, showing irregular interior surrounded by zoned rim. An exposed coesite inclusion is present at the outermost part of the grain (but invisible in a)). c) Fluid inclusion array within zircon from coesite-bearing gneiss YKC-3.

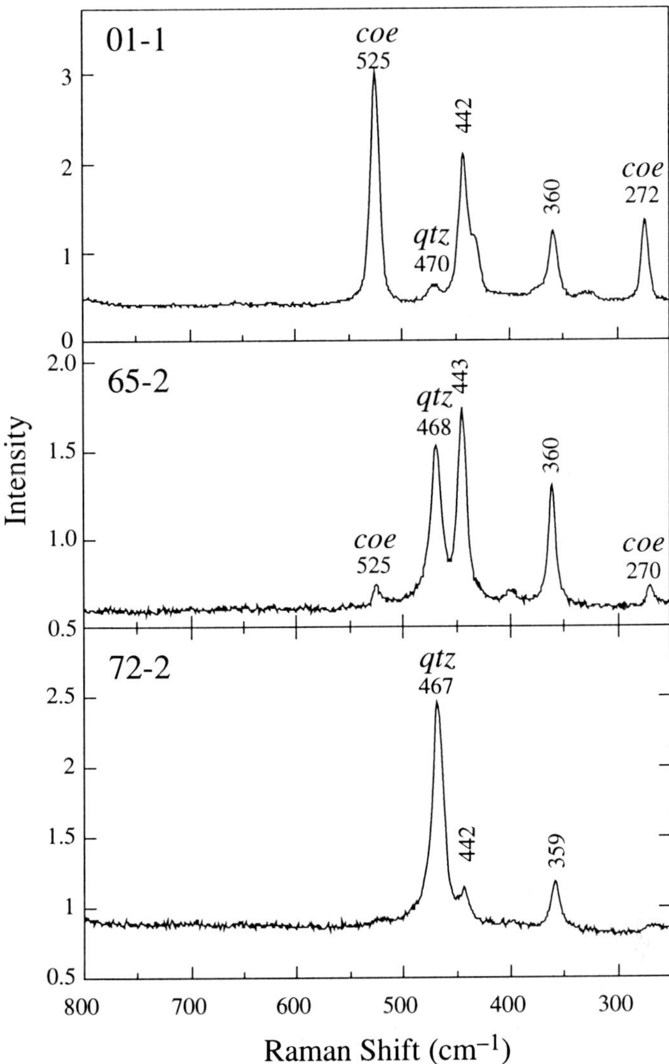

Figure 3. Representative Raman spectra obtained from three grains (no. 01, 65, and 72) from sample YKC-3. Inclusion spectra always contain zircon bands characterized by wave numbers of 442–443 cm^{-1} and 359–360 cm^{-1}. Note that bonding modes from both coesite and quartz appear in no. 01–1 and no. 65–2.

3. DISCUSSION

From U-Pb chronological investigations (Ames *et al.*, 1996; Rowley *et al.*, 1997; Hacker *et al.*, in press), zircons in gneisses are considered to have experienced at least two periods of crystal growth: original crystallization of the protoliths at

700–800 Ma and UHP metamorphism at 220 Ma. Cathodoluminescence images also show multiple growth of zircon grains (Fig. 2b) and our preliminary SHRIMP dating of zoned zircon from Wumiao confirmed a two-stage recrystallization (Liou *et al.*, 1996). The distribution of the coesite inclusions, however, has no relationship to the zoning patterns of the host zircons; coesite occurs not only in the outermost parts of zoned zircon but around the cores (compare Fig. 2a and b). From elastic models (Gillet *et al.*, 1984; van der Molen and van Roermund, 1986) and petrologic observations (Chopin *et al.*, 1991), it seems impossible that the coesite inclusions formed inside the host zircon crystal from quartz precursors. Thus some zircons that contain coesite inclusions certainly grew during UHP metamorphism. However, the occurrence of coesite in the core of zircon grains does not necessarily imply that it was trapped while the zircon core nucleated and grew. For example, Gebauer *et al.* (1997) demonstrated that coesite "inclusions" can form along cracks in older zircons.

The coexistence of coesite and quartz bands in the same inclusion (Fig. 3, 01–1 and 65–2) suggests that some coesite inclusions have converted to quartz. This phase transformation is possible when expansion of the coesite inclusion is allowed by cracking of the host zircon (Fig. 2c), and is conceivably enhanced by the influx of fluid during retrogression. The fluid inclusion arrays that we imaged around some quartz inclusions may be healed fractures that has behaved as fluid conduits. The conversion of coesite also indicates that some zircons were subjected to cracking in a way similar to other metamorphic minerals. The feasibility of zircon as a useful micro-container of UHP minerals may be related to the relative sizes of the inclusion and the host zircon. On the basis of an elastic inclusion model presented by van der Molen (1981), the internal pressure in the inclusion required to fracture the wall of the host mineral decreases as the inclusion size increases. The size of coesite inclusions we found is always less than 10 μm in diameter and very small compared with the host zircon grains (>250 μm in diameter). Thus there is a possibility that careful observation could find similarly sized coesite inclusions present in other minerals as well.

The size of an UHP slab is a significant constraint on the exhumation model of the slab. Previous researchers have found coesite inclusions in paragneisses, quartzites and marbles in the southern Dabie UHP terrane, but always from rocks closely associated with coesite-bearing eclogite blocks or layers. Our investigation shows that coesite inclusions occur at widespread localities in the amphibolite-facies gneisses of the coesite-eclogite zone. The domain including coesite-bearing lithologies extends laterally at least several kilometers, suggesting that a volumetrically substantial block of the upper crust was subducted to mantle depths. Recent structural and geochronological studies (Hacker *et al.*, 1995; Chavagnac and Jahn, 1996; Xue *et al.*, 1996; Rowley *et al.*, 1997) have proposed that UHP metamorphic rocks of the Dabie Mountains experienced a two-stage exhumation; rapid exhumation from mantle depths to lower crustal depths are followed by slower exhumation to the surface. However, it is still questionable whether the entire coesite-eclogite zone, tens of kilometers in scale, has retained complete coherency during subduction and exhumation. We plan to extend our technique to a broader area of the Dabie Shan to determine more precisely the areal extent of the UHP metamorphism.

ACKNOWLEDGMENTS

This manuscript benefited from input from several participants during the Qingdao workshop in August, 1997. We particularly appreciate Qingchen Wang, Dave Rowley and R.Y. Zhang for their suggestion with regard to the boundary between the northern and southern Dabie units. We also gratefully acknowledge the early reviews of Bradley Hacker and Gary Ernst. Constructive reviews of Christian Chopin and an anonymous reviewer have improved the manuscript. The study of UHP rocks was supported by research grants from Tokyo Institute of Technology and NSF EAR 95–06468.

REFERENCES

Ames, L., Zhou, G., and Xiong, B. (1996) Geochronology and geochemistry of ultrahigh-pressure metamorphism with implications for collision of the Sino–Korean and Yangtze cratons, central China, *Tectonics* **15**, 472–489.

Banno, S., Wallis, S.R., and Hirajima, T. (1997) From the depths of continental collision belt: Discovery and significance of UHP metamorphic rocks, *Kagaku* **67**, 39–47.

Boyer, H., Smith, D.C., Chopin, C., and Lasnier, B. (1985) Raman microprobe (RMP) determinations of natural and synthetic coesite, *Physics and Chemistry of Minerals* **12**, 45–48.

Carswell, D.A., O'Brien, P.J., Wilson, R.N., and Zhai, M. (1997) Thermobarometry of phengite-bearing eclogites in the Dabie Mountains of central China, *Journal of Metamorphic Geology* **15**, 239–252.

Chavagnac, V. and Jahn, B.-m. (1996) Coesite-bearing eclogites from the Bixiling Complex, Dabie Mountains, China; Sm-Nd ages, geochemical characteristics and tectonic implications, *Chemical Geology* **133**, 29–51.

Chopin, C. (1984) Coesite and pure pyrope in high-grade blueschists of the western Alps: a first record and some consequences, *Contributions to Mineralogy and Petrology* **86**, 107–118.

Chopin, C., Henry, C., and Michard, A. (1991) Geology and petrology of the coesite-bearing terrain, Dora Maira massif, western Alps, *European Journal of Mineralogy* **3**, 263–291.

Cong, B., Zhai, M., Carswell, T.A., Wilson, R.N., Wang, Q., Zhao, Z., and Windley, B.F. (1995) Petrogenesis of ultrahigh-pressure rocks and their country rocks at Shuanghe in Dabieshan, central China, *European Journal of Mineralogy* **7**, 119–138.

Dobrzhinetskaya, L.F., Eide, E.A., Larsen, R.B., Sturt, B.A., Tronnes, R.G., Smith, D.C., Taylor, W.R., Posukhova, T.V., and Posukhova, T.V. (1995) Microdiamond in high-grade metamorphic rocks of the Western Gneiss region, Norway, *Geology* **23**, 597–600.

Gebauer, D., Schertl, H.-P., Brix, M., and Schreyer, W. (1997) 35 Ma old ultrahigh-pressure metamorphism and evidence for very rapid exhumation in the Dora Maira massif, Western Alps, *Lithos* **41**, 5–24.

Gillet, P., Ingrin, J., and Chopin, C. (1984) Coesite in subducted continental crust: P-T history deduced from an elastic model, *Earth and Planetary Science Letters* **70**, 426–436.

Hacker, B.R., Ratschbacher, L., Webb, L., and Dong, S. (1995) What brought them up? Exhumation of the Dabie Shan ultrahigh-pressure rocks, *Geology* **23**, 743–746.

Hacker, B.R., Ratschbacher, L., Webb, L., Ireland, T., Walker, D., Calvert, A., and Dong, S.W. (1997) Exhumation of ultrahigh-pressure rocks, Dabie–Hong'an–Tongbai Shan, China, *Geological Society of America Abstracts with Programs* **28**.

Hacker, B.R., Ratschbacher, L., Webb, L., Ireland, T., Walker, D., and Dong, S. (in press) U/Pb zircon ages constrain the architecture of the ultrahigh-pressure Qinling–Dabie Orogen, China, *Earth and Planetary Science Letters.*

Hacker, B.R., Wang, X., Eide, E.A., and Ratschbacher, L. (1996) Qinling–Dabie ultrahigh-pressure collisional orogen, in A. Yin and T.M. Harrison (eds.), *The Tectonic Evolution of Asia*, Cambridge University Press, Cambridge, United Kingdom

Harley, S.L. and Carswell, D.A. (1995) Ultradeep crustal metamorphism: A prospective view, *Journal of Geophysical Research* **100**, 8367–8380.

Hu, K. and Liu, X. (1992) Diamond-bearing eclogites in central China: An example of ultra-high-pressure metamorphism of crustal rocks, *Proceedings of the 29th International Geological Congress* **2**, 599–600.

Liou, J.G., Maruyama, S., and Ernst, W.G. (1997a) Seeing a mountain in a grain of garnet, *Science* **276**, 48–49.

Liou, J.G. and Zhang, R.Y. (1996) Occurrence of intergranular coesite in ultrahigh-P rocks from the Sulu region, eastern China: Implications for lack of fluid during exhumation, *American Mineralogist* **81**, 1217–1221.

Liou, J.G., Zhang, R.Y., Eide, E.A., Maruyama, S., Wang, X., and Ernst, W.G. (1996) Metamorphism and tectonics of high-P and ultrahigh-P belts in Dabie-Sulu Regions, eastern central China, in A. Yin and T.M. Harrison (eds.), *The Tectonic Evolution of Asia*, Rubey IX, Cambridge University Press, Cambridge, United Kingdom, pp. 300–343.

Liou, J.G., Zhang, R.Y., and Jahn, B.M. (1997b) Petrogenesis of ultrahigh-pressure jadeite quartzite from the Dabie region, East- central China, *Lithos* **41**, 59–78.

Maruyama, S., Liou, J.G., and Zhang, R. (1994) Tectonic evolution of the ultrahigh-pressure (UHP) and high-pressure (HP) metamorphic belts from central China, *The Island Arc* **3**, 112–121.

Okay, A.I. (1993) Petrology of a diamond and coesite-bearing metamorphic terrain: Dabie Shan, China, *European Journal of Mineralogy* **5**, 659–675.

Okay, A.I. (1995) Paragonite eclogites from Dabie Shan, China: re-equilibration during exhumation?, *Journal of Metamorphic Geology* **13**, 449–460.

Okay, A.I., Sengör, A.M.C., and Satir, M. (1993) Tectonics of an ultrahigh-pressure metamorphic terrane: the Dabie Shan/Tongbai Shan orogen, China, *Tectonics* **12**, 1320–1334.

Okay, A.I., Xu, S.T., and Sengör, A.M.C. (1989) Coesite from the Dabie Shan eclogites, central China, *European Journal of Mineralogy* **1**, 595–598.

Roberts, S. and Beattie, I. (1995) Microprobe Techniques in the Earth Sciences, The Mineralogical Society Series, in P.J. Potts, F.W. Bowles, S.J.B. Reed, and M.R. Cave (eds.), *Micro-Raman spectroscopy in the earth sciences*, 6, Chapman and Hall, London, pp. 387–408.

Rowley, D.B., Xue, F., Tucker, R.D., Peng, Z.X., Baker, J., and Davis, A. (1997) Ages of ultrahigh pressure metamorphism and protolith orthogneisses from the eastern Dabie Shan: U/Pb zircon geochronology, *Earth and Planetary Science Letters* **151**, 191–203.

Schertl, H. and Schreyer, W. (1996) Mineral inclusions in heavy minerals of the ultrahigh-pressure metamorphic rocks of the Dora-Maira Massif and their bearing in the relative timing of the petrological events, in A. Basu and S. Hart (eds.), *Earth Processes: Reading the Isotopic Code, Geophysical Monograph*, **95**, American Geophysical Union, Washington, D.C., pp. 331–342.

Schertl, H.-P. and Okay, A.I. (1994) A coesite inclusion in dolomite in Dabie Shan, China: petrological and rheological significance, *European Journal of Mineralogy* **6**, 995–1000.

Smith, D.C. (1984) Coesite in clinopyroxene in the Caledonides and its implications for geodynamics, *Nature* **310**, 641–644.

Sobolev, N.V., Shatsky, V.S., and Shatskiy, V.S. (1990) Diamond inclusions in garnets from metamorphic rocks; a new environment for diamond formation, *Nature* **343**, 742–746.

Sobolev, N.V., Shatsky, V.S., and Vavilov, M.A. (1992) Inclusions of diamonds, coesite and coexisting minerals in zircons and garnets from metamorphic rocks of Kokchetav massif (northern Kazakhstan, USSR), *Proceedings of the 29th International Geological Congress* **2**, 599.

Sobolev, N.V., Shatsky, V.S., Vavilov, S.V., and Goryainov, S.V. (1994) Zircon from ultra high pressure metamorphic rocks of folded regions as an unique container of inclusions of diamond, coesite and coexisting minerals, *Doklady Akademii Nauk, USSR* **334**, 488–492.

Spear, F.S. (1993) *Metamorphic phase equilibria and pressure-temperature-time paths*, Mineralogical Society of America, Washington, D.C.

Tabata, H., Maruyama, S., and Shi, Z. (in press) 1998, *The Island Arc*.

van der Molen, I. (1981) The shift of the a-b transition temperature of quartz associated with the thermal expansion of granite at high pressure, *Tectonophysics* **73**, 323–342.

van der Molen, I. and van Roermund, H.L.M. (1986) The pressure path of solid inclusions in minerals: the retention of coesite inclusions during uplift, *Lithos* **19**, 317–324.

Wang, X. and Liou, J.G. (1991) Regional ultrahigh-pressure coesite-bearing eclogitic terrane in central China: evidence from country rocks, gneiss, marble, and metapelite, *Geology* **19**, 933–936.

Wang, X., Liou, J.G., and Mao, H.K. (1989) Coesite-bearing eclogite from the Dabie Mountains in central China, *Geology* **17**, 1085–1088.

Wang, X., Liou, J.G., and Maruyama, S. (1992) Coesite-bearing eclogite from the Dabie Mountains, central China: Petrology and P-T path, *Journal of Geology* **100**, 231–250.

Wang, X., Zhang, R.Y., and Liou, J.G. (1995) UHPM terrane in east central China, in R.G. Coleman and X. Wang (eds.), *Ultrahigh Pressure Metamorphism*, Cambridge University Press, Stanford, pp. 356–390.

Xu, S., Okay, A.I., Ji, S., Sengör, A.M.C., Su, W., Liu, Y., and Jiang, L. (1992) Diamond from the Dabie Shan metamorphic rocks and its implication for the tectonic setting, *Science* **256**, 80–82.

Xu, S. and Su, W. (1997) Raman determination on micro-diamond in eclogite from the Dabie Mountains, eastern China, *Chinese Science Bulletin* **42**, 87.

Xue, F., Rowley, D.B., and Baker, J. (1996) Refolded syn-ultrahigh-pressure thrust sheets in the south Dabie Complex, China; field evidence and tectonic implications, *Geology* **24**, 455–458.

Zhang, R.Y. and Liou, J.G. (1996) Significance of coesite inclusions in dolomite from eclogite in the southern Dabie mountains, China, *American Mineralogist* **80**, 181–186.

Zhang, R.Y., Liou, J.G., and Tsai, C.H. (1996) Petrogenesis of a high-temperature metamorphic terrane: a new tectonic interpretation for the northern Dabieshan, central China, *Journal of Metamorphic Geology* **14**, 319–333.

Chapter 11

H_2O Recycling During Continental Collision: Phase-Equilibrium and Kinetic Considerations

W. Gary Ernst
Department of Geological and Environmental Sciences, Stanford University, Stanford, CA 94305–2115, USA, ernst@geo.stanford.edu

Jed L. Mosenfelder
Bayerisches Geoinstitut, Universität Bayreuth, D-95440, Bayreuth, Germany

Mary L. Leech and Jun Liu
Department of Geological and Environmental Sciences, Stanford University, Stanford, CA 94305–2115, USA, mary@geo.stanford.edu and jun@geo.stanford.edu

Abstract: During the early stages of subduction of the lithosphere, anhydrous rocks become partially hydrated, and volatiles evolve from hydrous lithologies. Where present, aqueous fluid markedly enhances reaction rates. Phase-equilibrium studies demonstrate that, under typical subduction-zone P–T trajectories, clinoamphibole constitutes a major phase in deep-seated metamorphic rocks of MORB composition; other hydrous minerals are either absent or of relatively minor abundance. Clinoamphiboles dehydrate at pressures <2 GPa, so blueschists and amphibolites expel H_2O at great depth, and commonly achieve the stable eclogitic assemblage of garnet + omphacite + rutile ± phengite. Partly serpentinized mantle beneath the oceanic crust devolatilizes at comparable pressures. In contrast, for reasonable subduction-zone P–T gradients, white micas ± biotites remain stable to pressures substantially exceeding 3.5–4.0 GPa. Accordingly, when the micaceous lithologies that dominate the continental crust (granitic, pelitic, and quartzofeldspathic gneisses) are subducted to depths of ≥100 km, they fail to evolve significant H_2O, and may transform incompletely to the stable eclogitic assemblage of coesite + jadeite + K-spar + garnet + rutile + phengite. Thus, although all "juicy" rock types expel volatiles during compaction and shallow burial, the especially deep underflow of partly hydrated oceanic crust-capped lithosphere probably generates most of the ultrahigh-pressure volatile flux along and above a subduction zone prior to

B. R. Hacker and J. G. Liou (eds.),
When Continents Collide: Geodynamics and Geochemistry of Ultrahigh-Pressure Rocks, 275–295.
© 1998 *Kluwer Academic Publishers. Printed in the Netherlands.*

continental collision; as large volumes of sialic crust enter the convergent plate junction, volatile flux at deep levels severely diminishes.

Although rate-enhancing volatiles evolved during prograde metamorphism diffuse into the overlying, hanging-wall lithosphere (mantle wedge), aqueous fluids also migrate back up the subduction channel, with proportions depending on the relative permeabilities of the lithologic units involved. Tectonic slices of a collisional sialic massif may move back to shallow depths along the subduction zone, propelled chiefly by buoyancy. Rehydration attending exhumation and retrogression characteristically is incomplete; its extent reflects the coarse grain size and relative impermeability of the units undergoing exhumation, as well as declining temperature and general lack of aqueous fluid. Hence, in addition to phase-equilibrium constraints, retrograde and prograde reactions are sensitive functions of kinetics, reflecting the presence or absence of H_2O.

1. INTRODUCTION

During ultrahigh-pressure (UHP) metamorphism, continental crust is carried beneath the nonsubducted plate to depths exceeding 100 km (Schreyer *et al.*, 1987; Coleman and Wang, 1995). In the course of this descent, pressure within the slab rises, while temperature only gradually increases; volatiles—principally H_2O (and some CO_2)—may be added to relatively dry portions of the downgoing plate, and driven off from the more volatile-rich parts. Early, pre-eclogitic stages of these reactions are illustrated diagrammatically in Fig. 1.

Eventually, some of the subducted material, metamorphosed under high-pressure (HP) or UHP conditions (Harley and Carswell, 1995; Schreyer, 1995; Maruyama *et al.*, 1996) decouples from the descending slab and, driven largely by buoyancy, returns to more normal continental crustal levels (Ernst, 1970; England and Holland, 1979; Cloos, 1993; Chemenda *et al.*, 1995; 1996). During more-or-less adiabatic exhumation, temperature tends to remain relatively elevated, whereas if material ascends along an active subduction channel, it cools by thermal conduction across both upper and lower bounding surfaces (Hacker and Peacock, 1994; Hacker *et al.*, 1995; Ernst *et al.*, 1997). In either case, pressure declines with ascent. Volatiles gain access to the rising sialic mass, or diffuse away from it, depending on the composition, grain size, permeability, and deformation of the various constituent units. Some evolved fluids rise into and through the overlying mantle wedge, whereas the rest migrates back up the conduit provided by the subduction channel. The presence or absence of a separate aqueous phase during recrystallization is extremely important because this fluid has the capacity to catalyze reactions enormously (Rubie, 1986).

The present synthesis briefly summarizes experimental phase-equilibrium constraints on H_2O-bearing minerals stable in the lithosphere (e.g., Massonne, 1995), and qualitatively traces the complex recycling of H_2O attending

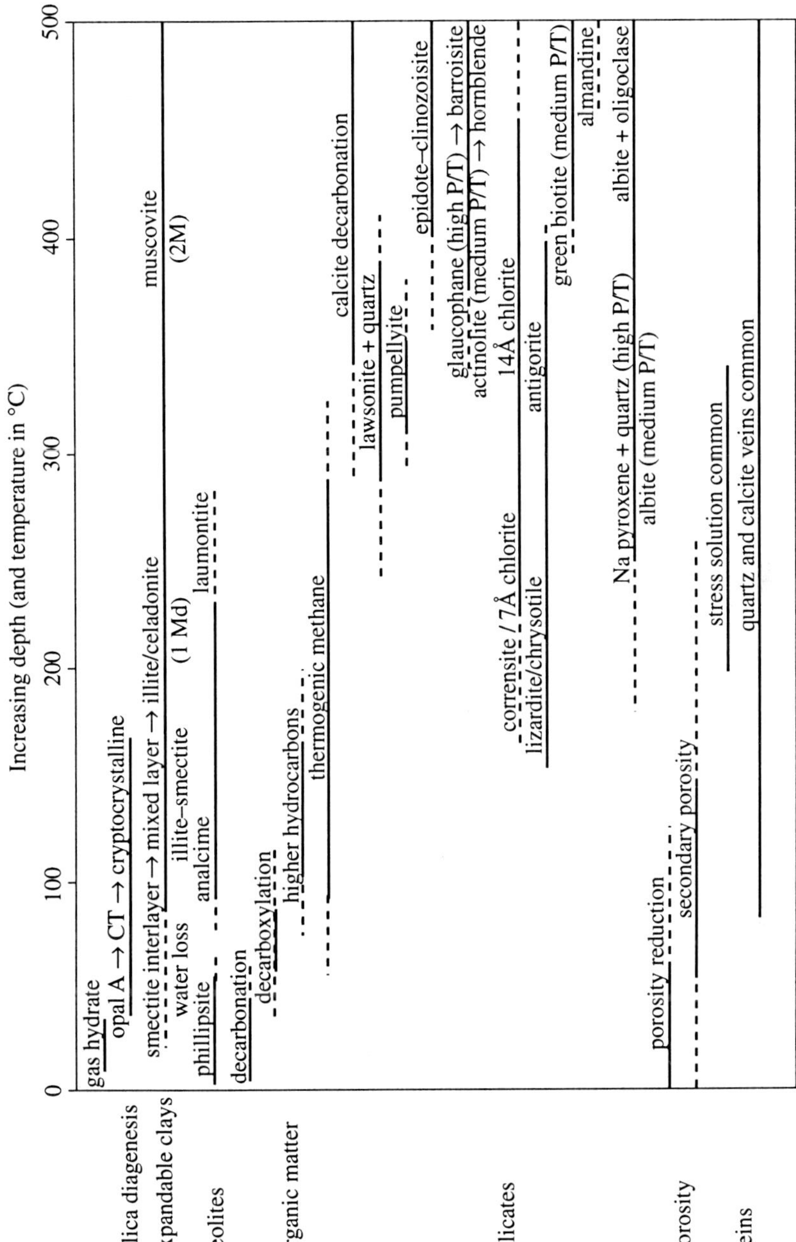

Figure 1. Schematic representation of hydration and fluid expulsion history of crustal rocks in early stages of the subduction of a lithospheric plate, generalized after Frey (1987) and Ernst (1990).

continental collision. Although carbon dioxide is important in certain subduction-zone histories, we do not consider the fate of CO_2 sequestered in carbonate strata, nor accessory carbonate minerals present in silicate-rich lithologies.

Study of natural UHP assemblages has documented the metastable persistence of low-pressure assemblages at surprisingly great depths, evidently due to the absence of an aqueous fluid phase (Austrheim, 1990; Jamtveit *et al.*, 1990; Hirajima *et al.*, 1993; Austrheim *et al.*, 1997; Zhang and Liou, 1997; Austrheim, this volume) as well as the metastable persistence of HP and UHP assemblages and mineral inclusions during exhumation (Chopin, 1984; Smith, 1984; Sobolev *et al.*, 1990; Liou and Zhang, 1996). Quantification of the relevant reaction rates is essential to an overall understanding of volatile recycling accompanying continental collision. It is clear from experimental studies that reaction rates are strongly influenced by the presence or absence of volatile constituents, especially H_2O (Rubie, 1990; Hacker, 1996). Volatile-rich rocks expel aqueous fluids during compaction and shallow-level underflow (Fig. 1), such that during the deep subduction of crustal lithologies, retained volatiles are sequestered exclusively in relatively refractory silicate phases (Poli and Schmidt, 1997). Under UHP conditions, aqueous fluid is probably not common as a separate phase (Rumble, this volume), and free H_2O becomes available only to the extent that such minerals undergo dehydration reactions.

In this chapter, we emphasize the special role played by pressure-limited clinoamphibole and antigorite in catalyzing the rate of attainment of equilibrium in deeply subducted mafic rocks. For quartzose felsic lithologies, H_2O is sequestered in micas that remain stable at UHP; accordingly, such rocks may not significantly devolatilize at depths of about 100 km, and therefore may fail to recrystallize to the stable UHP phase assemblage.

2. PHASE-EQUILIBRIUM CONSTRAINTS

In aggregate, four principal rock types constitute most of the noncarbonate, volatile-bearing subducted and recrystallized sialic crust: metabasalts and metagabbros; variably serpentinized mantle materials; pelitic schists, metagraywackes, and high-grade gneissic equivalents; and metagranitoids. Because the partial fusion of granitic rock types is treated in detail by Patiño Douce and McCarthy (this volume), our discussion concentrates on mafic, ultramafic, and felsic lithologies.

2.1 Metabasalts

Glaucophane–crossite, barroisite, and hornblende are important constituents in pre-eclogite mineral assemblages. Although common as low-temperature, low-pressure prograde and retrograde minerals, most other hydrous phases such as actinolites, micas, epidotes, chlorites, and chloritoid are uncommon, minor phases in UHP metabasalts of normal compositional ranges. The consistency of occurrence of this extremely limited subset of hydrous phases has been documented through a generation of petrologic studies world-wide (e.g., Banno, 1964; Ernst *et al.*, 1970; Cortesogno *et al.*, 1977; Dal Piaz and Ernst, 1978; Sobolev and Sobolev, 1980; Hirajima *et al.*, 1993; Beane *et al.*, 1995; Chopin and Sobolev, 1995; Dobretsov *et al.*, 1995; Krogh and Carswell, 1995; Liou *et al.*, 1996; Zhang *et al.*, 1997).

Various experimental studies have been performed on metabasaltic bulk compositions, encompassing the temperature and pressure range 300–1050°C and 0–3.5 GPa (Essene *et al.*, 1970; Helz, 1973; Liou *et al.*, 1974; Helz, 1979; Spear, 1981; Gilbert *et al.*, 1982; Apted and Liou, 1983; Moody *et al.*, 1983; Poli, 1993; Liu *et al.*, 1996; Poli and Schmidt, 1997). As documented by long-term experiments and reaction reversals for unaltered, chemically pristine mid-ocean ridge basalt (MORB), barroisitic to pargasitic amphibole is the major hydrous phase at temperatures >650°C. The Ca-amphibole stability field contracts drastically at high pressures, and disappears at ≤1.6–2.4 GPa, depending on temperature. The phase relations for unaltered MORB illustrated in Fig. 2 (Liu *et al.*, 1996) are generally applicable to unmetasomatized oceanic basalts and gabbros. Substantial amounts of lawsonite and sodic amphibole characterise metabasaltic rocks at moderate pressures and at temperatures <500–600°C (Maresch, 1977), but the glaucophane stability field is limited at elevated pressure, analogous to the P–T behavior of Ca-amphiboles (Koons, 1982; Carmen and Gilbert, 1983; Welch and Graham, 1992). Lawsonite remains stable in rocks of MORB composition as an important mineral under UHP conditions (Okamoto and Maruyama, 1997; Ono, 1997) , but only at low temperatures characterized by unnaturally low thermal gradients of ~2–5°C/km, as shown in Fig. 3.

Synthesis experiments involving mafic igneous rocks of more aluminous composition have been interpreted to suggest that epidote and chlorite possess broad stability ranges at temperatures <600–650°C (Pawley and Holloway, 1993; Schmidt and Poli, 1994; Poli and Schmidt, 1997). At these temperatures epidote and chlorite should coexist with actinolitic amphibole at low pressures, and with glaucophane–crossite at intermediate pressures. In contrast, in the senior author's experience with—and interpretation of—natural parageneses in rocks of MORB composition, the volume fractions of chlorite, actinolite, sodic amphibole, epidote, and in many cases, white mica, progressively decrease to zero as UHP conditions are approached. The lack of reaction reversal—hence demonstrated chemical equilibrium—in these earlier laboratory studies is the likely explanation for this discrepancy. In metabasalts more potassic than MORB, white mica is

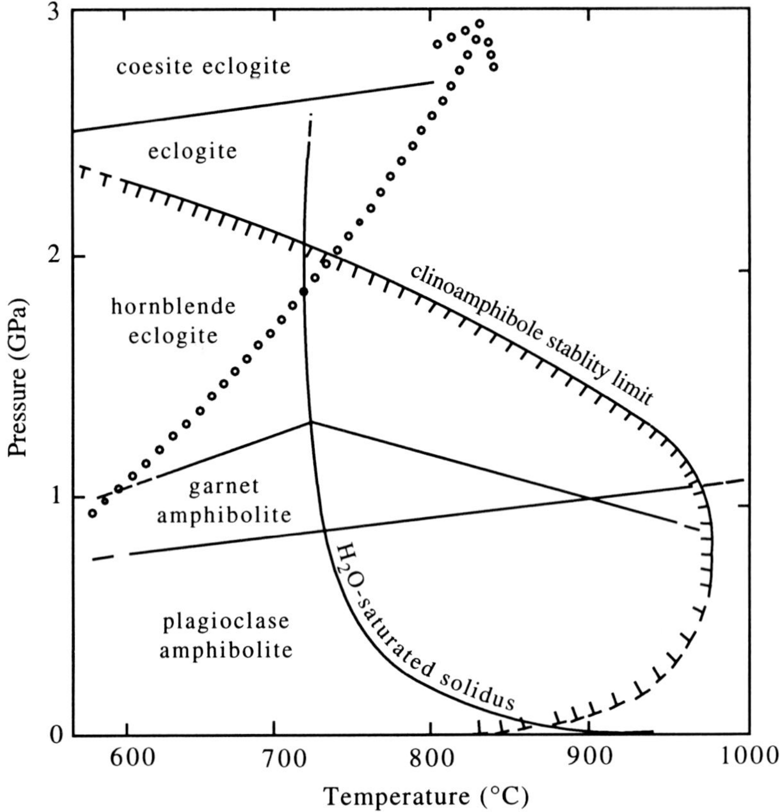

Figure 2. Petrogenetic grid for the amphibolite → eclogite transformation in the basalt–H_2O system with oxygen fugacity defined by the $FeSiO_4$–Fe_3O_4–magnetite–SiO_2 buffer, experimentally determined by Liu *et al.* (1996). Run lengths were up to 1630 hours at low temperatures; several reversals demonstrate a close approximation to chemical equilibrium. The illustrated P–T trajectory for a prograde subduction-zone thermal gradient (open dots) is ~10°C/km.

stabilized to relatively elevated pressures, and may be an associate of garnet, omphacite, and rutile in the HP/UHP paragenesis of altered mafic rocks. Although chloritoid has been reported in some synthesis experiments (Pawley and Holloway, 1993; Poli and Schmidt, 1997), there is little likelihood that this mineral represents an important phase in subducted MORBs because of its unusual Al_2O_3 + FeO-rich composition and restricted thermal stability.

Because calcic and sodic amphiboles break down at near-ultrahigh pressures, metabasaltic rocks generate at least modest amounts of H_2O by dehydration under conditions conducive to the production of eclogitic assemblages; accordingly, eclogitization is favored except in extremely dry, impermeable protoliths such as granulite-facies metagabbros (Austrheim, 1990; Hirajima *et al.*, 1993; Zhang and Liou, 1997).

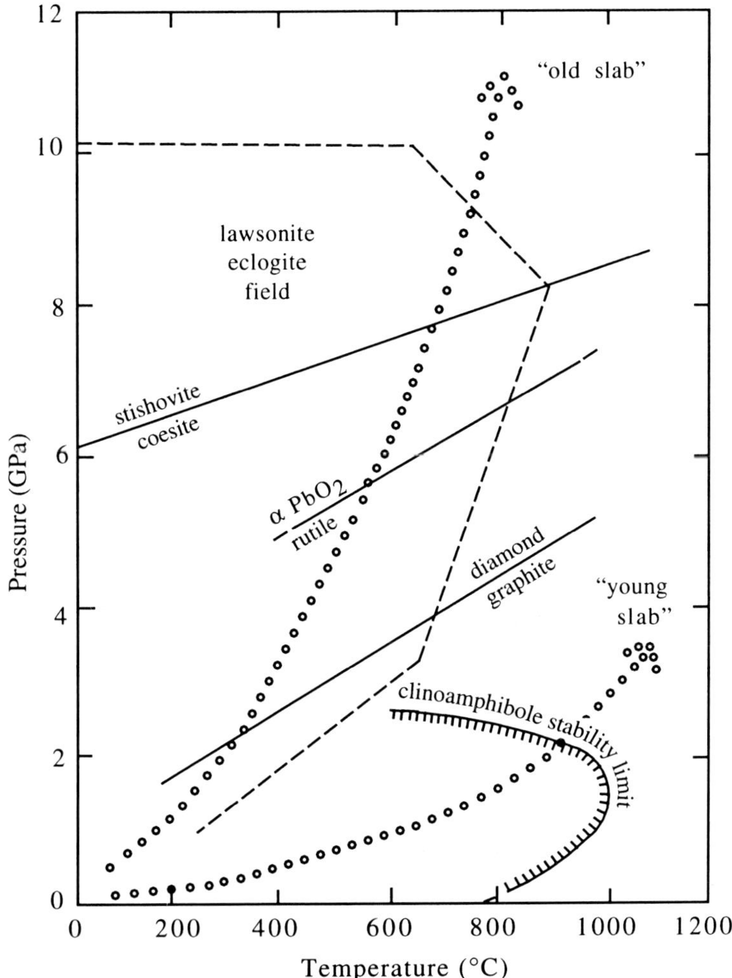

Figure 3. Synthesis fields and presumed stability relations of lawsonite in the unaltered basalt–H_2O system, after Okamoto and Maruyama (1997). The rutile–αPbO_2 transition is after Akaogi *et al.* (1992). White mica joins this assemblage for K_2O + Al_2O_3-rich bulk-rock compositions. Values illustrated for prograde subduction-zone thermal gradients (open dots) are: "old slab" ~2°C/km; "young slab" ~10°C/km.

2.2 Serpentinites

Antigoritic serpentinites dehydrate to orthopyroxene or talc + olivine, liberating large amounts of H_2O at approximately 550–600°C and 2–3 GPa (Johannes, 1975; Evans *et al.*, 1976; Ulmer *et al.*, 1994; Wunder and Schreyer, 1997). Thus, similar to clinoamphibole, the stability of antigorite is limited by pressure

(Fig. 4). How much of the suboceanic mantle is partly serpentinized is unknown, however, inasmuch as most sampling thus far has involved oceanic fracture zones, transform faults, and ophiolites—all sections characterized by unusually intense deformation and fluid circulation (e.g., Bonatti, 1976). Fully serpentinized mantle has a density of <2.5 g/cm^3, and should be virtually unsubductable. Even mantle lithosphere with >10–20% serpentinite is less dense than asthenosphere. This interpretation is supported by measured seismic P-wave speeds of 8.1 km/s for the oceanic mantle *vs.* a 5.5 km/s for serpentinite (Christensen and Salisbury, 1975). In any case, at least modest amounts of hydrated ultramafics are present in the uppermost parts of the lithosphere, so

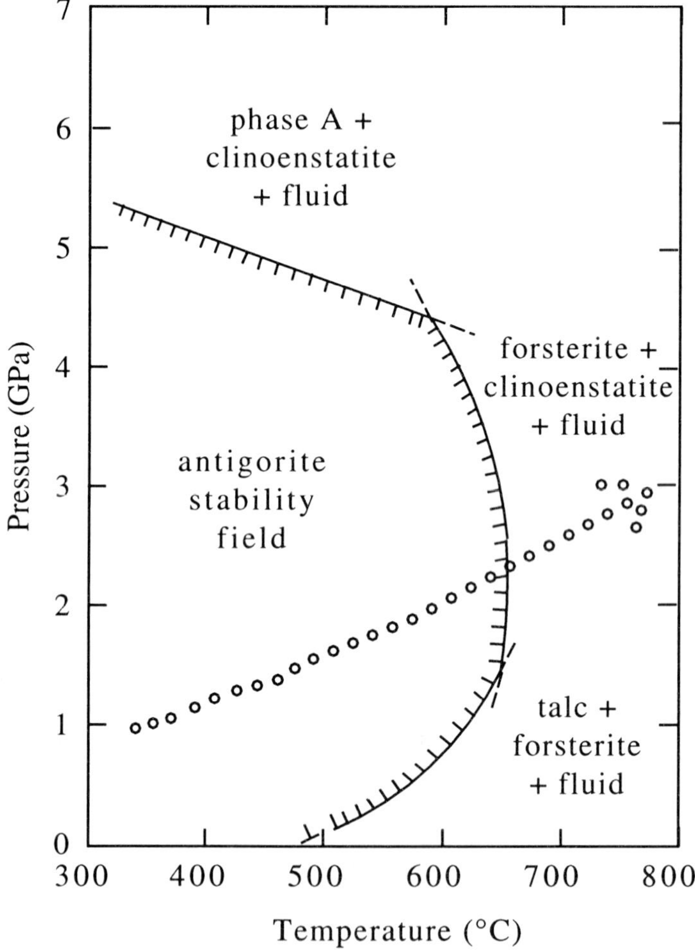

Figure 4. Typical phase diagram for the stability of antigorite, after Wunder and Schreyer (1997). See also Johannes (1975), Evans *et al.* (1976), and Ulmer and Trommsdorff (1994). The illustrated P–T trajectory for a prograde subduction-zone thermal gradient (open dots) is ~10°C/km.

their subduction and consequent devolatilization undoubtedly contribute to the aqueous fluid flux of a descending oceanic plate (Ulmer *et al.*, 1994) prior to continental collision.

2.3 Metapelites and Metagraywackes

Experimental studies of quartzofeldspathic lithologies over a wide range of laboratory conditions demonstrate that chlorites dehydrate above ~600°C, whereas potassic white micas and biotites remain stable to considerably higher temperatures—hence are present at greater subduction depths in metaclastics and orthogneisses (Le Breton and Thompson, 1988; Vielzeuf and Holloway, 1988; Vielzeuf and Montel, 1994; Gardien *et al.*, 1995; Patiño Douce and Beard, 1996; Skjerlie and Johnston, 1996; Luth, 1997; Massonne and Szpurka, 1997). A petrogenetic grid showing the phase relations for metagraywackes and pelitic schists to 2 GPa is illustrated in Fig. 5. This comprehensive treatment shows that, for peraluminous, quartz-bearing lithologies, biotite + kyanite are replaced by muscovite + garnet at pressures >1.7 GPa. For subduction-zone P–T gradients, muscovite–phengite solid solutions remain stable in gneissic units well into UHP conditions, at least to 3.5–4.0 GPa, and for metaluminous rocks, biotite persists to similar pressures, according to the various works referred to above. Because phengitic micas and biotites constitute the major, volumetrically important hydrous phases in equilibrium assemblages under UHP, quartzofeldspathic lithologies (pelitic schists + metagraywackes + gneisses) that dominate the continental crust will fail to evolve significant amounts of H_2O during UHP metamorphism at depths of 100 km or more; thus they may preserve metastable, lower pressure feldspar + quartz-bearing micaceous assemblages in the absence of a catalytic aqueous fluid, failing to develop the stable prograde mineralogic assemblage of coesite + jadeite = K-spar + garnet + rutile + phengite.

2.4 Evolution of H_2O at UHP

Although most downgoing low-grade metamorphic rocks lose important amounts of volatiles at shallow subduction depths (Fig. 1), by the time they reach HP/UHP conditions, micaceous quartzofeldspathic and metapelitic lithologies should be relatively deficient in H_2O-producing phases. Thus, the amphibole-bearing oceanic crust and serpentinized peridotite—not mica-bearing sialic crust—are the most significant dehydrating rock type at depths approaching or exceeding 100 km. If so, the closure of an ocean basin prior to continental collision provides the greatest proportion of deep-seated aqueous fluid rising through the subduction channel as well as diffusing into the hanging-wall mantle wedge. The fluid flux at UHP is probably curtailed or terminated at depth by the arrival, profound underflow, and suturing of continental crust.

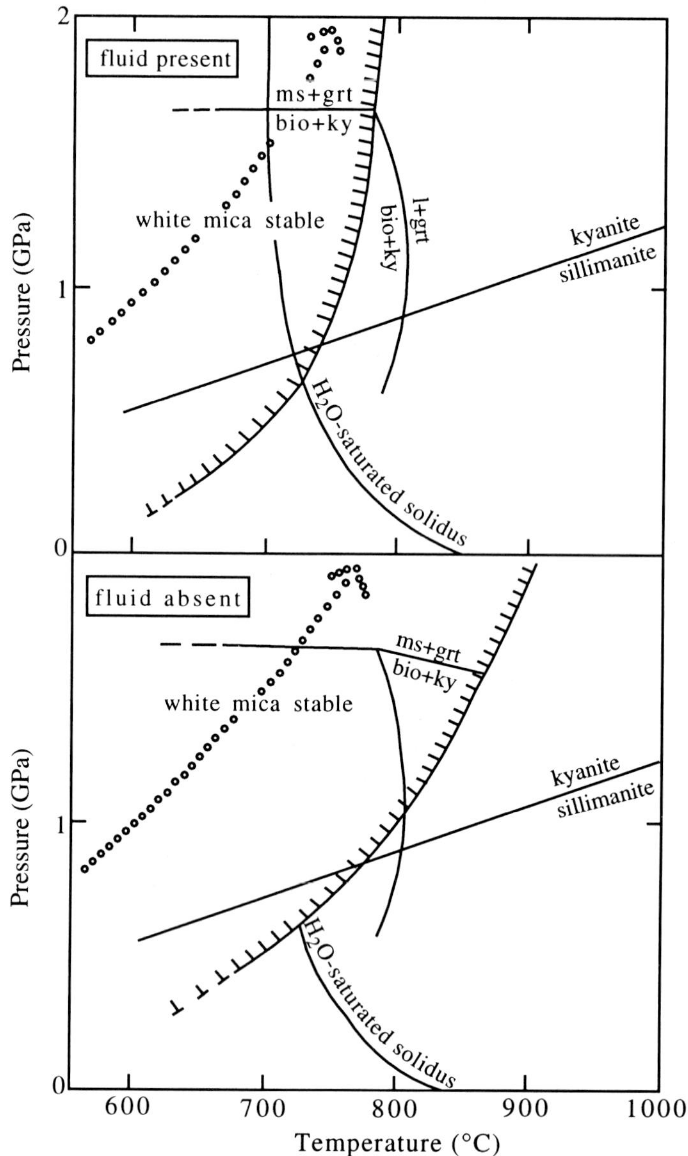

Figure 5. Experimentally determined and computed petrogenetic grid under fluid-present and fluid-absent conditions for quartzose metagraywackes + pelitic schists possessing intermediate bulk-rock Fe/Mg ratios, simplified after Vielzeuf and Holloway (1988). Abbreviations: bio, biotite; grt, garnet; ms, white mica. Phengitic micas are stable to pressures in excess of 3.5–4.0 GPa in quartzofeldspathic rock types (Le Breton and Thompson, 1988; Vielzeuf and Montel, 1994; Gardien *et al.*, 1995; Patiño Douce and Beard, 1996; Skjerlie and Johnston, 1996; Massonne and Szpurka, 1997). Biotite is stable for simplified metaluminous bulk compositions up to ~9 GPa (Luth, 1997). The illustrated P–T trajectory for a prograde subduction-zone thermal gradient (open dots) is ~10°C/km.

3. KINETIC CONSTRAINTS

The principal evidence testifying to the existence of UHP recrystallization of continental crust now exposed at Earth's surface resides in the fragmentary preservation of UHP silicate phases and yet rarer diamond. The persistence of such mineralogic relics is a reflection of initial prograde P–T conditions and even more importantly, sluggish rates of backreaction during ascent and decompression. What do we know about the kinetics of such transformations?

3.1 Rate of Graphitization of Diamond

Experiments on the anhydrous transformation of diamond → graphite show that the rate of graphitization can be estimated by the expression:

$$d\xi/dt = C\ \exp^{-(\Delta E + P\Delta V)/RT}$$

where ξ is the proportion converted, t time, C a constant, ΔE the activation energy, ΔV the activation volume, and R the universal gas constant (Davies and Evans, 1972). The low activation volume for graphitization (~10 $cm^3\ mol^{-1}$) indicates that the influence of pressure is small and that the rate of graphitization depends largely on temperature (Pearson *et al.*, 1995). Experimentally determined activation energies for graphitization of diamond are lowest for {110} faces, 760 kJ mol^{-1}, and highest for {111} faces, 1060 kJ mol^{-1} (Pearson and Nixon, 1996). Using the activation energy for the graphitization of {110} faces under anhydrous conditions, Pearson *et al.* (1995) calculated that the complete conversion of a diamond octahedron (~10 mm edge length) to graphite requires about 1 m.y. at 1200°C or 1 b.y. at 1000°C.

In the Kokchetav Complex of northern Kazakhstan, where inclusions of variably graphitized microdiamond are well documented (Sobolev and Sobolev, 1980; Sobolev *et al.*, 1990), the ascent to midcrustal levels seemingly took about 20 m.y., indicating an average exhumation rate of 5–6 mm/year (Ernst *et al.*, 1997). Ascent from mantle depths of the garnet lherzolite and eclogite complex at Alpe Arami, central Alps, took place at >20 mm/year, according to SHRIMP dating by Gebauer (1996). In the eclogites of the Western Gneiss Region of southwestern Norway (Dobrzhinetskaya *et al.*, 1995), and the Dabie Shan of east-central China (Xu *et al.*, 1992), exceedingly rare microdiamonds have been reported, but not generally confirmed; in these putative occurrences, the average exhumation rates to crustal depths are poorly constrained, but are comparable to the previously cited occurrences at approximately 1–4 mm/year (Ernst *et al.*, 1997). Kinetic studies on the graphitization of diamond require orders-of-magnitude slower exhumation rates and higher temperatures than are thought to have occurred in the petrotectonic evolution of these UHP complexes, suggesting that no diamonds in UHP metamorphic rocks should be altered to graphite. In fact, diamond is preserved in the Kokchetav Complex (and apparently in the Western Gneiss Region, and the Dabie Shan) under only very special conditions, chiefly as armored micro-inclusions in nonreactive, strong pressure vessels such as garnet or zircon. Re-equilibration that occurred during the exhumation history of these UHP rocks probably caused the complete graphitization of any diamond that was present in the matrix. Calculated rates of graphitization based on experiments are slower than nature probably because the dry graphitization experiments do not take into account the presence of aqueous fluids and other experimentally verified rate-enhancing constituents (Tagiri and Oba, 1986). The preferential preservation of these tiny inclusions of diamond relative to coesite micro-inclusions in northern Kazakhstan (Sobolev and Sobolev, 1980; Sobolev *et al.*, 1990) reflects the fact that at high temperatures (~900°C), coesite backreacts much more rapidly than diamond on decompression.

3.2 Rate of Transformation of Silicates

Evidence of widespread disequilibrium during eclogite-facies metamorphism (e.g., Rubie, 1990; Hacker, 1996; Austrheim, this volume) highlights the need for more quantitative experimental data on reaction rates. It has become increasingly clear from natural examples that availability of fluid, rather than P–T-induced thermodynamic changes, chiefly control the rate at which reactions occur. Although the catalytic effect of H_2O is well known and some specific catalytic mechanisms have been proposed (e.g., Rubie, 1986), little is known quantitatively from experiments about the dependence of reaction rates and mechanisms on fluid concentration in systems with low fluid:rock ratios—the most likely condition for large massifs of subducting continental crust.

Of primary concern in the present case are the kinetics of three types of transformation: polymorphic, multiphase solid–solid, and dehydration reactions. In particular, for quartzofeldspathic lithologies, quartz is expected to transform polymorphically to coesite, plagioclase should break down to form any of a series of denser phase assemblages, and phengite should dehydrate at high temperature. Below we consider the rather limited database on the kinetics of these three categories of reaction (cf. Hacker and Peacock, 1994).

3.3 Polymorphic Reactions

Preliminary results on quartz → coesite kinetics (Mosenfelder and Bohlen, 1995) as well as experimental data on the reverse transformation (Mosenfelder and Bohlen, 1997) suggest that this reaction is rapid in both directions at UHP conditions. This hypothesis is compatible with textural evidence from natural UHP metagranitoids. For instance, the undeformed UHP Brossasco metagranite

(Biino and Compagnoni, 1992) contains granoblastic quartz aggregates pseudomorphing original euhedral igneous quartz; the most probable explanation for this texture is a complete reversion from coesite. However, two factors have not been satisfactorily addressed in experimental studies. First, nucleation in laboratory experiments may be much faster than in nature because of the presence of cracks or other structural damage in starting materials. Second, trace amounts of H_2O in synthetic phases can further accelerate reaction (Mosenfelder and Bohlen, 1997).

Backreaction of UHP phases during exhumation presents a different problem because transformation may be inhibited by a number of factors. In particular, coesite and diamond included in other phases can be protected by an "overpressure" from the host phase until cracking occurs at lower temperatures (Gillet *et al.*, 1984). Furthermore, the effect of fluids, or lack thereof, may be especially important at lower temperatures during exhumation.

3.4 Multiphase Solid–Solid Reactions

At UHP conditions in a quartzofeldspathic system, plagioclase should ultimately break down via the discontinuous reaction plagioclase → jadeite + grossular + kyanite + quartz. Sufficient data to predict the kinetics of complex, multiphase, solid–solid reactions are unavailable. Such reactions, both prograde and retrograde, are expected to be considerably slower than polymorphic transformations because they are probably controlled by diffusion of several species, potentially along complex pathways, rather than by the diffusion of a single ion across an interface. A number of studies of natural rocks (e.g., Wayte *et al.*, 1989; Rubie, 1990; Biino and Compagnoni, 1992; Wain, 1997; Austrheim, this volume), have inferred that plagioclase breaks down only in the presence of an aqueous fluid, even at relatively high temperatures and large pressure oversteps. This suggests that extreme nucleation barriers exist for these reactions, a hypothesis which needs to be tested experimentally and theoretically.

3.5 Dehydration Reactions

It is commonly assumed that dehydration reactions either occur rapidly or are controlled by the rate of enthalpy production (Ridley, 1985). Most experiments, employing powdered starting materials, yield rates which are rapid when extrapolated to geologic conditions (e.g., Wegner and Ernst, 1983; Schramke *et al.*, 1987; Jové and Hacker, 1997). However, work on dehydration reactions employing solid rocks as starting materials yields slower rates and different mechanisms (Brearley, 1987; Rubie and Brearley, 1987; Brearley and Rubie, 1990; Hacker, 1990; Hacker and Christie, 1991). In these studies, metastable dehydration melting commonly occurs, which actually inhibits stable dehydration reactions. Extrapolation of these results to longer timescales and temperatures close to equilibrium is uncertain at present. Nevertheless, it seems clear that, because dehydration reactions generate a rate-enhancing constituent (H_2O), prograde transformations will proceed more rapidly than the equivalent backreaction. This phenomenon, and the fact that mafic lithologies carry substantial amounts of a pressure-limited hydrous phase, whereas felsic rocks are not so constrained, constitute the thrust of this paper.

4. DISCUSSION

The ease with which aqueous fluids gain access to—or egress from—subducting and ascending sialic masses is a complex function of rock composition, permeability, extent of deformation, grain size, and availability of volatiles. These quantities reflect the histories and ongoing dynamics of the lithotectonic units at the convergent plate junction. What we know reasonably well, based on computational thermal modelling (Peacock, 1992; Peacock, 1995; Ernst and Peacock, 1996) and laboratory phase-equilibrium studies (Vielzeuf and Holloway, 1988; Massonne, 1995; Schreyer, 1995; Liu *et al.*, 1996; Patiño Douce and McCarthy, this volume), is the evolving P–T structure of a continental collision zone and the P–T stability fields of the mineral assemblages for the major rock types. What we do not know with any degree of certainty is the rates at which mineralogic transformations take place, and therefore, the conditions and tectonic realms under which volatile constituents are evolved and consumed. Given the proper (anhydrous) circumstances, low-pressure assemblages can be retained under prograde UHP conditions, and UHP relics can be preserved under retrograde low-pressure conditions. Quantification of the kinetics for the controlling reactions is the key, and as briefly described above, these data are not yet available except for a few, chemically simple mineral systems.

5. CONCLUSIONS

The examination of UHP rocks exhumed from depths of >100 km, combined with experimental phase-equilibrium data, and available kinetic information, allows the following conclusions:

1) Subduction of oceanic and/or sialic massifs at moderate depths results in the partial hydration of relatively dry units and in the evolution of volatiles from "juicier" lithologies (Fig. 1).

2) Under typical deep-seated subduction-zone P–T trajectories, clinoamphiboles dehydrate at pressures <2 GPa (Fig. 2); amphibole-bearing rocks such as metabasalts expel aqueous fluids during eclogitization, and

recrystallize readily to the stable HP/UHP garnet + omphacite + rutile ± white mica (eclogite-facies) assemblage. Similar to clinoamphiboles, antigorite in serpentinized peridotites dehydrates at comparable to slightly higher pressures (Fig. 4).

3) For such P–T subduction paths, phengitic micas and biotites are stable to far greater pressures than characterise clinoamphiboles and antigorite (Fig. 5); accordingly, interlayered mica-rich lithologies (metagranitoids, gneisses, metagraywackes, and metashales) may fail to evolve significant amounts of H_2O under UHP conditions. Thus quartzose felsic units probably transform much less completely to eclogitic phase assemblages (coesite + jadeite = K-spar + garnet + rutile + phengite) than do spatially associated mafic rocks.

4) Volatile evolution is maximized during the subduction of amphibole-rich oceanic crust and associated serpentinized mantle prior to continental collision; once large volumes of micaceous quartzofeldspathic rock enter the subduction zone, dehydration may virtually cease at depths ≥100 km. If sialic crust is carried down substantially farther, mica-bearing units must eventually dehydrate due to higher pressures and temperatures.

5) Attending prograde evolution of volatiles, some H_2O diffuses into the overlying, nonsubducted lithospheric hanging wall, while the rest of the aqueous fluid migrates back up the subduction channel, depending on the permeabilities of the lithologic units involved.

6) As segments of a sialic massif are exhumed, propelled chiefly by buoyancy along the subduction zone acting as a stress guide, cooling takes place; the degree of retrograde rehydration depends on the nature of the units involved and on the availability of aqueous fluids.

7) During thermal relaxation and decompression (wet or dry), the rates of backreaction of silicates such as coesite exceed that of the diamond → graphite transition.

ACKNOWLEDGMENTS

Discussions with numerous researchers investigating intracontinental HP/UHP terranes helped lead us to the ideas expressed here. Our manuscript was reviewed and materially improved by John Holloway, Douglas Rumble III, and Werner Schreyer as well as the editors of this volume, B. R. Hacker and J. G. Liou. Sandy Keiser drafted the illustrations. We thank the above workers and both Stanford University and the Carnegie Institution of Washington for support.

REFERENCES

Akaogi, M., Kusaba, K., Susaki, J.-I., Yagi, T., Matsui, M., Kikegawa, T., Yusa, H., and Ito, E. (1992) High-pressure high-temperature stability of αPbO_2-type TiO_2 and $MgSiO_3$ majorite: calorimetric and in situ X-ray diffraction studies, in Y. Syono and M.H. Manghnani (eds.), *High Pressure Research: Applications to Earth and Planetary Sciences*, AGU, Washington, pp. 447–455.

Apted, M.J. and Liou, J.G. (1983) Phase relations among greenschist, epidote amphibolite, and amphibolite in a basaltic system, *American Journal of Science* **283–A**, 328–354.

Austrheim, H. (1990) The granulite–eclogite facies transition: a comparison of experimental work and a natural occurrence in the Bergen Arcs, western Norway, *Lithos* **25**, 163–169.

Austrheim, H. (this volume) The influence of fluid and deformation on metamorphism of the deep crust and consequences for geodynamics of collision zones, in B.R. Hacker and J.G. Liou (eds.), *When Continents Collide: Geodynamics and Geochemistry of Ultrahigh-Pressure Rocks*, Kluwer Academic Publishers, Dordrecht

Austrheim, H., Erambert, M., and Engvik, A.K. (1997) Processing of crust in the root of the Caledonian continental collision zone: the role of eclogitization, *Tectonophysics* **273**, 129–153.

Banno, S. (1964) Petrologic studies on Sanbagawa crystalline schists in the Bessi-Ino district, central Sikoku, Japan, *Journal of the Faculty of Science, University of Tokyo* **15**, 203–319.

Beane, R.J., Liou, J.G., Coleman, R.G., and Leech, M.L. (1995) Petrology and retrograde P-T path for eclogite of the Maksyutov Complex, Southern Ural Mountains, Russia, *The Island Arc* **4**, 254–266.

Biino, G. and Compagnoni, R. (1992) Very-high pressure metamorphism of the Brossasco coronite metagranite, southern Dora-Maira massif, Western Alps, *Schweizerische Mineralogische und Petrographische Mitteilungen* **72**, 347–363.

Bonatti, E. (1976) Serpentinite protrusions in the oceanic crust, *Earth and Planetary Science Letters* **32**, 107–113.

Brearley, A.J. (1987) An experimental and kinetic study of the breakdown of aluminous biotite at 800°C: reaction microstructures and mineral chemistry, *Bulletin de Mineralogie* **110**, 513–532.

Brearley, A.J. and Rubie, D.C. (1990) Effects of H2O on the disequilibrium breakdown of muscovite + quartz, *Journal of Petrology* **31**, 925–956.

Carmen, J.H. and Gilbert, M.C. (1983) Experimental studies of glaucophane stability, *American Journal of Science* **283A**, 414–437.

Chemenda, A.I., Mattauer, M., and Bokun, A.N. (1996) Continental subduction and a mechanism for exhumation of high-pressure metamorphic rocks: new modeling and field data from Oman, *Earth and Planetary Science Letters* **143**, 173–182.

Chemenda, A.I., Mattauer, M., Malavieille, J., and Bokun, A.N. (1995) A mechanism for syn-collisional rock exhumation and associated normal faulting: Results from physical modelling, *Earth and Planetary Science Letters* **132**, 225–232.

Chopin, C. (1984) Coesite and pure pyrope in high-grade blueschists of the western Alps: a first record and some consequences, *Contributions to Mineralogy and Petrology* **86**, 107–118.

Chopin, C. and Sobolev, N.V. (1995) Principal mineralogic indicators of UHP in crustal rocks, in R.G. Coleman and X. Wang (eds.), *Ultrahigh Pressure Metamorphism*, Cambridge University Press, Cambridge, pp. 96–131.

Christensen, N.I. and Salisbury, M.H. (1975) Structure and constitution of the lower oceanic crust, *Reviews of Geophysics and Space Physics* **13**, 57–86.

Cloos, M. (1993) Lithospheric buoyancy and collisional orogenesis: subduction of oceanic plateaus, continental margins, island arcs, spreading ridges, and seamounts, *Geological Society of America Bulletin* **105**, 715–737.

Coleman, R.G. and Wang, X. (1995) *Ultrahigh Pressure Metamorphism*, Cambridge University Press, New York.

Cortesogno, L., Ernst, W.G., Galli, M., Messiga, B., Pedemonte, G.M., and Piccardo, G.P. (1977) Chemical petrology of eclogitic lenses in serpentinite, Gruppo di Voltri, Ligurian Alps, *Journal of Geology* **85**, 255–277.

Dal Piaz, G.V. and Ernst, W.G. (1978) Areal geology and petrology of eclogites and associated metabasites of the Piemonte ophiolite nappe, Breuil-St. Jacques area, Italian western Alps, *Tectonophysics* **51**, 99–126.

Davies, G.R. and Evans, T. (1972) Graphitization of diamond at zero pressure and at a high pressure, *Proceedings of the Royal Society of London* **328**, 413–427.

Dobretsov, N.L., Sobolev, N.V., Shatsky, V.S., Coleman, R.G., and Ernst, W.G. (1995) Geotectonic evolution of diamondiferous paragneisses, Kokchetav Complex, northern Kazakhstan: The geologic enigma of ultrahigh-pressure crustal rocks within a Paleozoic foldbelt, *The Island Arc* **4**, 267–279.

Dobrzhinetskaya, L.F., Eide, E.A., Larsen, R.B., Sturt, B.A., Tronnes, R.G., Smith, D.C., Taylor, W.R., Posukhova, T.V., and Posukhova, T.V. (1995) Microdiamond in high-grade metamorphic rocks of the Western Gneiss region, Norway, *Geology* **23**, 597–600.

England, P.C. and Holland, T.J.B. (1979) Archimedes and the Tauern eclogites: the role of buoyancy in the preservation of exotic eclogite blocks, *Earth and Planetary Science Letters* **44**, 287–294.

Ernst, W.G. (1970) Tectonic contact between the Franciscan melange and the Great Valley sequence, crustal expression of a Late Mesozoic Benioff zone, *Journal of Geophysical Research* **75**, 886–901.

Ernst, W.G. (1990) Thermobarometric and fluid-expulsion history of subduction zones, *Journal of Geophysical Research* **95**, 9047–9053.

Ernst, W.G., Maruyama, S., and Wallis, S. (1997) Buoyancy-driven, rapid exhumation of ultrahigh-pressure metamorphosed continental crust, *Proceedings of the National Academy of Sciences* **94**, 9,532–9,537.

Ernst, W.G. and Peacock, S.M. (1996) A thermotectonic model for preservation of ultrahigh-pressure phases in metamorphosed continental crust, in G.E. Bebout, D.W. Scholl, S.H. Kirby, and J.P. Platt (eds.), *Subduction Top to Bottom*, American Geophysical Union, Washington, D.C., pp. 171–178.

Ernst, W.G., Seki, Y., Onuki, H., and Gilbert, M.C. (1970) Comparative Study of Low-grade Metamorphism in the California Coast Ranges and the Outer Metamorphic Belt of Japan, *Geological Society of America Memoir* **124**, 276.

Essene, E., Henson, B.J., and Green, D.H. (1970) Experimental study of amphibolite and eclogite stability, *Physics of Earth and Planetary Interiors* **3**, 378–384.

Evans, B.W., Johannes, W., Oterdoom, H., and Trommsdorff, V. (1976) Stability of chrysotile and antigorite in the serpentinite multisystem, *Schweizerisches Mineralogisches und Petrografisches Mittelungen* **56**, 79–93.

Frey, M. (1987) Low Temperature Metamorphism, p. 351. Blackie, Glasgow.

Gardien, V., Thompson, A.B., Grujic, D., and Ulmer, P. (1995) Experimental melting of biotite + plagioclase + quartz ± muscovite assemblages and implications for crustal melting, *Journal of Geophysical Research* **100**, 15,581–15,591.

Gebauer, D. (1996) A P-T-t path for an (ultra?-) high-pressure ultramafic/mafic rock association and its felsic country rocks based on SHRIMP dating of magmatic and metamorphic zircon domains. Example: Alpe Arami (central Swiss Alps), *Geophysical Monograph* **95**, 307–329.

Gilbert, M.C., Helz, R.T., Popp, R.K., and Spear, F.S. (1982) Experimental studies of amphibole stability, in D.R. Veblen and P.H. Ribbe (eds.), *Amphiboles: Petrology and Experimental Phase*

Relations, Reviews in Mineralogy, **9B**, Mineralogical Society of America, Washington, D. C., pp. 229–353.

Gillet, P., Ingrin, J., and Chopin, C. (1984) Coesite in subducted continental crust: P-T history deduced from an elastic model, *Earth and Planetary Science Letters* **70**, 426–436.

Hacker, B.R. (1990) Amphibolite-facies to granulite-facies reactions in experimentally deformed unpowdered amphibolite, *American Mineralogist* **75**, 1349–1361.

Hacker, B.R. (1996) Eclogite formation and the rheology, buoyancy, seismicity, and H2O content of oceanic crust, in G.E. Bebout, Scholl, D., Kirby, S.H., Platt, J.P. (ed.), *Dynamics of Subduction*, Monograph, American Geophysical Union, Washington, D.C., pp. 337–246.

Hacker, B.R. and Christie, J.M. (1991) Observational evidence for a possible new diffusion path, *Science* **251**, 67–70.

Hacker, B.R. and Peacock, S.M. (1994) Creation, preservation, and exhumation of coesite-bearing, ultrahigh-pressure metamorphic rocks, in R.G. Coleman and X. Wang (eds.), *Ultrahigh Pressure Metamorphism*, Cambridge University Press, Cambridge, United Kingdom

Hacker, B.R., Ratschbacher, L., Webb, L., and Dong, S. (1995) What brought them up? Exhumation of the Dabie Shan ultrahigh-pressure rocks, *Geology* **23**, 743–746.

Harley, S.L. and Carswell, D.A. (1995) Ultradeep crustal metamorphism: A prospective view, *Journal of Geophysical Research* **100**, 8367–8380.

Helz, R.T. (1973) Phase relations of basalts in their melting range at PH2O = 5 kb as a function of oxygen fugacity, *Journal of Petrology* **14**, 249–302.

Helz, R.T. (1979) Alkali exchange between hornblende and melt: a temperature-sensitive reaction, *American Mineralogist* **64**, 953–965.

Hirajima, T., Wallis, S.R., Zhai, M., and Ye, K. (1993) Eclogitized metagranitoid from the Su–Lu ultrahigh-pressure (UHP) province, eastern China, *Proceedings Of The Japan Academy. Series B* **69**, 249–254.

Jamtveit, B., Bucher-Nurminen, K., and Austrheim, H. (1990) Fluid controlled eclogitization of granulites in deep crustal shear zones, Bergen arcs, western Norway, *Contributions to Mineralogy and Petrology* **104**, 184–193.

Johannes, W. (1975) Zur Synthese und thermischen Stabilität von antigorit, *Fortschrit fur Mineralogie* **53**, 36.

Jové, C. and Hacker, B.R. (1997) Experimental investigation of laumontite → wairakite + H2O: A model diagenetic reaction, *American Mineralogist.* **82**, 781–789.

Koons, P.O. (1982) An experimental investigation of the behaviour of amphibole in the system Na2O-MgO-Al2O3–SiO2–H2O at high pressures, *Contributions to Mineralogy and Petrology* **79**, 258–267.

Krogh, E.J. and Carswell, D.A. (1995) HP and UHP eclogites and garnet peridotites in the Scandinavian Caledonides, in R.G. Coleman and X. Wang (eds.), *Ultrahigh Pressure Metamorphism*, Cambridge University Press, Stanford, pp. 244–298.

Le Breton, N. and Thompson, A.B. (1988) Fluid-absent (dehydration) melting of biotite in metapelites in the early stages of crustal anatexis, *Contributions to Mineralogy and Petrology* **99**, 226–237.

Liou, J.G., Kuniyoshi, S., and Ito, K. (1974) Experimental studies of the phase relations between greenschist and amphibolite in a basaltic system, *American Journal of Science* **274**, 613–632.

Liou, J.G. and Zhang, R.Y. (1996) Occurrence of intergranular coesite in ultrahigh-P rocks from the Sulu region, eastern China: Implications for lack of fluid during exhumation, *American Mineralogist* **81**, 1217–1221.

Liou, J.G., Zhang, R.Y., Eide, E.A., Maruyama, S., Wang, X., and Ernst, W.G. (1996) Metamorphism and tectonics of high-P and ultrahigh-P belts in Dabie-Sulu Regions, eastern central China, in A. Yin and T.M. Harrison (eds.), *The Tectonic Evolution of Asia*, Rubey IX, Cambridge University Press, Cambridge, United Kingdom, pp. 300–343.

Liu, J., Bohlen, S.R., and Ernst, W.G. (1996) Stability of hydrous phases in subducting oceanic crust, *Earth and Planetary Science Letters* **143**, 161–171.

Luth, R.W. (1997) Experimental study of the system phlogopite-diopside from 3.5 to 17 GPa, *American Mineralogist* **82**, 1198–1209.

Maresch, W.V. (1977) Experimental studies on glaucophane; an analysis of present knowledge, *Tectonophysics* **43**, 109–125.

Maruyama, S., Liou, J.G., and Terabayashi, M. (1996) Blueschists and eclogites of the world, and their exhumation, *International Geology Review* **38**, 485–594.

Massonne, H.J. (1995) Experimental and petrogenetic study of UHPM, in R.G. Coleman and X. Wang (eds.), *Ultrahigh Pressure Metamorphism*, Cambridge University Press, Cambridge, pp. 33–95.

Massonne, H.J. and Szpurka, Z. (1997) Thermodynamic properties of white micas on the basis of high-pressure experiments in the systems K_2O-MgO-Al_2O_3-SiO_2-H_2O and K_2O-FeO-Al_2O_3-SiO_2-H_2O, *Lithos* **41**, 229–250.

Moody, J.B., Meyer, D., and Jenkins, J.E. (1983) Experimental characterization of the greenschist/amphibolite boundary in mafic systems, *American Journal of Science* **283**, 48–92.

Mosenfelder, J.L. and Bohlen, S.R. (1995) Kinetics of the quartz to coesite transformation: Transactions of the American Geophysical Union, *Transactions of the American Geophysical Union, Eos* **76**, 531–532.

Mosenfelder, J.L. and Bohlen, S.R. (1997) Kinetics of the cœsite → quartz transformation, *Earth and Planetary Science Letters* **153**, 133–147.

Okamoto, K. and Maruyama, S. (1997) The high pressure stability limits of lawsonite in the MORB + H_2O system, *Tokyo Workshop on Kokchetav Deep-drilling Project, Waseda University*, 91–95.

Ono, S. (1997) Stability limits of hydrous minerals in sediment and MORB compositions, *Transactions of the American Geophysical Union, Eos* **78**, 825.

Patiño Douce, A.E. and Beard, J.S. (1996) Effects of P, f(O_2) and Mg/Fe ratio on dehydration-melting of model metagreywackes, *Journal of Petrology* **37**, 999–1024.

Patiño Douce, A.E. and McCarthy, T.C. (this volume) Melting of crustal rocks during continental collision and subduction

Pawley, A.R. and Holloway, J.R. (1993) Water sources for subduction zone volcanism; new experimental constraints, *Science* **260**, 664–667.

Peacock, S.M. (1992) Blueschist-facies metamorphism, shear heating, and P-T-t paths in subduction shear zones, *Journal of Geophysical Research* **97**, 17,693–17,707.

Peacock, S.M. (1995) Ultrahigh-pressure metamorphic rocks and the thermal evolution of continent collision belts, *The Island Arc* **4**, 376–383.

Pearson, D.G., Davies, G.R., and Nixon, P.H. (1995) Orogenic ultramafic rocks of UHP (diamond facies) origin, in R.G. Coleman and X. Wang (eds.), *Ultrahigh Pressure Metamorphism*, Cambridge University Press, New York

Pearson, D.G. and Nixon, P.H. (1996) Diamonds in young orogenic belts: graphitized diamond from Beni Bousera, N. Morocco, a comparison with kimberlite-derived diamond occurrences and implications for diamond genesis and exploration, *Africa Geoscience Review* **3**, 295–316.

Poli, S. (1993) The amphibolite-eclogite transformation; an experimental study on basalt, *American Journal of Science* **293**, 1061–1107.

Poli, S. and Schmidt, M.W. (1997) The high-pressure stability of hydrous phases in orogenic belts: an experimental approach on eclogite-forming processes, *Tectonophysics* **273**, 169–184.

Ridley, J. (1985) The effect of reaction enthalpy on the progress of a metamorphic reaction, in A.B. Thompson and D.C. Rubie (eds.), *Metamorphic reactions. Kinetics, textures, and deformation, Advances in Physical Geochemistry*, 4, Springer-Verlag, Berlin, pp. 80–97.

Rubie, D.C. (1986) The catalysis of mineral reactions by water and restrictions on the presence of aqueous fluid during metamorphism, *Mineralogical Magazine* **50**, 399.

Rubie, D.C. (1990) Role of kinetics in the formation and preservation of eclogites, in D.A. Carswell (ed.), *Eclogite Facies Rocks*, Blackie, Glasgow, pp. 111–140.

Rubie, D.C. and Brearley, A.J. (1987) Metastable melting during the breakdown of muscovite + quartz at 1 kbar, *Bulletin Mineralogie* **110**, 533–549.

Rumble, D. (this volume) Stable Isotope Geochemistry Of Ultrahigh-Pressure Rocks, in B.R. Hacker and J.G. Liou (eds.), *When Continents Collide: Geodynamics and Geochemistry of Ultrahigh-Pressure Rocks*, Kluwer Academic Publishers, Dordrecht

Schmidt, M.W. and Poli, S. (1994) The stability of lawsonite and zoisite at high pressures: experiments in CASH to 92 kbar and implications for the presence of hydrous phases in subducted lithosphere, *Earth and Planetary Science Letters* **124**, 105–118.

Schramke, J.A., Kerrick, D.M., and Lasaga, A.C. (1987) The reaction muscovite + quartz <--> andalusite + K-feldspar + water; Part 1, Growth kinetics and mechanism., *American Journal of Science* **287**, 517–559.

Schreyer, W. (1995) Ultradeep metamorphic rocks: The retrospective viewpoint, *Journal of Geophysical Research* **100**, 8353–8366.

Schreyer, W., Massonne, H.J., and Chopin, C. (1987) Continental crust subducted to depths near 100 km: implications for magma and fluid genesis in collision zones, in B.O. Mysen (ed.), *Magmatic Processes: Physiochemical Principles*, 1, Geochemical Society, University Park, pp. 155–163.

Skjerlie, K.P. and Johnston, A.D. (1996) Vapour-absent melting from 10 to 20 kbar of crustal rocks that contain multiple hydrous phases: Implications for anatexis in the deep to very deep continental crust and active continental margins, *Journal of Petrology* **37**, 661–691.

Smith, D.C. (1984) Coesite in clinopyroxene in the Caledonides and its implications for geodynamics, *Nature* **310**, 641–644.

Sobolev, N.V., Shatsky, V.S., and Shatskiy, V.S. (1990) Diamond inclusions in garnets from metamorphic rocks; a new environment for diamond formation, *Nature* **343**, 742–746.

Sobolev, V.S. and Sobolev, N.V. (1980) New proof on very deep subsidence of eclogitized crustal rocks, *Doklady Akademii Nauk* **250**, 683–685.

Spear, F.S. (1981) An experimental study of hornblende stability and compositional variability in amphibolite, *American Journal of Science* **281**, 697–734.

Tagiri, M. and Oba, T. (1986) Hydrothermal synthesis of graphite from bituminous coal at 0.5–5 kbar water vapor pressure and 300–600°C, *Journal of Japanese Association of Mineralogy, Petrology, and Economic Geology* **81**, 260–271.

Ulmer, P., Trommsdorff, V., and Reusser, E. (1994) Experimental investigation of antigorite stability to 80 kbar, *Mineralogical Magazine* **58A**, 918–919.

Vielzeuf, D. and Holloway, J.R. (1988) Experimental determination of the fluid-absent melting relations in the pelitic system. Consequences for crustal differentiation, *Contributions to Mineralogy and Petrology* **98**, 257–76.

Vielzeuf, D. and Montel, J.M. (1994) Partial melting of metagreywackes. 1. Fluid-absent experiments and phase relationships, *Contributions to Mineralogy and Petrology* **117**, 375–393.

Wain, A. (1997) New evidence for coesite in eclogite and gneisses; defining an ultrahigh-pressure province in the Western Gneiss region of Norway, *Geology* **25**, 927–930.

Wayte, G.J., Worden, R.H., Rubie, D.C., and Droop, G.T.R. (1989) A TEM study of disequilibrium plagioclase breakdown at high pressure: the role of infiltrating fluid, *Contributions to Mineralogy and Petrology* **101**, 426–437.

Wegner, W.W. and Ernst, W.G. (1983) Experimentally determined hydration and dehydration reaction rates in the system $MgO\text{-}SiO_2\text{-}H_2O$, *Amer. Jour. Sci.* **283–A**, 151–180.

Welch, M.D. and Graham, C.M. (1992) An experimental study of glaucophanic amphiboles in the system Na_2O-MgO-Al_2O_3-SiO_2-SiF_4(NMSF); some implications for glaucophane stability in natural and synthetic systems at high temperatures and pressures, *Contributions to Mineralogy and Petrology* **111**, 248–259.

Wunder, B. and Schreyer, W. (1997) Antigorite: high-pressure stability in the system MgO-SiO_2-H_2O (MSH), *Lithos* **41**, 213–227.

Xu, S., Okay, A.I., Ji, S., Sengör, A.M.C., Su, W., Liu, Y., and Jiang, L. (1992) Diamond from the Dabie Shan metamorphic rocks and its implication for the tectonic setting, *Science* **256**, 80–82.

Zhang, R.Y. and Liou, J.G. (1997) Partial transformation of gabbro to coesite-bearing eclogite from Yangkou, the Sulu Terrane, eastern China, *Journal of Metamorphic Geology* **15**, 183–202.

Zhang, R.Y., Liou, J.G., Ernst, W.G., Coleman, R.G., Sobolev, N.V., and Shatsky, V.S. (1997) Metamorphic evolution of diamond-bearing and associated rocks from the Kokchetav massif, northern Kazakhstan, *Journal of Metamorphic Geology* **15**, 479–496.

Chapter 12

Influence of Fluid and Deformation on Metamorphism of the Deep Crust and Consequences for the Geodynamics of Collision Zones

Håkon Austrheim
Mineralogisk-Geologisk Museum, Sarsgate 1, N-0562 Oslo, Norway. Present address: Dept. of Geology and Geophysics, University of Wisconsin, 1215 W. Dayton St. Madison, WI 53706, USA, hakon.austrheim@toyen.uio.no

Abstract: A survey of natural transitions from igneous rocks, amphibolite, and granulite, to eclogite demonstrates that the transitions may take place over cm-scale distances parallel to fluid fronts. Rocks ranging in composition from basaltic to granitic show incomplete reactions over the whole range of high pressure to ultrahigh-pressure conditions (500°C and 1.2 GPa to 800°C and >3.0 GPa), indicating overstepping of reaction boundaries of ~1 GPa. When fluid becomes available metastable crust may react forcefully and release earthquakes as indicated by occurrence of eclogite-facies pseudotachylite.

Eclogitization weakens the crust equivalent to a temperature increase of >100°C. At an eclogitization degree of ~40 % the crust loses coherence and pre-existing structure. Reaction of eclogites to granulites and amphibolites also depends on fluid availability. The rheology and density changes caused by fluid-induced eclogite formation and retrogression influence the evolution of collision zones by controlling the timing of collapse, the topography of a collision zone, the exhumation of deep crustal sections, and the amount of material returned to the mantle. The evolution of collision and subduction zones depends not only on temperature and pressure evolution, but also on the fluid budget.

1. INTRODUCTION

Rocks buried in subduction zones and crustal root zones are exposed to changing pressures, temperatures and stresses that may cause eclogitization. In the basaltic system, plagioclase feldspar is replaced during eclogitization by denser phases such as omphacite, zoisite, kyanite and garnet. These mineralogical changes lead

B. R. Hacker and J. G. Liou (eds.),
When Continents Collide: Geodynamics and Geochemistry of Ultrahigh-Pressure Rocks, 297–323.
© 1998 *Kluwer Academic Publishers. Printed in the Netherlands.*

to drastic changes in petrophysical properties. The density changes may be greater than those at most lithological boundaries, including the Moho (Green and Ringwood, 1967). During the heating of thickened crust and during the exhumation of deep crustal rocks, eclogites may enter granulite- and amphibolite-facies P–T conditions. If the eclogites re-equilibrate mineralogically, their density is reduced through the production of feldspar. Metamorphic reactions influence the rheology of the crust instantaneously through reaction-enhanced ductility (including grain-size reduction, increased fluid pressure through metamorphic reactions (Hacker, 1997) and transformation plasticity) and more permanently through new mineral paragenesis. Importantly, eclogitization reduces the strength of the crust (Rubie, 1983; Klaper, 1990; Austrheim, 1991).

Mineralogical changes during eclogite formation also influence the geophysical signature (seismic velocity, reflectivity, and gravity anomalies) of the lithosphere. Sapin and Hirn (1997) found by combining wide angle and vertical reflection seismic data that the Moho beneath the Himalaya is a heterogeneous transition zone >10 km thick that they interpreted to represent the granulite → eclogite transition. Geophysical properties form one basis for our view of the composition and structure of the deep crust and mantle and are frequently used to constrain geodynamic models. It is therefore important to explore whether rocks that are exposed to changing P–T conditions equilibrate to the new conditions or remain metastable.

The changes in petrophysical properties related to metamorphism, and eclogitization in particular, also directly affect the geodynamics of a collision or a subduction zone. The density contrast between basalt and eclogite is at least three time greater than the density contrast resulting from cooling of the basalt during drift away from the mid-ocean ridge. This density contrast may dictate whether a subduction zone becomes arrested (Cloos, 1993). Ahrens and Schubert (1975b) found that the downward body force on a descending slab due to eclogitization is an important component of the driving force for plate motion. Whether the crust becomes eclogitized therefore has a profound influence on the geodynamics of collision and subduction zones. Mafic eclogites have densities higher than typical mantle values of 3.3 g/cm^3 and, if present in large amounts, may sink into the mantle. Such a disappearance of material from the crust has been suggested for the Alps (Laubscher, 1988) and the Himalaya (Le Pichon *et al.*, 1992; 1997) on the basis of mass-balance calculations. Eclogitization can also help produce crustal thickness in excess of 100 km (required to stabilize coesite) without forming impossibly high mountains (Dewey *et al.*, 1993).

The rheological changes associated with eclogitization, either temporary or permanent, may also influence the geodynamics of collision zones. Ruff and Kanamori (1983) suggested that the basalt → eclogite transition in the downgoing oceanic crust is responsible for the decoupling of subduction zones below 40 km. Ryan and Dewey (1997) pointed to the importance of strength reduction during eclogitization for the opening and closing of oceans (the Wilson cycle). Backreaction of eclogites to granulites and amphibolites affect the exhumation of deep-seated rocks (Dewey *et al.*, 1993; Le Pichon *et al.*, 1997).

Whether a metamorphic reaction takes place is not only a question of mineral stability but also reaction kinetics. The concentration of H_2O in grain boundaries is one of the factors controlling diffusivity and metamorphic reaction rates (Rubie, 1986). From this Rubie concluded that the absence of a free aqueous fluid prevented parts of the Sesia Zone eclogites from retrogressing during exhumation. In accordance with this, Henry *et al.* (1997) explained the 75–km deep root zone of the Himalaya as metastable granulite prevented from transforming to eclogite by sluggish reaction kinetics caused by insufficient fluid. Ahrens and Schubert (1975a) used diffusion rates in minerals to calculate that the gabbro → eclogite transition is only possible within geological time if the temperature exceeds 600–800°C or if a fluid phase is present to promote element mobility.

However, during regional metamorphism a close interplay between mineral reactions and deformation occurs. Metamorphism is a dynamic process that can cause or influence deformation. The expansion or contraction of reacting rocks produces forces (Wheeler, 1987) that may result in deformation that can happen at a catastrophic rate and cause earthquakes (Austrheim and Boundy, 1994; Kirby *et al.*, 1996). Rheological changes associated with phase transitions may aid in the release of stress. Reaction rates are therefore likely to be controlled by an interplay between temperature, deformation and fluid activity. It is impossible to assess theoretically or experimentally the relative importance of these parameters and to calculate reaction rates with confidence (Hacker and Peacock, 1994; Ernst *et al.*, this volume). Exhumed high P terrains are alternative sources of information for understanding metamorphic processes in the deep crust. Where the metamorphic reactions have been arrested, incompletely transformed rocks provide a laboratory for the study of reaction kinetics. This chapter presents eclogite formation and retrogression in various exposed deep crustal segments to unravel the controlling parameters and the kinetics behind these important metamorphic transitions. The significance of the observations is discussed in relation to existing geodynamic models.

2. FIELD EXAMPLES OF INCOMPLETE ECLOGITIZATION

Eclogite-facies rocks form from rocks ranging in composition from ultramafic to granitic. Rocks subducted into the eclogite-facies field have variable mineralogies due to differences in composition and metamorphic history. Since the amount of fluid in a rock may determine reaction rate it is useful to divide protoliths according to the amount of hydrous phases. Granulite-facies rocks typically form in the lower and middle crust (Fountain and Salisbury, 1981) and contain anhydrous phases, mainly feldspars, pyroxenes and garnet. As such rocks are frequently recycled into root zones it is important to unravel their fate. Plutonic rocks (including gabbros) that contain a limited amount of hydrous phases in the form of amphibole and mica also represent potential starting material for the eclogitization process. Amphibolites and greenschist-facies rocks with abundant hydrous phases commonly experience eclogite-facies conditions in subduction zones. Incomplete transformation has been reported for all these rock types from a number of different HP and UHP localities (Table 1; Fig. 1). Exposed transitions from granulite → eclogite facies are known from the Bergen Arcs (Boundy *et al.*, 1992; Austrheim and Engvik, 1997), the Sunnfjord and Nordfjord regions of the Western Gneiss Region, Norway (Austrheim *et al.*, 1997; Wain, 1997), the Austroalpine Mt. Emilius Klippe (Pennacchioni, 1996) and from the Musgrave range of Australia (Ellis and Maboko, 1992). Studies of the partial conversion of plutonic rocks have focused mainly on gabbros, including Yangkou, China (Zhang and Liou, 1997), the Western Gneiss Region, Norway (Mørk, 1985) and the Allalin gabbro, Western Alps (Meyer, 1983), but partially transformed granitoids are known from Monte Mucrone (Koons *et al.*, 1987) Dora Maira (Biino and Compagnoni, 1992), NW Spain (Gil Ibarguchi, 1995) and the Sulu region in China (Wallis *et al.*, 1997). Exposures of the amphibolite → eclogite facies transition are more scarce, but exist in the Marunkeu complex of Polar Urals (Lennykh *et al.*, 1997) and in the Bergen Arcs of Western Norway as described below.

2.1 Bergen Arcs: Granulite → Eclogite and Amphibolite → Eclogite

In the Bergen Arcs, eclogite-facies and amphibolite-facies metamorphism were superimposed on Grenvillian granulites (T: 800–850°C and P: <1.0 GPa). The granulite-facies rocks of the Bergen Arcs comprise an "anorthosite complex" ranging from pure anorthosite to gabbro, with various members of the mangerite–charnockite suite as well. The eclogites formed when these granulites were subducted during the Caledonian orogeny and metamorphosed at 650–700°C and pressures >1.5 GPa. The eclogitization of the anorthosite complex, best studied on the island of Holsnøy (Austrheim and Engvik, 1997) occurred in different structural settings: along metasomatic veins (Fig. 1a), related to shear zones, in pseudotachylite "veins" and in "eclogite breccias". Metasomatic veins typically contain up to 5–cm long hydrous phases (phengite, clinozoisite and amphibole) in addition to omphacite, garnet, kyanite, plagioclase and quartz. The formation of abundant hydrous phases from anhydrous granulites required the introduction of an aqueous fluid. The rock is eclogitized as far as

Table 1. Field examples of incomplete metamorphic transitions in deep crustal settings

	P–T of protolith	P–T of product	Reference
Granulite to eclogite facies			
Bergen Arcs, W. Norway	<1.0 GPa 800 °C	>1.5 GPa 650–700 °C	Boundy et al. (1992)
Mt Emelius Klippe, Alps		1.1–1.3 GPa 450–550 °C	Pennacchioni (1996)
Sunnfjord, WGR, W. Norway		1.2–1.5 GPa 525–635 °C	Krogh (1980)
Nordfjord, WGR, W. Norway	1.0–1.2 GPa 800 °C	2.0–2.2 (> 2.8?) GPa 700–800 °C	Wain (1997)
Musgrave Range, Australia	1.2 GPa 850–900 °C	1.2 GPa 650 °C	Ellis and Maboko (1992)
Gabbro to eclogite facies			
Yangkou, E. China		>3.0 GPa 800–850 °C	Zhang and Liou (1997)
Nordfjord, WGR, Norway		2.0–2.2, (>2.8?) GPa 700–800 °C	Wain (1997)
Sunnfjord, WGR, Norway		1.2–1.5 GPa 525–635 °C	Krogh (1980)
Flemsøy, WGR, Norway		1.5–2.0 GPa 750 ± 60 °C	Mørk (1985)
Allalin, W. Alps		>2.0 GPa 600 °C	Meyer (1983)
Lanzo and Rocciavre, W. Alps		1.2–1.3 GPa 450–500 °C	Pognante and Kienast (1987)
Marianske Lazne, Czech Republic		1.4–1.6 GPa 625–730 °C	Beard et al. (1995)
Granitoid to eclogite facies			
Malpica–Tuy allochthon, NW Spain		1.6 ± 0.1 GPa 675 ± 30 °C	Gil Ibarguchi (1995)
Lago Mucrone, W. Alps		>1.4–1.6 GPa 500–560 °C	Koons et al. (1987)
Brossasco, Dora Maira, W. Alps		>2.8 GPa 700–750 °C	Biino and Compagnoni (1992)
Yangkou, E. China		>3.0 GPa 800–850 °C	Zhang and Liou (1997) Wallis et al. (1997)
Amphibolite to eclogite facies			
Marun-keu, Polar Urals		1.0–1.1 GPa 500–650 °C	Udovkina (1971)
Eclogite to granulite facies			
Hareidland, W. Norway	2.5 GPa 800 °C	0.6–0.8 GPa 800 °C	Jamtveit (1987) Straume (1997)
Eclogite to amphibolite facies			
Hareidland, W. Norway		2.5 GPa 800 °C	Jamtveit (1987)
Trescolmen, Adula Nappe, Alps	>1.5 GPa 550 °C	0.4–0.6 GPa 525 °C	Heinrich (1982)

WGR: Western Gneiss Region

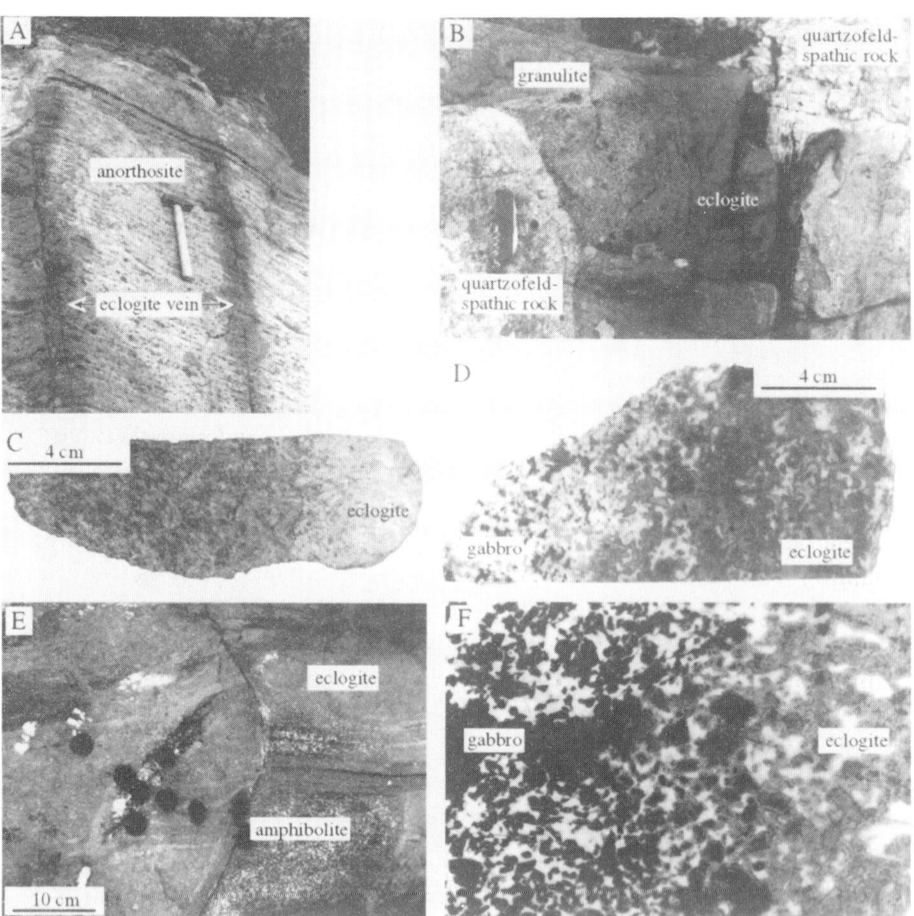

Figure 1. Examples of incomplete eclogite formation from granulites, gabbros and amphibolites. In all the examples the transformation takes place over a distance of cm. A) Eclogite veins cutting granulite-facies anorthosite, Bergen Arcs, W. Norway (Austrheim *et al.*, 1997). B) Mafic two-pyroxene granulite layer bounded by quartzofeldspathic layers. Part of the mafic layer is converted to eclogite. The eclogitization front is a well-defined vertical line. Banded granulite facies complex, Bårdsholmen, Sunnfjord, Western Gneiss Region, Norway (Engvik *et al.*, in preparation). C) Amphibolite transformed to eclogite. Marun-keu complex, Polar Urals. D) Transition from troctolitic gabbro to eclogite, Marun-keu complex, Polar Urals. E) Ghostly lenses of amphibolite surrounded by eclogite (partly retrogressed). Bergen Arcs, Norway. F) Transition from gabbro to eclogite. Allalin gabbro, Alps.

20 cm from the veins. The granulite protolith only a few cm away from the vein is medium grained with a typical granoblastic texture of equant and subhedral garnet grains. Near the veins, omphacite rims developed between garnet and feldspar, and diopsidic pyroxene is replaced by omphacite (Jd_{50}). The shear zones vary from cm to several hundred meters across and range from completely

eclogitized to clasts of the granulite protolith enclosed within an eclogite matrix. "Eclogite breccias" (Austrheim and Mørk, 1988) refer to rotated angular blocks of granulites embedded in foliated eclogite. Such mixtures of eclogite and granulite crop out in patches of over 1000 m^2 and are a characteristic and dominant lithology on northern Holsnøy. The breccias, containing about 40% eclogite, are often found along the borders of major eclogite-facies shear zones. Within the breccias, the old granulite complex has lost coherence and blocks of granulite float in a matrix of eclogite. The granulite blocks are typically angular and range in size from decimeters to meters. The texture of the granulite blocks is similar to that of the granulites described above. Erambert and Austrheim (1993) noted that the garnets in the eclogites were inherited from the granulite-facies event and contain fractures where the garnet has reequilibrated to the eclogite facies. The degree of reequilibration depends on the density of fractures that formed before of during the eclogite facies metamorphism, and on the availability of fluid.

Pseudotachylites are dark aphanitic veins showing intrusive relationships and containing clasts and crystals of the host rocks. They are interpreted as having formed by frictional melting during faulting. In support of this interpretation, Weiss and Wenk (1983) formed pseudotachylites in gabbros during experiments at high pressures (2–4 GPa) and high strain rates. Syn-eclogite-facies pseudotachylites (Fig. 2a) occur in the partially eclogitized granulites of the Bergen Arcs (Austrheim and Boundy, 1994). Outcrops of up to 100 m^2 are shattered by pseudotachylites that are often associated with cataclasites and ultramylonites. Austrheim *et al.* (1996; 1997) illustrated how seismic faulting promoted eclogitization of the crust by fracturing the granulite-facies minerals in the host rock. During eclogitization of the pseudotachylite-bearing rocks by subsequent fluid infiltration (Fig. 2a), the fractured minerals reequilibrated along these fractures (Fig. 2b).

The granulite-facies complex of the Bergen Arcs is also locally transformed to amphibolite facies. The amphibolitization occurred along structures similar to those along which eclogitization occurred, although amphibolite-facies pseudotachylites have not been described. The amphibolites of the Bergen Arcs have been interpreted as due to hydration during exhumation. However, newly discovered field relationships suggest that an early stage of amphibolitization may have predated eclogite formation because ghostly lenses of amphibolites are bordered by retrogressed eclogites (Fig. 1e). The P–T conditions and age of the amphibolites are not known.

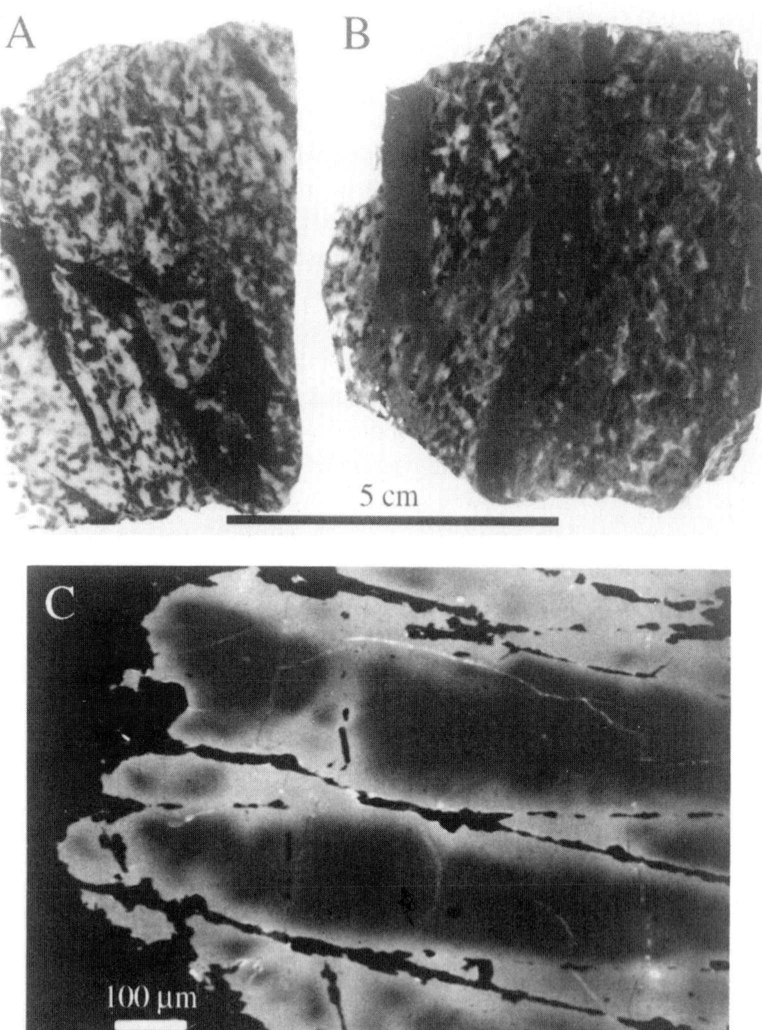

Figure 2. Pseudotachylite-bearing samples from Holsnøy, Bergen Arcs. A) Dark glassy pseudotachylite veins cutting granulite (Sample ABE 9). B) Eclogitized pseudotachylite-bearing rock from a small eclogite shear zone (Sample ABE 11). Pseudotachylites are fine-grained green veins (dark gray). Eclogitization of the wall rock is only partial as dark spots of diopside can still be seen (Austrheim *et al.*, 1997). C) Back-scattered electron image of garnet from the side wall of a pseudotachylite that has been eclogitized. The dark parts of the garnets are remnants that preserve the granulite-facies chemistry. The garnet has equilibrated to a more almandine-rich composition along fractures. These fractures contain inclusions of eclogite-facies minerals (omphacite, amphibole and phengite). The fractures formed during seismic faulting of the deep crust and the photo illustrates the importance of this fracturing in the metamorphism of the deep crust.

2.2 The Austroalpine Mt. Emilius Klippe: Granulite → Eclogite

The pre-Alpine mafic granulites of the Austroalpine Mt. Emilius Klippe underwent Alpine subduction-related eclogite-facies metamorphism (Pennacchioni, 1996) at temperatures of 450–550°C and minimum pressures of 1.1–1.3 GPa. As in the Bergen Arcs, the granulites are incompletely transformed to eclogites along shear zones and fluid-enriched veins. Pennacchioni (1996) found that the fluid-enriched eclogite veins controlled the nucleation of the shear zones. Fluid-influenced crystal plastic flow in matrix plagioclase and amphibole promoted fracturing of harder minerals (garnets and clinopyroxene). Reequilibration of garnet and clinopyroxene took place along these fractures. Thus the fluid not only accelerated the crystal plastic processes but also promoted brittle deformation of the strong phases. Pennacchioni (1996) concluded that the extent of metamorphic reequilibration was closely related to the density of the fluid pathways.

2.3 Western Gneiss Region: Granulite → Eclogite and Gabbro → Eclogite

Eclogites are found over an area of 45,000 km^2 in western Norway referred to as the Western Gneiss Region. A gradient in estimated temperature and pressure over 100 km ranging from 600°C and 1.2 GPa inland to 750–800°C and a minimum pressure of 2.0 GPa in the outer coastal area was reported by Krogh (1977). In the high-temperature coastal areas, coesite-bearing eclogites were first described by Smith (1984), suggesting pressures >2.8 GPa and crustal metamorphism at depths >100 km. In the low-temperature Sunnfjord region, Austrheim and Engvik (1997) and Austrheim *et al.* (1997) described a transition from granulite → eclogite facies. The partially eclogitized granulite in the Sunnfjord region is Grenvillian in age and the P–T path followed by these rocks was similar to that of the Bergen Arcs, however, the maximum pressure (1.2–1.5 GPa) and temperature (525–635°C) recorded during the Caledonian eclogite-facies event in Sunnfjord were lower (Krogh, 1980). The protolith consists of alternating quartzofeldspathic and mafic layers. The transformation may be localized (Fig. 1b) and follows fractures and shear zones. Where complete eclogitization occurred, a melange-like rock developed in which cm- to m-scale blocks of eclogite derived from the mafic layers float in a phengite matrix derived by transformation of the quartzofeldspathic layers. The banded structure of the granulite is destroyed by this stage of eclogitization. Like eclogitization in the Bergen Arcs, the transformation of the granulite in the Sunnfjord region was fluid controlled.

Dolerites occur throughout the Western Gneiss Region and are partly transformed to eclogites (Mørk, 1985). The dolerites are also found where UHP rocks occur (Austrheim and Engvik, 1997; Wain, 1997). There is no obvious correlation in this terrain between the degree of reaction and P–T conditions. In the low temperature area of Sunnfjord and in the UHP area of Nordfjord, both completely reacted and incompletely transformed dolerites occur.

The Western Gneiss Region is dominated by quartzofeldspathic (granodioritic) rocks; mafic and ultramafic rocks constitute less than 5%. The fate of these felsic rocks during the UHP metamorphism is difficult to ascertain and disputed. In most cases they were totally reworked and reequilibrated during subsequent amphibolite-facies conditions. There are, however, low strain areas like the Flatraket body (Wain, 1997), where the feldspar survived UHP metamorphism without reaction to eclogite-facies minerals. In the Nordfjord area, UHP metamorphism was superimposed on a suite of rocks ranging in composition from gabbroic to anorthositic to granitic. The pre-UHP metamorphism of these rocks is less well known, but Wain (1997) calculated metamorphic conditions of 1.0–1.2 GPa at 800°C for the granulites. It is not clear if the granulite assemblages formed during an earlier event or equilibrated during the Caledonian collision. The plagioclase must have survived pressures of >2.0–2.2 GPa and possibly >2.8 GPa (Wain, 1997).

2.4 Yangkou, Eastern China: Gabbro → Coesite-Bearing Eclogite

Zhang and Liou (1997) reported partial transformation of gabbro via coronitic stages to coesite-bearing eclogite. The prograde P–T path calculated for this complex passes from 540 ± 50°C at ~1.3 GPa, 600–800°C at ≥1.5–2.5 GPa and 800–850°C at >3.0 GPa. This is in accordance with the finding from the Western Gneiss Region that even during UHPM, reaction can be incomplete. Zhang and Liou (1997) argued that the primary low P phases survived untransformed due to sluggish reaction resulting from a lack of fluid during the prograde and retrograde P–T evolution.

2.5 Marun-keu, Polar Urals: Gabbro → Eclogite and Amphibolite → Eclogite

The Marun-keu complex of the Polar Urals, Russia, includes well exposed transitions of troctolitic gabbro → eclogite as well as the more rarely observed transition between amphibolite and eclogite (Lennykh *et al.*, 1997). The eclogite-facies event of the Marun-keu complex took place at 500–650°C and 1.0–1.1 GPa (Udovkina, 1971). The transformation of the troctolitic gabbro → eclogite occurred in veins (Fig. 1d) and in connection with shear zones.

The reactions transforming amphibolite to eclogite are dehydration reactions and are generally assumed to be less hindered by kinetics (Bousquet *et al.*, 1997). The occurrences in the Polar Urals (Fig. 1e) as well as in the Bergen Arcs (Fig. 1f) suggest that this transition may also be driven by fluid rather than by temperature and pressure alone. In the Marun-keu complex of Polar Urals, lenses of amphibolite are surrounded by eclogite, and locally by veins of garnet and omphacite. This relationship resembles the Bergen Arcs and takes place over centimeters (Fig. 1c and 1e). Across the transition the modal composition changes from dominantly amphibole with minor quartz, albite and garnet, to omphacite (Jd_{40}) + garnet + phengite. It is unknown whether the amphibolite was metastable or whether the amphibolite and eclogite were stable at the same P and T but at different fluid compositions. Again, however, the eclogites and their amphibolite-facies protoliths occur in the same outcrop and the transition takes place over the scale of centimeters.

3. FIELD EXAMPLES OF ECLOGITE → GRANULITE AND AMPHIBOLITE → GRANULITE

Provided subduction stops, the temperature of thickened crust will increase due to conduction and eclogite-facies conditions may be replaced by granulite-facies conditions. Since the albite → jadeite + quartz reaction boundary has a slight positive dP/dT slope, isothermal decompression leads to stabilization of plagioclase at the expense of pyroxene and the formation of granulite assemblages after eclogite. Eclogite-facies rocks from many terrains are variably reacted to granulites and amphibolite/greenschist-facies assemblages. The reactions converting eclogite to granulite are solid–solid or dehydration reactions. The granulite-forming reactions do not require fluid. In contrast, the amphibolitization of mafic eclogite cannot proceed without the addition of fluid as the reactions are hydration reactions. A major factor that controls the rate of equilibration of mafic eclogites to amphibolite is the rate of supply of H_2O (Rubie, 1990).

Most work on the retrogression of eclogite has focused on defining the P–T evolution (e.g., O'Brien, 1993), and only a few investigations have been conducted to outline the kinetics behind these important reactions. Heinrich (1982) dealt with retrogression of the Adula Nappe, Alps, and Straume (1997) studied retrogression of eclogites at Hareidland, Western Gneiss Region. Both works describe textures, the process of retrogression, and the kinetics of metamorphic reactions in general.

3.1 Western Gneiss Region: Eclogite → Granulite and Eclogite → Amphibolite

The eclogites of the Western Gneiss Region are extensively retrogressed (Dunn and Medaris, 1989), which has been explained as initial exhumation at relatively high temperature followed by increasing temperature during exhumation (Jamtveit, 1987). However, the association of retrogressed eclogites with preserved eclogites suggests that fluid availability was also important (Rubie, 1990). The rocks studied by Straume (1997) had peak conditions of 2.5 GPa and 800°C and then passed through the granulite- and amphibolite-facies fields (Jamtveit, 1987; Dunn and Medaris, 1989). The eclogite-facies assemblage in mafic rocks consists of omphacite (Jd_{30}), garnet, zoisite, quartz and kyanite. Straume (1997) recognized two stages in the process of conversion of the eclogite to granulite; an initial stage where the reactions were closely connected to fractures and a second, symplectic stage. En echelon sigmoidal fractures 1 cm long and 100 mm thick in garnet are filled with anorthite, spinel and orthopyroxene. Reaction at grain boundaries was due to fracturing. Orthopyroxene and anorthite formed along quartz inclusions in garnet along only that part of the grain boundary that was fractured. The rest of the grain boundary is without reaction products (Fig. 3A). Similar fractures along kyanite/garnet grain boundaries contain the reaction products sapphirine, anorthite and orthopyroxene (Fig. 3B). Unfractured kyanite/garnet boundaries show no reaction. Straume (1997) demonstrated that the fractures were channels of mass transfer and critical for the reaction to occur. She also noted that the fractures terminate in garnets as a trail of minute amphibole inclusions, suggesting that the apparent solid–solid reaction was enhanced by fluid. It was assumed that the symplectitic stage where the backreaction was more advanced also started by fracturing. Temperatures calculated for the granulite-facies paragenesis based on orthopyroxene-garnet pairs are 750–800°C (Straume, 1997), illustrating that even at 800°C retrogression to granulites was limited and depended on trace fluids and fracturing. The granulite-facies symplectite grades spatially into amphibolite. The amphibolitization of the eclogites at Hareidland is also connected to pegmatites and fracture systems (Straume, 1997). The fractures are parallel dark zones where the symplectitic eclogites replaced a rock with abundant hydrous phases—amphibole, chlorite and margarite. The transition from symplectitic eclogite to amphibolite is discrete and occurs over a width of 1 cm. Based on density measurements and whole-rock chemistry Straume (1997) found that the amphibolitization was associated with a volume increase of 8–18% for the most garnet-rich eclogites. Also at the microscale it was demonstrated that fracturing was important for the hydration to occur. Relict garnets along the hydration front were faulted, and pull-apart domains are filled with amphibole. Zoning developed in the garnets along these microfaults. It is unclear whether the microfracturing with the granulite-facies assemblages formed as a result of the volume expansion during amphibolitization or for other reasons. If the cracking occurred during the amphibolitization it must mean that the granulite and amphibolite represent the same P–T conditions but different fluid activities.

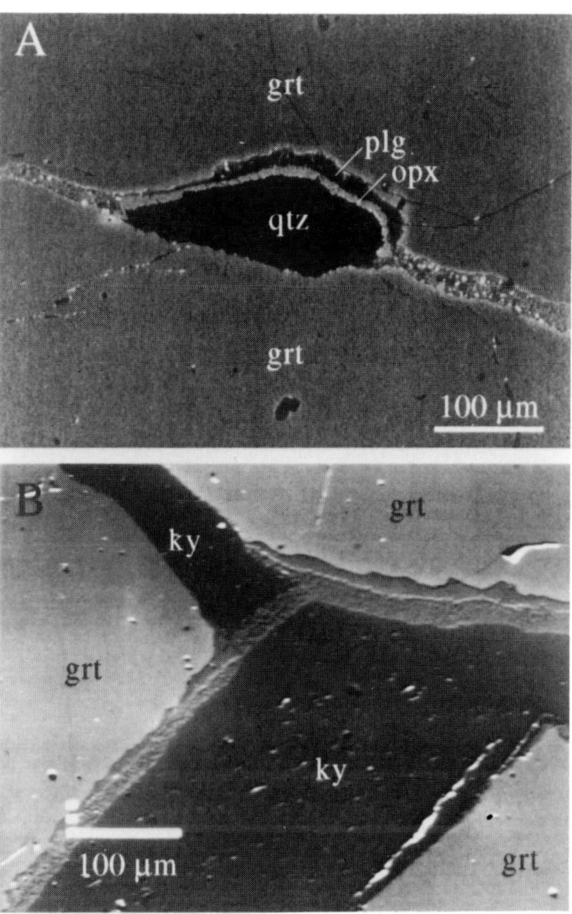

Figure 3. Eclogite → granulite facies transition, Hareidland, Western Gneiss Region, Norway (Straume, 1997). A) Reaction between garnet and a quartz inclusion produced the granulite-facies assemblage orthopyroxene + plagioclase along the top of the inclusion where a fracture parallels the grain boundary. On the lower edge where the fracture is absent the reaction products are missing. B) Fractures filled with the granulite-facies assemblage plagioclase + sapphirine + orthopyroxene at grain boundaries between eclogite-facies garnet and kyanite. The granulite-facies assemblage is present only along the interphase fractures. Fracture that are wholly within kyanite are filled with plagioclase + sapphirine. This implies that the fracture is necessary for the formation of the granulite assemblage from the eclogite-facies assemblage.

3.2 Sesia Zone: Eclogite → Granulite and Eclogite → Amphibolite

Amphibolitization of eclogite in the Sesia zone was described by Heinrich (1982). Lenses of eclogites surrounded by metapelites were hydrated from the rim inward and concentric zones developed, consisting of a symplectite of diopside + albite (inner rim) and amphibolite (outer rim) separating the pristine eclogite core from the metapelite. Heinrich (1982) showed that amphibolitization of the eclogite was caused by water liberated from the breakdown of phengite in the metapelite. Importantly the solid–solid reaction (omphacite + quartz → diopside + plagioclase) forming the symplectite requires fluid as a catalyst. Rubie (1990) interpreted this to mean that the amphibolite and the diopside–plagioclase assemblages formed at different f_{H2O}. According to Rubie (1990) the diopside–plagioclase symplectite needed some H_2O to catalyze the reaction. In both the Sesia Zone and the Western Gneiss Region the retrogression to eclogite depended on fluid even when dry parageneses formed. The rocks investigated by Straume (1997) were mafic and contained minor mica. The information may therefore only apply to mafic eclogites. The work by Heinrich (1982) suggests that phengite-rich eclogite-facies rocks may behave differently and be more easily retrogressed. However, as phengite is found in many high P areas, its breakdown during exhumation is not ubiquitous.

4. The Role of Deformation and Fluid in Metamorphism of the Deep Crust

The described field examples show that the formation of eclogites and backreaction to granulites and amphibolites can be incomplete. Although it may be argued that incompletely eclogitized crust is more easily be exhumed and thus be over-represented, the processes seen in these rocks do take place in the deep crust. The examples discussed here illustrate eclogites formed at temperatures varying from 450–800°C at pressures of 1.1–3.0 GPa. Fig. 4 shows the P–T conditions calculated for these various terrains compared to the albite = jadeite + quartz and quartz = coesite reactions. The degree of overstepping of the reaction boundary is difficult to evaluate, because mafic/gabbroic rock may react to eclogite well below the albite-out reaction, but with respect to this boundary the pressure overstepping is largest for the areas with the highest temperature, i.e. >1.0 GPa for Yangkou. There does not seem to be a principally different evolution for different temperatures within each of the different protolith groups. For the low temperature eclogites of Mt. Emilius, both unreacted and completely reacted rocks are present. The same is true for the Sunnfjord and Bergen Arcs granulites that were eclogitized at higher temperature. Even in the Nordfjord area which enjoyed UHP metamorphism at 700–800°C, unreacted and completely reacted rocks are mixed. Considering only gabbroic protoliths, coronas are common even in the least-reacted case. However, the coronitic gabbros and the completely eclogitized gabbro are mixed in all the 450–500°C eclogites from Lanzo, Rocciavre, and Yangkou. The metagranitoids also behaved similarly regardless of metamorphic conditions. The four cases summarized in Table 1 were metamorphosed at widely different conditions, but igneous textures and minerals including K-feldspar are found in each. In all cases, however, albite was partially replaced by aggregates including jadeitic pyroxene. The expected negative correlation between degree of reaction and temperature cannot be read from these data. Metastable protoliths and reacted rocks are often found in the

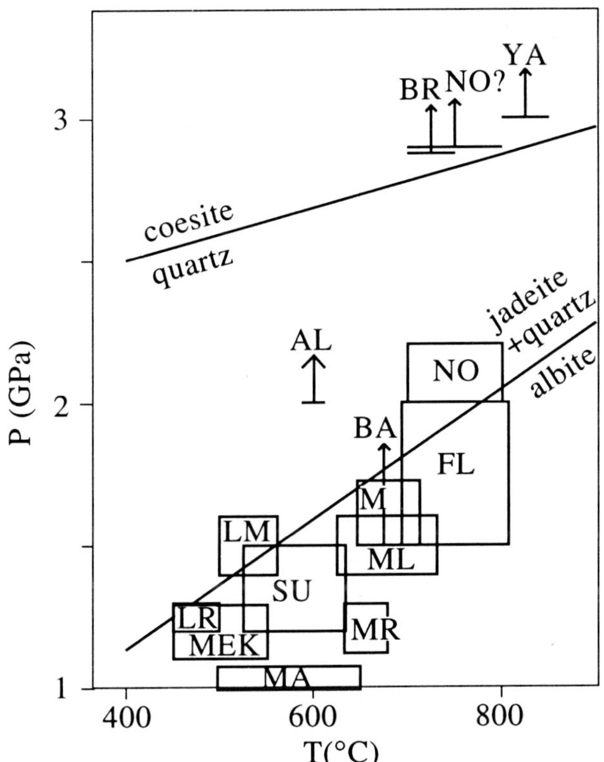

Figure 4. Calculated P T conditions for the areas showing incomplete eclogitization listed in Table 1 shown in relation to albite = jadeite + quartz (Holland, 1980) and quartz–coesite (Bohlen and Boettcher, 1982). Where only minimum pressures are given the area is represented by an arrow. BA: Bergen Arcs; MEK: Mt. Emilius Klippe; SU: Sunnfjord, Western Gneiss Region; NO: Nordfjord, Western Gneiss Region; MR: Musgrave Range, Australia; YA: Yangkou, China; FL: Flemsøy, Western Gneiss Region; AL: Allalin, Alps; LR: Lanzo and Rocciavre, Alps; ML: Marianske Lazne, Czech Republic; M: Malpica-Tuy, Spain; LM: Lago Mucrone, Alps; BR: Brossasco, Alps; MA: Marun-keu, Polar Ural.

same outcrop and passed through the same regional P–T history. The difference in degree of reaction must therefore be found in parameters other than time, pressure and temperature. This points to fluid and deformation as critical factors controlling the transformations. These parameters must overrule temperature and pressure within the whole range of physical conditions for which data exist. It is difficult to ascertain the respective roles of fluid and deformation; for example is fluid necessary for deformation?

However, Pennacchioni (1996) observed that ductile shear zones developed from the hydrated granulites. In the same way Austrheim *et al.* (1997) concluded that hydration is a prerequisite for ductile deformation. Where granulite remained dry and unreacted, it deformed by faulting. The fluid therefore is the primary cause of metamorphism and the shear zones developed as a consequence of the crust having been weakened by reaction.

4.1 Deformation and Eclogite Formation

As shown by the examples from the Bergen Arcs, the Western Gneiss Region and the Mt. Emilius Klippe, nearly complete eclogitization occurs in high strain zones. It has long been recognized that eclogites may develop in high strain zones and the role of ductile deformation on reaction kinetics, outlined by Rubie (1990), will not be repeated here.

The deep crust is generally assumed to deform by ductile flow. However, observations of deep crustal sections suggest that the overall deformation of the deep crust depends on the presence or absence fluid and that the deep crust can fracture on the micro and outcrop scale. Such fracturing is important during both eclogitization and the backreaction of eclogites. Garnets in eclogites from the Bergen Arcs were mainly inherited from the granulite protolith and many have preserved their granulite-facies composition, but are transected by veins with higher iron content established during the eclogite-facies overprint. The reequilibrated parts of the garnets locally contain inclusions of eclogite-facies minerals such as omphacite. Erambert and Austrheim (1993) interpreted these veins as traces of fluid channels and concluded that fluid and deformation were fundamental to the reequilibration of the garnet.

These fractures enhanced mobility and allowed equilibration of garnet as well as mass transfer between various reaction domains. By comparing the length of fractures to the length of diffusion on grain contacts it can be suggested that the mobility along the fracture is infinitely faster than volume diffusion. In addition to causing reequilibration of existing minerals, the formation of granulite-facies minerals at the expense of eclogite parageneses starts along fractures, and where these fractures are absent the granulite-facies minerals are also missing (Straume, 1997). From the UHP Dora Maira eclogites, Chopin (1997) also described fractures with granulite-facies minerals in eclogite-facies garnets. Fluid is the primary feature that caused reaction, because fractured rocks otherwise show no sign of reaction (Austrheim *et al.*, 1997). The importance of fracturing is also demonstrated by oxygen isotope ratios in granulites from the Adirondacks (Valley and Graham, 1993). Volume diffusion of oxygen through the crystal structure of magnetite contributed to isotope exchange in the outer 10 mm rim of small grains, but larger grains were cut by healed cracks several 1000 mm long where the isotope exchange took place.

4.2 Fluid and Metamorphism of the Deep Crust

There is ample evidence that fluids are temporarily present and play important roles in the deep crust, although stable isotope studies on gabbro → eclogite (Barnicoat and Cartwright, 1997) and granulite → eclogite (Van Wyck *et al.*, 1996) transformations do not support the involvement of a large fluid flux. The preservation of very low pre-subduction $\delta^{18}O$ values through UHP metamorphism (Yui *et al.*, 1995; Baker *et al.*, 1997; Yui *et al.*, 1997; Rumble, this volume) also support this. Fluids are part of many metamorphic reactions, they act as catalysts in solid–solid reactions, they may lead to changes in rock composition (metasomatism), and they may control rheology (Kirby, 1984). This is true when granulite reacts to eclogite and when eclogite and granulite transform to amphibolite. The transformation of amphibolite to eclogite, as observed in the Marun-keu complex of the Polar Urals and in the Bergen Arcs, indicate that fluid composition may play a role in the stability of eclogite versus amphibolite. Since the rock in both cases contained sufficient H_2O in the form of abundant amphibole and the reactions involved were dehydration reactions, there is no need to call upon fluid as an added component, or for kinetic reasons. This suggests that the amphibolites and eclogites were stable with different fluids.

4.3 Amount of Eclogite as a Function of Fluid Content

The degree of eclogitization depends on the amount of fluid available. Le Pichon *et al.* (1997) and Ryan and Dewey (1997) suggested that subducted crust contains enough fluid to allow complete eclogitization. Eclogitization of granitic rocks requires fluid because phengitic micas will formed at the expense of K-feldspar. Even if the granite contains biotite there will be insufficient H_2O to convert K-feldspar to phengite. The metamorphosed quartz diorite of the Sesia zone contains between 1.16 and 1.51 wt% H_2O, while the incompletely transformed quartz diorite has H_2O values around 0.85 wt%. Both rocks have moderate K_2O values of 1–2.73 wt%. Average plutonic rocks exposed at the surface of Earth world-wide contain only 0.6 wt% H_2O and <3.4 wt% K_2O (Wedepohl, 1970). If this potassium reacts to form phengitic mica it will require ~1.5 wt% H_2O.

Additional water will then be needed if zoisite, paragonite and amphibole are to form. Such rocks will be a fluid sink and may dry out the crust.

4.4 Fluid-Induced Eclogitization—A Forceful and Rapid Process

Even though gabbro and granulite can remain metastable at eclogite-facies conditions, eclogitization is a forceful process that may be catastrophically fast provided that fluid is available (Austrheim and Engvik, 1997). There is ample evidence that fluid can penetrate the deep crust and cause eclogitization. The volume changes that follow this transformation are likely to produce stresses that induce fracturing. This stress may be released rapidly to form pseudotachylite. The fractures that form may act as channels for fluid-aided eclogite formation. Thus Austrheim *et al.* (1997) viewed eclogitization as a process initiated by fluid and spread through the rock as a wave. The rheology changes and volume changes that result from this transformation can cause seismic failure in the unreacted rock. In addition to slip along the fault surface, the seismic energy released may cause deformation away from the fault plane. Because earthquakes are frequent on a geological time scale, seismicity is an integrated part of the eclogitization process and may play an important role in the fracturing and fluid movement in the subducting slab.

5. Fluid-Controlled Metamorphism of the Deep Crust and Geodynamics of Collision and Subduction Zones

Recent geodynamic models (Andersen *et al.*, 1991; Austrheim, 1991; Dewey *et al.*, 1993; Bousquet *et al.*, 1997; Henry *et al.*, 1997; Le Pichon *et al.*, 1997; Ryan and Dewey, 1997) have highlighted the importance of eclogite formation and retrogression in the evolution of continent collision zones. Similarly, eclogite formation is important in the behavior of subduction zones (Hacker, 1996; Kirby *et al.*, 1996). Most of these models presented are based on a temperature-controlled reaction kinetics, however eclogite formation, its back reaction to granulites and amphibolites, and the rheology of the deep crust, are to a large extent controlled by fluid.

5.1 Crustal Elevation and Crustal Thickness

The reviewed field examples indicate that dry rock will not react even at temperatures as high as 800°C and at large degrees of overstepping. Experiments on the reaction albite → jadeite + quartz were carried out by Hacker *et al.* (1993)

on water impregnated and dry samples at an overstepping of 0.5 GPa. The water-impregnated samples produced reaction products at temperatures as low as 600°C. Vacuum-dried samples heated in the temperature range of 900–1200°C all gave reaction products. This suggests that reaction should proceed at equivalent conditions in Earth. However, further experiments at dry conditions at 600–900°C are required to localize the "kinetic boundary" for this reaction and to test the findings from the field examples. Substantial overstepping of reaction boundaries is in accordance with the existence of a 75 km thick crust below the Himalaya and with theoretical considerations (Ahrens and Schubert, 1975a). Thermal modelling by Henry *et al.* (1997) shows that a large part of the Himalayan crust is under eclogite-facies conditions. Because the degree of eclogitization depends on the availability of fluid in the rock being subducted, it is possible that different collision zones may behave differently depending on the mix of rock types and the amount of fluid added from outside. Eclogitization of 30 km of lower crust with a density increase of 0.3 g/cm^3 will reduce the elevation of a mountain range 3 km (Le Pichon *et al.*, 1997). Thus Henry *et al.* (1997) described the difference in elevation along the Alpine–Himalayan system, where the elevation increases abruptly from 2.5 km west of Zagros to 5 km in the High Himalaya as due to eclogitization at 55 km under the Alps and at 75 km under the Himalaya. These different levels of eclogitization were proposed to result from different conversion rates and colder temperatures under the Himalaya. An alternative explanation in accordance with field observation presented here is that the lower crust under the Alps may be wetter.

5.2 Timing and Fluid Source of Metamorphism

If the fluid to flux eclogitization derives from the subducted slab, the development of the collision zone will be a function of the subducted material and its thermal evolution. If the fluid is added from a source external to the slab that is independent of the temperature evolution of the slab, the evolution of the collision zone may be disconnected from the thermal evolution of the subducted slab. In this case eclogitization might start at various temperatures and be related to deeper processes in the mantle and less to the thermal evolution of the subducting slab. The moment of fluid infiltration from the mantle will be an important moment in the evolution of the orogen. It is commonly inferred that incomplete eclogitization reflects rapid exhumation and short residence time at depth. In light of the findings that reactions are fluid and deformation controlled this is correct only in so far as the probability for addition of fluid increases with time. At an extreme limit, this means that a dry granulite or a gabbro can stay at depth for many millions of years unless fluid is added and that granulites may have been exposed to eclogite-facies conditions and returned to the surface without any record in the rock. Support for this view can be found in the Urals where a Palaeozoic mountain range has preserved its deep root zone for more than 300 m.y. (Steer *et al.*, 1995; Carbonell *et al.*, 1996). The low geothermal gradient across the region indicates that the root zone be at eclogite-facies conditions, yet its crustal signature suggests that it never converted to eclogite-facies rocks.

5.3 Weakening of the Crust and Orogenic Collapse

It has been argued (Mooney and Meissner, 1991) that crustal root zones disappear over time and that Palaeozoic mountain ranges have lost their root zones by lateral ductile flow in an isostatic adjustment to former high topography. The existence of a root zone under the Palaeozoic Urals (Steer *et al.*, 1995; Carbonell *et al.*, 1996), however, shows that this may not be universally true. However, the presence of a root zone may be explained if the crust below the Urals was dry and did not become eclogitized and weakened. The collapse of thickened crust depends on rheology and is presently viewed as a temperature increase caused by removal of a thermal boundary layer (Dewey *et al.*, 1993). However the Mt. Emilius Klippe, the Bergen Arcs and the Sunnfjord examples all show that crust is weakened by fluid-induced eclogitization and that existing structures are completely disrupted during the eclogitization process. This weakening was not caused by a temperature increase because the stiff granulites and the weak eclogites experienced the same temperature evolution. Since this weakening can take place at different temperatures, ranging from 450–550°C in the Mt. Emilius Klippe to 525–635°C in the Sunnfjord and 650–700°C in the Bergen Arcs, fluid may have a larger influence on the strength of crust and may be more efficient than a temperature increase of >100°C. Although it cannot be quantified, this will strongly influence the deformation of thickened crust and may be the factor controlling the collapse of orogenic belts. Furthermore, eclogitization not only weakens the crust, but markedly changes its density as well. A partly eclogitized crust produces an ideal situation for vertical movements. Austrheim *et al.* (1997) presented a model for the fractionation of root zones wherein less dense, unreacted and felsic material rises and dense eclogite descends into the mantle. Such a model explains both the overall felsic and unreacted nature of exposed deep crustal terranes like the Western Gneiss Region and is in accordance with mass balance calculations suggesting crustal loss in collision zones (Laubscher, 1988).

5.4 Retrogression, Uplift and Exhumation of Deep-Seated Terranes

The backreaction of eclogites to granulites and amphibolites profoundly influences exhumation of deep-seated rocks. Dewey *et al.* (1993) explained that advective thinning of the lithosphere combined with the resultant heating and the conversion of eclogite to granulite/amphibolite causes a surface uplift of about 2 km. Similarly Le Pichon *et al.* (1997) proposed that the uplift of the Tibetan plateau is not due to tectonic shortening, but to the transformation of eclogite to granulite in the lower crust. They argued that the eclogite → granulite reaction is much less likely to be metastable because it releases H_2O, occurs as a result of a slow temperature increase over millions of years, and occurs at temperatures >700°C. The discovery that the transformation of eclogite to granulite is controlled by fracturing and fluid even at temperature as high as 750–800°C (Straume, 1997) and that retrogression of eclogites in the Sesia Zone (Heinrich, 1982) and Western Gneiss Region (Straume, 1997) was fluid controlled suggests that this transition cannot be modeled solely as a function of P and T. If the fluid is derived by dehydration reactions (as in the Sesia zone), retrogression of the mafic eclogites will be controlled by fluid transport. The Marun-keu, Polar Urals, and Bergen Arcs rocks, all suggest that amphibolitization and eclogitization may occur at the same crustal level. If so, the introduction of an amphibolitizing fluid may be an effective way of exhuming eclogitized crust from continental root zones and from subduction zones, independent of temperature increases.

5.5 Eclogitization of Metastable Crust and Basin Subsidence

Steep geothermal gradients in rift zones and in some sedimentary basins have been used as an argument against eclogitization as a driving force for basin subsidence. Realizing that the eclogite transition is kinetically controlled removes this obstacle. If the reaction boundary is overstepped—as suggested by the many examples listed above—eclogite formation will be enhanced by rising temperature or fluid infiltration even in the absence of subduction. From a detailed analysis of the foreland basin of the Carpathians, Artyushkov *et al.* (1996) concluded that this basin cannot have formed by flexure of the lithosphere due to loading by nappes as commonly thought; however, it might have formed by eclogitization of a metastable crust during fluid infiltration.

5.6 Eclogitization in Subduction Zones

The observations outlined above are mainly derived from eclogitization in continent collision zones, but they must also apply to subduction zones; exposures of eclogites as well as seismological data support this (Hacker, 1996). For example, Barnicoat and Cartwright (1997) found that only those parts of ophiolitic gabbros that were hydrothermally altered in the ocean floor transformed to eclogite; they estimated that the eclogitization reactions were overstepped by 0.6–1.1 GPa. Seismological studies show that the subducting oceanic crust of the Philippine Sea and Pacific Plates beneath Japan may remain as untransformed gabbro down to 60 and 70–80 km depth, respectively (Hori, 1990; Hurukawa and Imoto, 1992; Hurukawa, 1993). Earthquakes at depths of 90–150 km in the Nazca plate being subducted beneath northern Chile are interpreted to result from eclogitization of the oceanic crust (Comte and Suárez, 1994; Kirby *et al.*, 1996). These examples show that overstepping of reaction boundaries in subduction zones may be large. Hurukawa and Imoto (1993) ascribed the different eclogitization level inferred for the two plates under Japan to a more rapid convergence rate and to an older age of the Pacific plate compared to the Philippine Sea plate. In accord with this, Kirby *et al.* (1996) suggested that eclogitization reactions are highly overstepped in mature slabs and take place at depths of 100–170 km, while earthquakes and eclogitization are limited to 50–100 km depth in young lithosphere because it is warmer. These predictions agree with field data showing extensive overstepping of reaction boundaries, and imply that eclogitization takes place over a considerable range of temperature and pressure and that the reactions must be incomplete. The spread of eclogitization over depth intervals of at least 50 km must mean that temperature cannot be controlling the kinetics of these reactions. Kirby *et al.* (1996) explained the earthquakes as caused by the release of fluid by dehydration reactions in the slab and noted that this should happen at a shallower level in a warm slab. As discussed above, the timing of eclogitization will depend on the minerals undergoing hydration as well as the transport of fluid from the site of dehydration to the site of reaction. As oceanic crust is variably hydrated (see summary by Hacker, 1996) it is to be expected that subduction zones, like continent collision zones, may develop in different ways depending on the fluid budget. The degree of eclogitization must be limited in that part where earthquakes occur. Following the information from the Bergen Arcs, the slab will disintegrate and deform by ductile flow when the degree of eclogitization reaches ~40 %. The suggestion, derived from the Bergen Arcs and the Marun-keu complex, that eclogite and amphibolite may coexist with different fluids, may have relevance to the evolution of subduction zones and the exhumation of eclogite-bearing melanges.

6. CONCLUDING REMARKS

Field observations and seismological data indicate that metamorphic reactions in the deep crust are controlled by fluid and that metamorphic reactions in a dry crust can be widely overstepped at temperatures up to 800°C. Since metamorphic reactions must strongly influence the evolution of collision zones, such zones cannot simply be to modeled function of P and T alone, but must be viewed as a function of the fluid budget, including both amount and composition. Many features of collisional orogens such as uplift, exhumation of deep crust, orogenic collapse (removal of root zones) and basin subsidence, may depend on the fluid budget at depth. Different collision zones and subduction zones may have behaved differently solely because of fluid availability. If fluid is available from rocks undergoing dehydration, the rate of metamorphism depends on the transport from one rock to another and on the spatial distribution of rocks at depth. However if the fluid is added from the underlying mantle, the orogenic evolution is controlled from outside the subducting slab. In this case the key to understanding the evolution of collisional and subduction zones is understanding fluid movement in the deeper parts of the lithosphere. A better understanding of collisional evolution demands an answer to many problems connected to fluid in the deeper parts of the lithosphere, such as its amount, source, transport and its control on the minerals formed. Earth scientists have barely started to investigate complicated relationships and only through a joint interdisciplinary effort can we hope to find answers.

ACKNOWLEDGMENTS

Support from the Norwegian Research Council to project 107603/410, *Fluid-Induced metamorphism and Geodynamic Processes*, is gratefully acknowledged. An expedition to the Polar Urals to study the Marun-keu complex were possible by support from the Nansen foundation and Europrobe. P. Philippot is thanked for providing Fig. 1d. Åse Straume and Johannes Glodny generously allowed the use of unpublished illustrations. I thank Muriel Erambert for many interesting discussions on various aspects of eclogite formations. This chapter was written while the author was at sabbatical leave at the Department of Geology and Geophysics, University of Wisconsin. J. Valley is thanked for providing stimulating working conditions and for discussions on mobility on fractures. Constructive reviews and editorial help from J. Mosenfelder and B. Hacker are gratefully acknowledged.

REFERENCES

Ahrens, T.J. and Schubert, G. (1975a) Gabbro-eclogite reaction rate and its geophysical significance, *Reviews of Geophysics and Space Physics* **13**, 383–400.

Ahrens, T.J. and Schubert, G. (1975b) Rapid formation of eclogite in a slightly wet mantle, *Earth and Planetary Science Letters* **27**, 90–94.

Andersen, T.B., Jamtveit, B., Dewey, J.F., and Swensson, E. (1991) Subduction and eduction of continental crust: major mechanism during continent-continent collision and orogenic extensional collapse, a model based on the south Caledonides, *Terra Nova* **3**, 303–310.

Artyushkov, E.A., Baer, M.A., and Mørner, N.-A. (1996) The East Carpathians: indications of phase transitions, lithospheric failure and decoupled evolution of thrust belts and its foreland, *Tectonophysics* **262**, 101–133.

Austrheim, H. (1991) Eclogite formation and dynamics of crustal roots under continental collision zones, *Terra Nova* **3**, 492–499.

Austrheim, H. and Boundy, T.M. (1994) Pseudotachylytes generated during seismic faulting and eclogitization of deep crust, *Science* **265**, 82–83.

Austrheim, H. and Engvik, A.K. (1997) Fluid transport, deformation and metamorphism at depth in a collision zone, in B. Jamtveit and B.W.D. Yardley (eds.), *Fluid flow and transport in rocks: mechanisms and effects*, Chapman and Hall, London, pp. 123–137.

Austrheim, H., Erambert, M., and Boundy, T.M. (1996) Garnet recording deep crustal earthquakes, *Earth and Planetary Science Letters* **139**, 223–238.

Austrheim, H., Erambert, M., and Engvik, A.K. (1997) Processing of crust in the root of the Caledonian continental collision zone: the role of eclogitization, *Tectonophysics* **273**, 129–153.

Austrheim, H. and Mørk, M.B.E. (1988) The lower continental crust of the Caledonian mountain chain: evidence from former deep crustal sections in western Norway, in Y. Kristofferson (ed.), *Progress in Studies of the Lithosphere in Norway*, **3**, Norges Geologiske Undersøkelse, Trondheim, pp. 102–113.

Baker, J., Matthews, A., Mattey, D., Rowley, D., and Xue, F. (1997) Fluid-rock interactions during ultra-high pressure metamorphism, Dabie Shan, China, *Geochimica Cosmochimica et Acta* **61**, 1685–1696.

Barnicoat, A.C. and Cartwright, I. (1997) The gabbro-eclogite transformation: an oxygen isotope and petrographic study of west Alpine ophiolites, *Journal of Metamorphic Geology* **15**, 93–104.

Biino, G. and Compagnoni, R. (1992) Very-high pressure metamorphism of the Brossasco coronite metagranite, southern Dora-Maira massif, Western Alps, *Schweizerische Mineralogische und Petrographische Mitteilungen* **72**, 347–363.

Bohlen, S.R. and Boettcher, A.L. (1982) The quartz-coesite transformation: a precise determination and the effects of other components, *Journal of Geophysical Research* **97**, 7073–7078.

Boundy, T.M., Fountain, D.M., and Austrheim, H. (1992) Structural development and petrofabrics of eclogite facies shear zones, Bergen arcs, western Norway; implications for deep crustal deformational processes, *Journal of Metamorphic Geology* **10**, 127–146.

Bousquet, R., Goffe, B., Henry, P., Le Pichon, X., and C., C. (1997) Kinematic, thermal and petrological model of the Central Alps: Lepontine metamorphism in the upper crust and eclogitisation of the lower crust, *Tectonophysics* **273**, 105–127.

Carbonell, R., Perez-Estaun, A., Gallart, J., Diaz, J., Kashubin, S., Mechie, J., Stadtlander, R., Schulze, A., Knapp, J.H., and Morozov, A. (1996) Crustal root beneath the Urals; wide-angle seismic evidence, *Science* **274**, 222–224.

Chopin, C., Simon, G., and Schenk, V. (1997) Granulite-facies overprint in Ultrahigh-pressure rocks (Dora-Maira massif): Evidence for low temperature granulites?, *Terra Nova* **9**, 6.

Cloos, M. (1993) Lithospheric buoyancy and collisional orogenesis: subduction of oceanic plateaus, continental margins, island arcs, spreading ridges, and seamounts, *Geological Society of America Bulletin* **105**, 715–737.

Comte, D. and Suárez, G. (1994) An inverted double seismic zone in Chile; evidence of phase transformation in the subducted slab, *Science* **263**, 212–215.

Dewey, J.F., Ryan, P.D., and Andersen, T.B. (1993) Orogenic uplift and collapse, crustal thickness, fabrics and metamorphic phase changes; the role of eclogites, *Geological Society Special Publications* **76**, 325–343.

Dunn, S.R. and Medaris, J., L.G. (1989) Retrograded eclogites in the Western Gneiss Region, Norway, and thermal evolution of a portion of the Scandinavian Caledonides, *Lithos* **22**, 229–245.

Ellis, D.J. and Maboko, M.A.H. (1992) Precambrian tectonics and the physicochemical evolution of the continental crust. I. The gabbro–eclogite transition revisited, *Precambrian Research* **55**, 491–506.

Engvik, A.K., Austrheim, H., and Andersen, T.B. (in preparation) Structural, mineralogical and petrophysical changes in Proterozoic crust recycled through the root zone of the Caledonian mountain chain and consequences for collisional geodynamics

Erambert, M. and Austrheim, H. (1993) The effects of fluid and deformation on zoning and inclusion pattern in poly-metamorphic garnets, *Contributions to Mineralogy and Petrology* **115**, 204–214.

Ernst, W.G., Mosenfelder, J.L., Leech, M.L., and Liu, J. (this volume) H_2O recycling during continental collision: phase-equilibrium and kinetic considerations

Fountain, D.M. and Salisbury, M.H. (1981) Exposed cross-sections through the continental crust: implications for crustal structures, petrology and evolution, *Earth and Planetary Science Letters* **56**, 267–277.

Gil Ibarguchi, J.I. (1995) Petrology of jadeite metagranite and associated orthogneisses from the Malpica-Yuy allochthon (Northwest Spain), *European Journal of Mineralogy* **7**, 403–415.

Green, D.H. and Ringwood, A.E. (1967) An experimental investigation of the gabbro-eclogite transformation and its petrological implications, *Geochimica Cosmochimica et Acta* **31**, 767–833.

Hacker, B.R. (1996) Eclogite formation and the rheology, buoyancy, seismicity, and H_2O content of oceanic crust, in G.E. Bebout, Scholl, D., Kirby, S.H., Platt, J.P. (ed.), *Dynamics of Subduction*, Monograph, American Geophysical Union, Washington, D.C., pp. 337–246.

Hacker, B.R. (1997) Diagenesis and the fault-valve seismicity of crustal faults, *Journal of Geophysical Research* **102**, 24,459–24,467.

Hacker, B.R., Bohlen, S.R., and Kirby, S.H. (1993) Albite $\rightarrow$ jadeite + quartz transformation in albitite, *Eos, Transactions American Geophysical Union* **74**, 611.

Hacker, B.R. and Peacock, S.M. (1994) Creation, preservation, and exhumation of coesite-bearing, ultrahigh-pressure metamorphic rocks, in R.G. Coleman and X. Wang (eds.), *Ultrahigh Pressure Metamorphism*, Cambridge University Press, Cambridge, United Kingdom

Heinrich, C.H. (1982) Kyanite-Eclogite to Amphibolite facies evolution of hydrous mafic and pelitic rocks, Adula nappe, central Alps, *Contributions to Mineralogy and Petrology* **81**, 30–38.

Henry, P., Le Pichon, X., and Goffe, B. (1997) Kinematics, thermal and petrological model of the Himalayas: constraints related to metamorphism within the underthrust Indian crust and topographic elevation, *Tectonophysics* **273**, 31–56.

Holland, T.J.B. (1980) The reaction albite = jadeite + quartz determined experimentally in the range 600–1200°C, *American Mineralogist* **65**, 129–134.

Hori, S. (1990) Seismic waves guided by untransformed oceanic crust subducted into the mantle: The case of the Kanto district, central, Japan, *Tectonophysics* **176**, 355–376.

Hurukawa, N. and Imoto, M. (1992) Subducting oceanic crusts of the Philippine Sea and Pacific plates and weak-zone-normal compression in the Kanto District, Japan, *Geophysical Journal International* **109**, 639–652.

Hurukawa, N.a.I., M. (1993) A non double-couple earthquake in the subducting oceanic crust of the Philippine Sea plate, *Journal of the Physics of the Earth* **41**, 257–269.

Jamtveit, B. (1987) Magmatic and metamorphic controls on the chemical variations within the Eiksunddal eclogite complex, Sunnmøre, western Norway, *Lithos* **20**, 369–389.

Kirby, S.H. (1984) Introduction and digest to the special issue on chemical effects of water on the deformation and strength of rocks, *Journal of Geophysical Research* **89**, 3991–3995.

Kirby, S.H., Engdahl, E.R., and Denlinger, R. (1996) Intermediate-depth intraslab earthquakes and arc volcanism as physical expressions of crustal and uppermost mantle metamorphism in subducting slabs, in G.E. Bebout, D. Scholl, and S. Kirby (eds.), *Subduction top to bottom, Geophysical Monograph*, 96, AGU, Washington, D.C., pp. 195–214.

Klaper, E.M. (1990) Reaction-enhanced formation of eclogite-facies shear zones in granulite-facies anorthosites, in R.J. Knipe and E.H. Rutter (eds.), *Deformation Mechanisms, Rheology and Tectonics, Geological Society of London Special Publication*, **54**, London, pp. 167–173.

Koons, P.O., Rubie, D.C., and Frueh-Green, G. (1987) The effects of disequilibrium and deformation on the mineralogical evolution of quartz diorite during metamorphism in the eclogite facies, *Journal of Petrology* **28**, 679–700.

Krogh, E.J. (1977) Evidence of Precambrian continent-continent collision in Western Norway, *Nature* **267**, 17–19.

Krogh, E.J. (1980) Geochemistry and petrology of glaucophane-bearing eclogites and associated rocks from Sunnfjord, Western Norway, *Lithos* **13**, 355–380.

Laubscher, H. (1988) Material balance in alpine orogeny, *Geological Society of America Bulletin* **100**, 1313–1328.

Le Pichon, X., Fournier, M., and Jolivet, L. (1992) Kinematics, topography, shortening, and extrusion in the India–Eurasia collision, *Tectonics* **11**, 1085–1098.

Le Pichon, X., Henry, P., and Goffe, B. (1997) Uplift of Tibet: from eclogite to granulites - implications for the Andean Plateau and Variscan belt, *Tectonophysics* **273**, 57–76.

Lennykh, V.I., Valizer, P., and Schulte, B.A. (1997) Eclogites and blue schists of the Urals: Evolution and Geodynamics, *Terra Nova* **9**, 17.

Meyer, J. (1983) Mineralogie und Petrologie des Allalin gabbros. Basel, p. 331. University of Basel, Switzerland.

Mooney, W.D. and Meissner, R. (1991) Continental crustal evolution observations, *Transactions of the American Geophysical Union, Eos* **72**, 537–538.

Mørk, M.B.E. (1985) A gabbro to eclogite transition on Flemsøy, Sunnmøre, western Norway, *Chemical Geology* **50**, 283–310.

O'Brien, P.J. (1993) Partially retrograded eclogites of the Muenchberg Massif, Germany; records of a multi-stage Variscan uplift history in the Bohemian Massif, *Journal of Metamorphic Geology* **11**, 241–260.

Pennacchioni, G. (1996) Progressive eclogitization under fluid-present conditions of pre-Alpine mafic granulites in the Austroalpine Mt Emilius Klippe (Italian western Alps), *Journal of Structural Geology* **18**, 549–561.

Rubie, D.C. (1983) Reaction-enhanced ductility: the role of solid-solid univariant reactions in deformation of the crust and mantle, *Tectonophysics* **96**, 331–352.

Rubie, D.C. (1986) The catalysis of mineral reactions by water and restrictions on the presence of aqueous fluid during metamorphism, *Mineralogical Magazine* **50**, 399.

Rubie, D.C. (1990) Role of kinetics in the formation and preservation of eclogites, in D.A. Carswell (ed.), *Eclogite Facies Rocks*, Blackie, Glasgow, pp. 111–140.

Ruff, L. and Kanamori, H. (1983) Seismic coupling and uncoupling at subduction zones, *Tectonophysics* **99**, 99–117.

Rumble, D. (this volume) Stable Isotope Geochemistry Of Ultrahigh-Pressure Rocks, in B.R. Hacker and J.G. Liou (eds.), *When Continents Collide: Geodynamics and Geochemistry of Ultrahigh-Pressure Rocks*, Kluwer Academic Publishers, Dordrecht

Ryan, P.D. and Dewey, J.F. (1997) Continental eclogites and Wilson Cycle, *Journal of the Geological Society of London* **154**, 437–442.

Sapin, M. and Hirn, A. (1997) Seismic structure and evidence for eclogitization during the Himalaya convergence, *Tectonophysics* **273**, 1–16.

Smith, D.C. (1984) Coesite in clinopyroxene in the Caledonides and its implications for geodynamics, *Nature* **310**, 641–644.

Steer, D.N., Knapp, J.H., Brown, L.D., Rybalka, A.V., and Sololov, V.B. (1995) Crustal structure of the Middle Urals based on reprocessing of Russian seismic reflection data, *Geophysical Journal International* **123**, 673–683.

Straume, Å.K. (1997) Retrograde metamorphism of eclogites from Haddal-Ulsteinvik, Sunnmøre. Oslo, p. 150. University of Oslo, Norway.

Udovkina, N.G. (1971) *Eklogity Poliarnogo Urala (Eclogites of the Polar Urals)*, Nauka, Moscow.

Valley, J.W. and Graham, C.M. (1993) Cryptic grain-scale heterogeneity of oxygen isotope ratios in metamorphic magnetite, *Science* **259**, 1729–1733.

Van Wyck, N., Valley, J.W., and Austrheim, H. (1996) Oxygen and carbon isotopic constraints on the development of eclogites, Holsnøy, *Lithos* **38**, 129–147.

Wain, A. (1997) New evidence for coesite in eclogite and gneisses; defining an ultrahigh-pressure province in the Western Gneiss region of Norway, *Geology* **25**, 927–930.

Wallis, S.R., Ishiwatari, A., Hirajima, T., Ye, K., Guo, J., Nakamura, D., Kato, K., Zhai, M., Enami, M., Cong, B., and Banno, S. (1997) Occurrence and field relationships of ultrahigh-pressure metagranitoid and coesite eclogite in the Su-Lu terrane, eastern China, *Journal of the Geological Society of London* **154**, 45–54.

Wedepohl, K.H. (1970) *Geochemistry*, Holt, Rinehart and Winston, New York.

Weiss, L.E. and Wenk, H.R. (1983) Experimentally produced pseudotachylite-like veins in gabbro, *Tectonophysics* **96**, 299–310.

Wheeler, J. (1987) The significance of grain-scale stresses in kinetics of metamorphism, *Contributions to Mineralogy Petrology* **97**, 397–404.

Yui, T.F., Rumble, D., and Lo, C.H. (1995) Unusually low 18O ultrahigh-pressure metamorphic rocks from the Sulu terrain, eastern China, *Geochimica Cosmochimica et Acta* **59**, 2859–2864.

Yui, T.Z., Rumble, D., Chen, C.H., and Lo, C.H. (1997) Stable isotope characteristics of eclogites from the ultra-high-pressure metamorphic terrain, east-central China, *Chemical Geology* **137**, 135–147.

Zhang, R.Y. and Liou, J.G. (1997) Partial transformation of gabbro to coesite-bearing eclogite from Yangkou, the Sulu Terrane, eastern China, *Journal of Metamorphic Geology* **15**, 183–202.

CARDIFF UNIVERSITY ★ PRIFYSGOL CAERDYDD ★

Petrology and Structural Geology

1. J. P. Bard: *Microtextures of Igneous and Metamorphic Rocks*. 1986
ISBN Hb: 90-277-2220-X; ISBN Pb: 90-277-2313-3
2. A. Nicolas: *Principles of Rock Deformation*. 1987
ISBN Hb: 90-277-2368-0; ISBN Pb: 90-277-2369-9
3. J. D. Macdougall (ed.): *Continental Flood Basalts*. 1988 ISBN 90-277-2806-2
4. A. Nicolas: *Structures of Ophiolites and Dynamics of Oceanic Lithosphere*. 1989
ISBN 0-7923-0255-9
5. Tj. Peters, A. Nicolas and R.G. Coleman (eds.): *Ophiolite Genesis and Evolution of the Oceanic Lithosphere*. Proceedings of the Ophiolite Conference (Muscat, Oman, January 1990). 1991 ISBN 0-7923-1176-0
6. R.L.M. Vissers and A. Nicolas (eds.): *Mantle and Lower Crust Exposed in Oceanic Ridges and in Ophiolites*. Contributions to a Specialized Symposium of the VII EUG Meeting, Strasbourg, Spring 1993. 1995 ISBN 0-7923-3491-4
7. C. Bolin (ed.): *Ultrahigh-Pressure Metamorphic Rocks in the Dabieshan-Sulu Region of China*. 1997 ISBN 0-7923-4163-5
8. J.L. Bouchez, D.H.W. Hutton and W.E. Stephens (eds.): *Granite: From Segregation of Melt to Emplacement Fabrics*. 1997 ISBN 0-7923-4460-X
9. O. Merle: *Emplacement Mechanisms of Nappes and Thrust Sheets*. 1998
ISBN 0-7923-4879-6
10. B.R. Hacker and J.G. Liou (eds.): *When Continents Collide: Geodynamics and Geochemistry of Ultrahigh-Pressure Rocks*. 1998 ISBN 0-4128-2420-5

KLUWER ACADEMIC PUBLISHERS – DORDRECHT / BOSTON / LONDON